AF675316

Nutritive Value or Bioutilization Technology of Alternative Forages for Ruminant Nutrition

Nutritive Value or Bioutilization Technology of Alternative Forages for Ruminant Nutrition

Guest Editors

Dingfu Xiao

Tiantian Meng

MDPI

Basel • Beijing • Wuhan • Barcelona • Belgrade • Novi Sad • Cluj • Manchester

Guest Editors
Dingfu Xiao
Hunan Agricultural University
Changsha
China

Tiantian Meng
Xinyang Normal University
Xinyang
China

Editorial Office
MDPI AG
Grosspeteranlage 5
4052 Basel, Switzerland

This is a reprint of the Special Issue, published open access by the journal *Animals* (ISSN 2076-2615), freely accessible at: https://www.mdpi.com/journal/animals/special_issues/3BI24N6R4R.

For citation purposes, cite each article independently as indicated on the article page online and as indicated below:

Lastname, A.A.; Lastname, B.B. Article Title. *Journal Name* **Year**, *Volume Number*, Page Range.

ISBN 978-3-7258-2689-6 (Hbk)
ISBN 978-3-7258-2690-2 (PDF)
https://doi.org/10.3390/books978-3-7258-2690-2

© 2024 by the authors. Articles in this book are Open Access and distributed under the Creative Commons Attribution (CC BY) license. The book as a whole is distributed by MDPI under the terms and conditions of the Creative Commons Attribution-NonCommercial-NoDerivs (CC BY-NC-ND) license (https://creativecommons.org/licenses/by-nc-nd/4.0/).

Contents

animals

Editorial

Nutritional Value Evaluation and Processing Technology of Feed and Nutrition Regulation Measures for Ruminants

Dingfu Xiao [1] and Tiantian Meng [2,*]

1 Yuelushan Laboratory, College of Animal Science and Technology, Hunan Agricultural University, Changsha 410128, China; xiaodingfu2001@hunau.edu.cn
2 College of Life Science, Xinyang Normal University, Xinyang 464000, China
* Correspondence: meng-tiantian@foxmail.com

1. Introduction

In the context of modern livestock farming, particularly in the ruminant industry, feed is a crucial factor that affects both the efficiency of production and animal health. Over the last decade, rising feed costs, driven by factors such as increased demand, raw material shortages, and global crises such as COVID-19, have posed significant challenges to the cattle and sheep farming industries. Feed expenses make up the largest part of production costs, and the price of essential feed ingredients such as corn has surged, placing increased economic pressure on farmers [1]. Thus, understanding the nutritional value of alternative feed sources and improving feed processing technologies is essential in order to enhance the productivity, product quality, and sustainability of livestock.

High-concentrate diets are often rich in carbohydrates. When ruminants consume these diets, volatile fatty acids (VFAs) such as acetic, propionic, and butyric acids are produced in the rumen through microbial fermentation. However, the absorption capacity of the rumen wall is limited, leading to the accumulation of VFAs, which lowers the pH of the rumen and triggers subacute ruminal acidosis (SARA) [2]. The prolonged consumption of high-concentrate diets increases the risk of SARA, disrupting rumen metabolism and impairing normal rumen function, leading to a reduction in production. Therefore, it is crucial to mitigate SARA through strategies such as the bioconversion of unconventional feeds and the use of plant extracts to prevent the negative effects associated with high-concentrate diets.

Citation: Xiao, D.; Meng, T. Nutritional Value Evaluation and Processing Technology of Feed and Nutrition Regulation Measures for Ruminants. *Animals* **2024**, *14*, 3153. https://doi.org/10.3390/ani14213153

Received: 13 September 2024
Accepted: 30 October 2024
Published: 2 November 2024

Copyright: © 2024 by the authors. Licensee MDPI, Basel, Switzerland. This article is an open access article distributed under the terms and conditions of the Creative Commons Attribution (CC BY) license (https://creativecommons.org/licenses/by/4.0/).

2. Feed Ingredients and Nutritional Value

The nutritional value of feed varies significantly depending on its source, variety, and processing methods. The quality of forage, including its fiber content, digestibility, and nutrient availability, plays a critical role in the overall health and productivity of ruminants [3]. High-quality forage promotes enhanced digestion and nutrient absorption, ensuring that livestock can maintain optimal health and performance levels. On the other hand, poor-quality feed leads to digestive problems and reduced productivity, which can result in significant economic losses [4]. In contrast, feeding overly mature grasses to ruminants increases the production of methane, which is strongly correlated with the carbon-to-nitrogen ratio in grasses [5,6]. Additionally, the nutrient content of different forages can affect the production of methane post-feeding [7]. Leguminous forages produce less methane than grasses, as the condensed tannins in legumes significantly impact the degradation of fiber and dry matter [8–11]. Not only does the quality of forage affect methane emissions, but different processing methods can also influence the production of methane in ruminants. For example, treatments such as ammonization, crushing, pelletizing, and ensiling have been shown to reduce methane emissions to varying degrees [12]. This reduction is likely due to an increase in the levels of non-fibrous carbohydrate (NFC) after ensiling, which lowers the NDF content and promotes the production of propionic acid in the rumen while reducing acetic acid; thus, methane synthesis is decreased [13,14]. Furthermore, high-concentrate diets, which are commonly used in intensive ruminant production, are rich in carbohydrates. Therefore, evaluating the nutritional value of feed ingredients and formulating balanced diets that prevent digestive disturbances are crucial to ensuring the well-being of ruminants.

3. Feed Processing Technology

The processing of feed plays a pivotal role in maintaining or even enhancing its nutritional quality. Technologies such as ensiling, pelleting, and steam-flaking can help preserve nutrients and enhance the efficiency of feed [15]. For instance, ensiling is a method used to preserve forage crops via fermentation, which not only extends the shelf life of feed but also enhances its digestibility by breaking down fiber components [16]. Advances in feed processing have enabled the enhanced utilization of alternative feed ingredients, thus mitigating the high costs associated with conventional feed materials.

Moreover, bioconversion techniques, including the microbial fermentation of unconventional feedstuffs, have gained attention due to their ability to enhance the value of feed while reducing dependence on traditional, high-cost ingredients [17]. This method can transform low-nutrient materials into more digestible and nutritionally valuable feed through the action of microbes, making it an attractive option for sustainable livestock farming.

4. Nutrition Regulation Measures

To address the challenges posed by high-concentrate diets, various nutritional regulation strategies have been explored. One approach involves the use of plant extracts and essential oils as natural additives to prevent the negative effects of rumen acidosis. These additives have been shown to modulate rumen fermentation patterns, promoting the production of beneficial VFAs while reducing methane emissions [18]. This not only enhances the health of the rumen, but also contributes to environmental sustainability by mitigating the greenhouse gas emitted from livestock [19].

Another promising nutritional regulation strategy is the integration of precision feeding techniques. This involves the use of real-time monitoring systems to adjust the composition and delivery of feed based on the requirements of individual animals [20]. Precision feeding can optimize the utilization of nutrients, improve the performance of animals, and reduce waste, leading to a more efficient and sustainable production system.

4.1. Unconventional Feeds for Ruminants

Unconventional feed resources possess unique nutritional qualities, but are associated with various challenges; these include processing difficulties, high transportation costs, susceptibility to deterioration, poor palatability, the presence of anti-nutritional factors, strong seasonality, and limited geographical availability. To maximize their benefits, these resources must be developed rationally and used in a scientifically sound manner. For example, the dietary supplementation of DHA-rich microalgae improved the antioxidant status, increased the deposition of DHA and enhanced the characteristic flavor of beef [21]. Silva found that partially substituting corn silage with whole soybean silage or black oat silage reduced the intake of feed while increasing rumination and chewing activities in cows, without compromising growth performance [22]. Similarly, Zhang et al. reported that sweet straw and wheat straw could serve as more affordable alternatives to whole-plant corn silage (WPCS) for fattening beef cattle, without significantly diminishing the economic returns [23]. Additionally, the inclusion of whole pomegranate byproducts in sheep diets, replacing traditional grain feed, improved the antioxidant capacity of the meat, likely due to the presence of vitamin E in the pomegranate byproducts [24].

4.2. Applied Plant Extract

A recent study using rumen simulation technology evaluated the effects of chito-oligosaccharides (COS) (0.02%, 0.04%, 0.08%) on the diets of beef cattle. The results showed that COS increased the disappearance rates of dry matter DM, crude protein (CP), ether extract, and acid detergent fiber, while raising the pH, microbial CP, and concentrations of propionate and butyrate. It also reduced the production of gas, NH3-N, the A/P ratio, and the acetate, isobutyrate, valerate, and protozoa counts, shifting fermentation from an acetate to a propionate mode. COS did not change the overall diversity of the microbiota but increased the relative abundance of *Methanosphaera*, *Ruminococcus*, *Endomycobium*, and *Eubacterium* [25].

In vitro experiments were conducted by adding mixtures of Firmwood tannin and chestnut tannin to forage and hay. The addition of these tannin mixtures to hay significantly reduced the acetic acid-to-propionic acid ratio; after 24 h, the CO_2/CH_4 ratio increased notably with 30 g/kg of mixed tannins. While the addition of mixed tannins to hay significantly lowered the production of ammonia, no such effect was observed in forage. Mixed tannins also protected against the degradation of protein in the rumen, particularly under low-protein conditions; this was possibly due to the expansion of the binding spectrum [26]. Furthermore, the addition of 1% and 2% Neolamarckia cadamba leaf extracts significantly reduced the production of CH_4 and increased the digestibility of dry matter, although it did not affect the concentrations of VFAs. While the microbial community of the rumen was largely unaffected, the extracts significantly decreased the relative abundance of *Methanobacterium*. This reduction in methane-producing bacteria was positively correlated with a reduction in the production of CH_4 in the rumen [27].

5. Conclusions

The rising cost of feed and the challenges associated with maintaining ruminant health in intensive production systems necessitate a comprehensive approach to evaluating the nutritional value of feed and processing technologies. By leveraging bioconversion methods, improving the quality of feed, and implementing effective nutrition regulation measures, the ruminant industry can achieve enhanced productivity and sustainability. Further research is needed to develop innovative feed solutions and nutritional strategies that not only enhance animal performance but also reduce the environmental impact of livestock farming.

Funding: This research received no external funding.

Conflicts of Interest: The authors declare no conflict of interest.

References

1. Jones, A.; Smith, R. The impact of feed costs on livestock production. *Anim. Feed Sci. Technol.* **2022**, *265*, 114–125.
2. Zebeli, Q.; Dijkstra, J.; Tafaj, M.; Steingass, H.; Ametaj, B.; Drochner, W. Modeling the Adequacy of Dietary Fiber in Dairy Cows Based on the Responses of Ruminal pH and Milk Fat Production to Composition of the Diet. *J. Dairy Sci.* **2008**, *91*, 2046–2066. [CrossRef] [PubMed]
3. De Brito, G.F.; McGrath, S.R.; Holman, B.W.; Friend, M.A.; Fowler, S.M.; van de Ven, R.J.; Hopkins, D.L. The effect of forage type on lamb carcass traits, meat quality and sensory traits. *Meat Sci.* **2016**, *119*, 95–101.
4. Wilson, T.; Moore, C. Forage quality and its implications on livestock performance. *Livest. Prod. Rev.* **2020**, *45*, 210–218.
5. Himanshu, H.; Murphy, J.; Grant, J.; O'Kiely, P. Synergies from co-digesting grass or clover silages with cattle slurry in in vitro batch anaerobic digestion. *Renew. Energy* **2018**, *127*, 474–480. [CrossRef]
6. Cole, N.; Meyer, B.; Parker, D.; Neel, J.; Turner, K.; Northup, B.; Jennings, T.; Jennings, J. Effects of diet quality on energy metabolism and methane production by beef steers fed a warm-season grass-based hay diet. *Appl. Anim. Sci.* **2020**, *36*, 652–667. [CrossRef]
7. Eugène, M.; Klumpp, K.; Sauvant, D. Methane mitigating options with forages fed to ruminants. *Grass Forage Sci.* **2021**, *76*, 196–204. [CrossRef]
8. Waghorn, G.C.; McNabb, W.C. Consequences of plant phenolic compounds for productivity and health of ruminants. *Proc. Nutr. Soc.* **2003**, *62*, 383–392. [CrossRef]
9. Patra, A.K. A meta-analysis on effects of supplementing diets with tannins on methane production and nitrogen excretion in ruminants. *J. Anim. Physiol. Anim. Nutr.* **2014**, *98*, 102–116.
10. Beauchemin, K.A.; Kreuzer, M.; O'Mara, F.; McAllister, T.A. Nutritional management for enteric methane abatement: A review. *Aust. J. Exp. Agric.* **2008**, *48*, 21–27. [CrossRef]

11. Tiemann, T.T.; Lascano, C.E.; Wettstein, H.-R.; Mayer, A.C.; Kreuzer, M.; Hess, H.D. Effect of the tropical tannin-rich shrub legumes Calliandra calothyrsus and Flemingia macrophylla on methane emission and nitrogen and energy balance in growing lambs. *Animal* **2008**, *2*, 790–799. [CrossRef]
12. Boadi, D.A.; Wittenberg, K.M.; Scott, S.L.; Burton, D.; Buckley, K.; Small, J.A.; Ominski, K.H. Effect of low and high forage diet on enteric and manure pack greenhouse gas emissions from a feedlot. *Can. J. Anim. Sci.* **2004**, *84*, 445–453. [CrossRef]
13. Martin, C.; Ferlay, A.; Mosoni, P.; Rochette, Y.; Chilliard, Y.; Doreau, M. Increasing linseed supply in dairy cow diets based on hay or corn silage: Effect on enteric methane emission, rumen microbial fermentation, and digestion. *J. Dairy Sci.* **2016**, *99*, 3445–3456. [CrossRef] [PubMed]
14. Hammond, K.; Jones, A.; Humphries, D.; Crompton, L.; Reynolds, C. Effects of diet forage source and neutral detergent fiber content on milk production of dairy cattle and methane emissions determined using GreenFeed and respiration chamber techniques. *J. Dairy Sci.* **2016**, *99*, 7904–7917. [CrossRef] [PubMed]
15. Zhang, L.; Liu, H. Advances in feed processing technologies for ruminants. *Anim. Nutr.* **2018**, *4*, 137–144.
16. Miller, R.; Davis, P. The effects of silage processing on feed quality. *J. Agric. Res.* **2021**, *18*, 123–135.
17. Green, J.; Miller, S. Bioconversion of unconventional feedstuffs in livestock diets. *Sustain. Agric.* **2019**, *23*, 67–75.
18. Arowolo, M.A.; He, J. Use of probiotics and botanical extracts to improve ruminant production in the tropics: A review. *Anim. Nutr.* **2018**, *4*, 241–249.
19. Smith, G.; Brown, H. Methane mitigation strategies in ruminant livestock. *Environ. Sustain. Agric.* **2020**, *32*, 45–58.
20. Carter, A.; Thomas, E. Precision feeding in ruminant livestock: A review. *Anim. Feed Sci. Technol.* **2021**, *269*, 116–127.
21. Xu, C.; Zhang, S.; Sun, B.; Xie, P.; Liu, X.; Chang, L.; Lu, F.; Zhang, S. Dietary Supplementation with Microalgae (*Schizochytrium* sp.) Improves the Antioxidant Status, Fatty Acids Profiles and Volatile Compounds of Beef. *Animals* **2021**, *11*, 3517. [CrossRef]
22. Silva, L.H.; Assis, D.E.; Estrada, M.M.; Assis, G.J.; Zamudio, G.D.; Carneiro, G.B.; Valadares Filho, S.C.; Paulino, M.F.; Chizzotti, M.L. Carcass and meat quality traits of Nellore young bulls and steers throughout fattening. *Livest. Sci.* **2019**, *229*, 28–36. [CrossRef]
23. Zhang, H.; Zhang, L.; Xue, X.; Zhang, X.; Wang, H.; Gao, T.; Phillips, C. Effect of feeding a diet comprised of various corn silages inclusion with peanut vine or wheat straw on performance, digestion, serum parameters and meat nutrients in finishing beef cattle. *Anim. Biosci.* **2022**, *35*, 29–38. [CrossRef] [PubMed]
24. Natalello, A.; Priolo, A.; Valenti, B.; Codini, M.; Mattioli, S.; Pauselli, M.; Puccio, M.; Lanza, M.; Stergiadis, S.; Luciano, G. Dietary pomegranate by-product improves oxidative stability of lamb meat. *Meat Sci.* **2020**, *162*, 108037. [CrossRef] [PubMed]
25. He, J.; Li, J.; Gao, Q.; Shen, W.; Liu, W.; Xia, M.; Xiao, H.; Xiao, D. In Vitro Evaluation of Chito-Oligosaccharides on Disappearance Rate of Nutrients, Rumen Fermentation Parameters, and Micro-Flora of Beef Cattle. *Animals* **2024**, *14*, 1657. [CrossRef]
26. Menci, R.; Coppa, M.; Torrent, A.; Natalello, A.; Valenti, B.; Luciano, G.; Priolo, A.; Niderkorn, V. Effects of two tannin extracts at different doses in interaction with a green or dry forage substrate on rumen fermentation and biohydrogenation-an in vitro study. *Anim. Feed Sci. Technol.* **2021**, *278*, 114977. [CrossRef]
27. Zhang, Q.; Wu, S.; Zou, X.; Ruan, S.; Kholif, A.E.; Hu, L.; Chen, X.; Zhou, W. Effects of Neolamarckia cadamba leaves extract on methanogenesis, microbial community in the rumen and digestibility of stylo silage. *J. Clean. Prod.* **2022**, *369*, 133338. [CrossRef]

Disclaimer/Publisher's Note: The statements, opinions and data contained in all publications are solely those of the individual author(s) and contributor(s) and not of MDPI and/or the editor(s). MDPI and/or the editor(s) disclaim responsibility for any injury to people or property resulting from any ideas, methods, instructions or products referred to in the content.

animals

MDPI

Article

In Vitro Evaluation of Chito-Oligosaccharides on Disappearance Rate of Nutrients, Rumen Fermentation Parameters, and Micro-Flora of Beef Cattle

Jianfu He [1,2], Jing Li [1], Qian Gao [1], Weijun Shen [1,2], Wenchang Liu [1], Min Xia [1], Haixiang Xiao [1] and Dingfu Xiao [1,2,*]

1 College of Animal Science and Technology, Hunan Agricultural University, Changsha 410128, China; hejianfu807@sina.com (J.H.); jing1173442331@outlook.com (J.L.); gaoqian996@sina.com (Q.G.); shenweijun@hunau.edu.cn (W.S.); bgmusic79@163.com (W.L.); xmin1017001@163.com (M.X.); xhx248260@sina.com (H.X.)

2 Yuelushan Laboratory, Changsha 410128, China

* Correspondence: xiaodingfu2001@163.com

Simple Summary: This study investigated whether COS in the diet can improve the digestion and rumen health of beef cattle under in vitro fermentation conditions. A total of 24 fermentation tanks were randomly allocated into four groups. Each treatment had six replicates, with one tank per replicate. Each group received a different concentration of 0.02% COS, 0.04% COS, 0.08% COS, and a control group (CON). The results demonstrated that the addition of COS facilitates a shift in rumen fermentation from the acetate model to the propionate model. This is achieved by influencing the microbial population, resulting in alterations to the fermentation level, nutrient disappearance rate, and gas production. These changes have the potential to enhance rumen health.

Abstract: The study aimed to investigate the effect of dietary chitosan oligosaccharides (COS) meal levels on the nutrient disappearance rate, rumen fermentation, and microflora of beef cattle in vitro. A total of 24 fermentation tanks were randomly divided into four treatments containing 0% COS (CON), 0.02% COS, 0.04% COS, and 0.08% COS for an 8-day experiment period, with each treatment comprising six replicates. The disappear rates of DM, CP, EE, and total gas production were quadratically increased with increasing COS levels. The disappear rates of DM, CP, EE, and ADF were greatest, whereas the total gas production was lowest in the 0.08% COS group. The pH, NH_3-N, MCP, the content of propionate, isobutyrate, butyrate, valerate, and the A/P were quadratically increased with increasing COS levels, while the A/P were linearly decreased. The pH, MCP, and the content of propionate, and butyrate were highest, whereas the NH_3-N and the content of acetate, isobutyrate, valerate, and the A/P were lowest in the 0.08% COS group. Microbiomics analysis showed that the rumen microbial diversity was not altered between the CON and the 0.08% COS group. However, the relative abundance of *Methanosphaera*, *Ruminococcus*, *Endomicrobium*, and *Eubacterium* groups was increased, and the relative abundance of pathogenic bacteria *Dorea* and *Escherichia-Shigella* showed a decrease in the 0.08% COS group. Overall, the 0.08% COS was the most effective among the three addition levels, resulting in an increase in the disappearance rate of in vitro fermented nutrients and improvements in rumen fermentation indexes and microbial communities. This, in turn, led to the maintenance of rumen health.

Keywords: beef cattle; chitosan oligosaccharides; RUSITEC system; rumen fermentation; microorganisms

Citation: He, J.; Li, J.; Gao, Q.; Shen, W.; Liu, W.; Xia, M.; Xiao, H.; Xiao, D. In Vitro Evaluation of Chito-Oligosaccharides on Disappearance Rate of Nutrients, Rumen Fermentation Parameters, and Micro-Flora of Beef Cattle. *Animals* **2024**, *14*, 1657. https://doi.org/10.3390/ani14111657

Academic Editor: Donald C Beitz

Received: 6 May 2024
Revised: 27 May 2024
Accepted: 30 May 2024
Published: 31 May 2024

Copyright: © 2024 by the authors. Licensee MDPI, Basel, Switzerland. This article is an open access article distributed under the terms and conditions of the Creative Commons Attribution (CC BY) license (https://creativecommons.org/licenses/by/4.0/).

1. Introduction

A number of challenges have constrained the rapid development of beef cattle farming, including long-distance transport, stress caused by climate change, high feed prices, and low nutrient utilisation [1]. These factors have resulted in suboptimal growth performance, meat quality, and reduced economic benefits. In the previous studies, the rumen fermentation indicators and microbial community distribution are important indices to understand the health status of the rumen [2–4]. However, due to the special multi-stomach structure of ruminants, the rumen health of ruminants has always been a major concern in ruminant breeding [5]. However, the rumen has complex physiological functions and digestive characteristics [6], and the mechanism of these physiological functions has not been completely clarified so far.

In previous studies, people have tried almost all feed substitutes and achieved remarkable results, as well as tried different types of additives, which have been proven to improve animal growth performance improving growth and enhancing immunity [7–10]. Chitin is a natural polymer with high production and biodegradability in nature, second only to cellulose. It can be deacetylated to obtain chitosan [11], and COS can be prepared by enzymatic or chemical hydrolysis of chitosan [12]. The COS was an additive with multiple biological activities and functions, which was prepared by enzymatic or chemical hydrolysis of chitosan. Compared to chitosan, COS exhibits low molecular weight, high degree of deacetylation, high polymerisation, and low viscosity, and the molecular weight of COS is in the range of 0.5–2.5 kDA [13,14]. Therefore, COS not only has the biological effects of chitosan, but its effect is even stronger than that of chitosan [13]. The physicochemical properties of COS give it significant biological properties, including anti-oxidant, anti-inflammatory, drug, and *DeoxyriboNucleic Acid* (DNA) delivery capabilities [15]. Consistently, research has shown that COS can enhance animal growth performance, facilitate the development of immune organs, reinforce the small intestinal mucosal barrier function [16], and mitigate the inflammatory response of the intestinal tract [17], regulating gastrointestinal function [18]. However, the effect of COS on the rumen fermentation of beef cattle is not yet clear.

The rumen simulation technology (RUSITEC) system is an in vitro fermentation device that mimics the physiological functions of the rumen with the aim of reducing the limitations of in vivo experiments, such as inconsistencies in the genetic background and physiological state of individual animals [19]. The RUSITEC system plays an important role in the study of rumen microorganisms and the mechanism of rumen fermentation [20]. Therefore, in this study, the RUSITEC system was used to investigate the effects of different levels of COS added to diets on rumen fermentation of beef cattle and to evaluate whether COS can be used as a safe and effective feed additive, with particular attention on nutrient disappearance rates, rumen fermentation parameters and rumen microbial community.

2. Materials and Methods

2.1. Experimental Animals, Feeding Management, and Experimental Design

Three *Xiangxi* yellow cattle (*Bos taurus;* a native breed in Hunan Province, China) possessing rumen fistulas were enlisted as donors of rumen fluid, the weight of the experimental animals is 385 ± 28.7 (mean ± standard deviation) kg, with an age of 3 ± 0.5 years. Rumen fluid collection was followed by its deployment in the RUSITEC system for in vitro fermentation assays. Formulation of the diet and feeding amount for rumen fluid donors adhered to the nutritional benchmarks delineated in the Chinese Beef Cattle Feeding Standards (NY/T815-2004) [21]. The dietary regimen for the cattle was comprised of a blend of wheat straw and concentrates in a 1:1 ratio (Table 1). Feeding sessions were conducted at 08:00 and 18:00, restricted dietary feeding. The study utilised a single-factor randomised trial design, dividing 24 fermentation tanks into four groups, each group with six replicates. The substrate was the basal diets with supplemented 0.02%, 0.04%, and 0.08% COS, respectively, COS are provided by Zhongke Rongxin Biotechnology Co., Ltd. (Suzhou, China) with a purity of ≥85%, mixed well with the fermentation substrate by mixing step by step. The adaptation period lasted for five days and was then followed by a three-day sampling period. The experiment was conducted at the Animal Science and Technology Experimental Building and Animal Training Center of Hunan Agricultural University.

Table 1. Composition and nutrient levels of basal diets and fermentation substrates (DM basic).

Items	Content (%)
Ingredients	
Wheat straw	50
Corn	12.5
Soybean meal	4.35
Unhusked rice	14
Wheat bran	7
Oil bran	3.5
Sprayed corn husk	1
Soybean germ powder	1.5
Brow rice	1.5
Soybean husk	1
Rumen undegradable fat powder	0.15
Expanded urea	0.4
Unified bran	0.625
Premix [a]	2.475
Total	100
Nutrient levels [b]	
DM	91.89
CP	10.99
EE	3.00
NDF	69.95
ADF	32.26
Ash	10.69
Ca	0.81
P	0.25

[a] Every 1 kg of premix contained 250,000 IU of vitamin A, 50,000 IU of vitamin D3, 800 IU of vitamin E, 0.9 g of $CuSO_4$, 12 g of $FeSO_4$, 14 g of $MnSO_4$, 10 g of $ZnSO_4$, 0.03 g of Na_2SeO_4, 0.02 g of KI, 0.02 g of $CoCl_2$, 55 g of $MgSO_4$. [b] The nutritional levels of the diets in the table are all measured values. DM, dry matter; CP, crude protein; EE, ether extract; NDF, neutral detergent fibres; ADF, acid detergent fibres; Ca, calcium; P, phosphorus.

2.2. Rumen Simulation Technique Fermentation

The construction and operation programme of the RUSITEC system utilised in this experiment is detailed in the study by Adebayo [22]. In brief, the RUSITEC system is a dual-flow continuous culture system that simulates rumen fermentation in vitro. It primarily comprises a simulated fermentation device, a constant temperature device, an information processing section, and a fermentation parameter control section. The simulated fermentation device comprises a fermentation tank, stirrer, buffer container, water-cooled overflow bottle, gas collection bag, and gas, pressure, temperature, and pH sensors, which are primarily employed for the collection of pertinent parameters. The constant temperature device encompasses a constant temperature heating device and a reflux pipe, which are utilised to ensure that the fermentation process maintains a constant temperature. The information processing section comprises four modules: gas, pressure, temperature, and pH, which are utilised for the detection of relevant parameters. The fermentation parameter control section comprises an intelligent operating system, hardware motherboard, and chips, which are employed for the overall control and operation of the device. The fermentation temperature of the RUSITEC system's fermentation tank can be maintained at 39 ± 0.5 °C. The McDougall [23] buffer solution is continuously injected into the fermentation tank at a set rate through a peristaltic pump, and the overflow liquid and undegraded solid components of each fermentation tank are automatically and continuously discharged into a water-cooled overflow bottle in a 4 °C water bath. The water-cooled overflow bottle is maintained at a low temperature to prevent further fermentation, thereby achieving a more accurate simulation of rumen fermentation in living animals.

Prior to the formal test, it is necessary to ascertain the airtightness of the RUSITEC system and conduct a trial run. During the formal experiment, the thermos was preheated to 39 °C and filled with CO_2 in advance. Prior to the morning feeding, rumen contents were collected from three *Xiangxi* yellow cattle in a rumen fistula tube, filtered through four layers of sterilised gauze, and returned to the laboratory. A constant N_2 airflow was then introduced into the device to maintain anaerobic conditions, while 500 mL of preheated filtered rumen fluid and 500 mL of McDougall buffer were introduced into each fermentation tank. Once the device has commenced operation, a constant N_2 airflow should be introduced, after which 20 g of fermentation substrate (DM basis) should be added to the fermentation tank. This should be performed at 08:00 and 18:00 each day. The composition of the fermentation substrate should be consistent with that of the rumen fluid donor cow feed. Wheat straw and concentrate supplement are dried to a constant weight in a 65 °C constant temperature drying oven and then ground by a grinder through a 1 mm aperture sieve. They are mixed in a 1:1 ratio. The constant temperature device maintains a temperature of 39 ± 0.5 °C in the fermentation tank. The intelligent operating system controls the stepper motor to stir the contents of the fermentation tank at a rate of 25 r/min. The McDougall buffer is introduced into the fermentation tank via a pressure pump at a rate of 6% per hour. The overflow bottle conduit is monitored to ensure that the overflow liquid and undegraded solid phase fermentation substrate in each fermentation tank can be collected in a timely manner and terminated in the overflow bottle.

2.3. Sample Collection

Prior to the addition of substrate each morning, a 5 mL sample of fermentation broth was collected from the liquid phase sample collection port of the fermenter by filtering the collection through four layers of sterile gauze, which was then placed in a 50 mL centrifuge tube and added to 15 mL of methyl green staining solution. The methyl green staining solution was prepared by completely dissolving 6 g of methyl green and 8 g of sodium chloride in 1 L of 35% formaldehyde solution. The solution is then mixed well and placed in a dark location to shake and stored for protozoa counting. During the sampling period (days 6 to 8 of the experiment), the collected material in the water-cooled overflow bottle is filtered through a nylon bag before the addition of substrates. A solid phase sample is then collected. After cleaning, drying, crushing, and storage in a self-sealing bag, the sample is used to determine nutrient content and calculate the nutrient disappearance rate. Prior to the addition of substrates each morning during the sampling period, 15 mL of fermentation broth is collected from the liquid phase sample collection port of the fermentation tanks. The pH is then measured using a portable pH meter (PHS-3C, Shanghai Yidian Scientific Instrument Co., Ltd., Shanghai, China), after which 1.5 mL of the collected fermentation broth is divided into sterile and enzyme-free 2 mL centrifuge tubes. A total of 300 μL of 25% metaphosphate is added to the samples of two centrifuge tubes, which are then mixed and acidified uniformly. These samples are subsequently used for the determination of volatile fatty acids (VFA) and NH_3-N concentrations, respectively. Prior to measurement, the samples were stored at a temperature of −20 °C. Two tubes of samples were used for microbial crude protein (MCP) determination and microbial omics analysis and were stored at −20 °C prior to measurement. The remaining samples were stored in a −80 °C refrigerator for future use. On the morning of the sampling period, after adding substrates, a 3 L gas bag was installed through the gas phase sample collection port to collect the discharged gas. The volume of gas collected in the bag is measured using a graduated syringe, and the total volume of gas collected is recorded.

2.4. Chemical and Biological Analysis

The determination of DM, ash, EE, and CP content in both substrate and fermentation residue followed the methods outlined by AOAC [24]. Moreover, the calcium (Ca) and phosphorus (P) content in the feed were assessed using the AOAC [24] method. The calculation method for organic matter (OM) is: $\text{OM}(\%) = 1 - \text{ash}(\%)$. The NDF and

ADF content in both feed and fermentation residue were determined as described by Van Soest [25]. And the calculation method for the disappearance rate of nutrients is:

$$x = [(a \times b - c \times d)/(a \times b)] \times 100\%$$

where x: disappearance rate of nutrients; a: mass of substrate added every day, dry matter basis; b: measured value of a certain nutrient component in the substrate; c: mass of residue collected every day during the sampling period, dry matter basis; d: measured value of a certain nutrient component in the residue).

Protozoa enumeration was conducted using the Sedgewick Rafter counting plate. A proportional mixture of fermentation broth and methyl green staining solution was prepared, with 1 mL of the resultant solution transferred onto the counting plate. The plate was examined under an optical microscope according to Kisidayov's methodology [26]. VFA concentration in the fermentation broth was determined via gas chromatography [27], NH_3-N content was quantified using the phenol hypochlorite method [28], and MCP content was assessed following the method outlined by Makkar [29].

2.5. Microbiological Analysis of Fermentation Broth

Microbiological analysis of fermentation broth samples was conducted by Beijing Ovison Gene Technology Co., Ltd. (Beijing, China). The detailed purification, quantification, sequencing steps and analytical methods are consistent with those in our previous studies [30], as described in detail below.

Microbial genomic DNA was extracted from the fermentation liquid samples using the E.Z.N.A. Soil DNA Kit (Omega Bio-tek, Inc., Norcross, GA, USA), following the manufacturer's instructions. The purity and concentration of the extracted DNA were determined using the TBS-380 and NanoDrop2000 spectrophotometers (Thermo Fisher Scientific, Waltham, MA, USA), respectively. Furthermore, the integrity of the extracted DNA was verified through 1% agarose gel electrophoresis. The DNA samples were subsequently stored at −80°C for subsequent experiments.

The rumen microbial community structure can be obtained by sequencing the region of the bacterial 16S rRNA gene and the eukaryotic 18S rRNA gene. The V3 and V4 hypervariable regions of the bacterial 16S rRNA gene were amplified using universal primers: 338F (5′-ACTCCTACGGGAGGCAGCAG-3′) and 806R (5′-GGACTACNNGGGTATCTAAT-3′). The V4 hypervariable region of the eukaryotic 18S rRNA gene was amplified using the primers 573F (5′-CGCGGTAATTCCAGCTCCA-3′) and 951R (5′-TTGGYRAATGCTTTCGC-3′). The purification, quantification, and sequencing of the 16S rRNA gene and 18S rRNA gene were conducted on the Illumina MiSeq/NovaSeq (Illumina, Inc., Albany, NY, USA) platform at Beijing Allwegene Technology Co., Ltd. (Beijing, China).

Based on previous studies [31,32], the original sequencing reads of the 16S rRNA gene and the 18S rRNA gene were demultiplexed, quality filtered, and merged. The qualified sequences were clustered into operational taxonomic units (OTUs) with 97% similarity using the Uparse algorithm of the Vsearch (v2.7.1) software and all OTU representative sequences were classified into different taxonomic groups against the Silva database using the BLAST tool. The α-diversity indices (Chao1, Observed_species, PD_whole_tree, and Shannon) were calculated based on the OTU information with the QIIME software (v1.8.0). The difference test of these indices between the two groups was conducted by the Wilcoxon rank sum test. A principal coordinates analysis (PCoA) was conducted based on the Bray–Curtis distance at the OTU level in order to assess β-diversity. Furthermore, an Adonis (PERMANOVA) analysis was conducted to assess significant differences in β-diversity of bacteria and protozoa between the two groups. The taxonomic annotation and relative abundance of microbial species at the phylum and genus levels were visualised as barplot diagrams using the R software (v3.6.0). The Linear Discriminant Analysis Effect Size (LEfSe) analysis was conducted using the Python (v2.7) software to identify the signature microbiota between the two groups.

2.6. Statistical Analysis of Data

Data analysis was performed using SPSS 20.0 software (SPSS, Inc., Chicago, IL, USA), which was employed for analysing data nutrient disappearance rate, VFA, NH_3-N, MCP, gas production, and pH using single-factor random design. Results are presented as mean ± Standard Error of Mean. The Duncan multiple range test was utilised to determine the significance of differences between treatment groups. A *p*-value of less than 0.05 indicates significant difference, and less than 0.01 indicates extremely significant differences, while *p*-value between 0.05 and 0.1 indicates a trend. Additionally, GraphPad Prism 8.0 software (Origin, CA, USA) was utilised for data visualisation and plotting. Figure 1 was drawn using GraphPad Prism (9.5.0.730), all images (Figures 1 and 2) are grouped and spliced using the PDF editing function in WPS Office (12.1.0.16729) and enhanced in pixel and resolution using Adobe Illustrator 2023.

3. Results

3.1. Rumen Protozoa Count

The number of rumen protozoa (Figure 1) in the four groups showed a rapid decline on day 2 after inoculation into the fermenters, followed by a slower decline on days 2–4 and a gradual stabilisation of the numbers after day 5. As shown in Table 2, during the initial five days of the experiment, when compared with the CON group, the inclusion of COS in the diet had no significant effect on the rumen protozoa count ($p > 0.05$). However, the 0.08% COS group witnessed a significant decline in protozoa count on the 6th and 8th day ($p < 0.05$), and there is a declining trend on the 7th day.

Table 2. Effects of dietary different levels of COS on the rumen protozoa count of beef cattle.

Items	Dietary COS Level, Feeding Basis				SEM	*p*-Value
	CON	0.02% COS	0.04% COS	0.08% COS		
	Number of Rumen Protozoa on Days 1–8 (10^5/mL)					
Day 1 [A]	7.86	7.86	7.86	7.86	0	/
Day 2	2.82	2.84	2.85	2.73	0.072	0.374
Day 3	1.13	1.22	1.21	1.15	0.057	0.357
Day 4	0.74	0.73	0.74	0.72	0.023	0.783
Day 5	0.67	0.68	0.68	0.64	0.029	0.441
Day 6	0.57 [a]	0.58 [a]	0.54 [ab]	0.50 [b]	0.024	0.015
Day 7	0.49 [ab]	0.51 [a]	0.46 [ab]	0.44 [b]	0.028	0.065
Day 8	0.47 [a]	0.46 [ab]	0.44 [ab]	0.42 [b]	0.020	0.144

In the table, the superscript "a–b" represents the differences in data between the treatment groups. Different letters indicate differences between the groups ($p < 0.05$), while the same or no letters indicate no significant differences ($p > 0.05$); SEM, standard error of the mean. [A] On the first day, all fermentation tanks were filled with a solution of fermentation broth and RUSITEC system buffer for rumen protozoa counting. This solution was evenly distributed throughout the tanks, ensuring that the protozoa count on the first day was consistent.

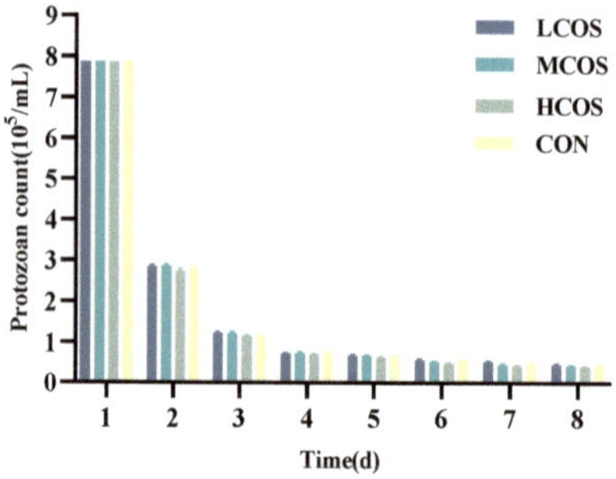

Figure 1. Dynamic changes in the number of rumen protozoa in four treatment groups on days 1–8.

3.2. Nutrient Disappearance Rates and Total Gas Production

As shown in Table 3, the disappear rates of DM, OM, CP, and EE, and the total gas production of day 6, day 7, and day 8 were quadratically decreased ($p < 0.05$) with an increasing COS level. With the increased level of COS, the total gas production of day 6, day 7, and day 8 also linearly decreased ($p < 0.05$). The disappearance rate of DM, OM, CP, EE, and ADF was greatest ($p < 0.05$), whereas the total gas production was least ($p < 0.05$) in the 0.08% COS group. There were no differences ($p > 0.05$) in the disappearance rate of OM and NDF among the COS groups.

Table 3. Effects of adding different levels of COS to the diet on nutrient disappearance rates and total gas production of rumen nutrients in beef cattle.

Items	Dietary COS Level, Feeding Basis				SEM	*p*-Value		
	CON	0.02% COS	0.04% COS	0.08% COS		Main Effect	Linear	Quadratic
Nutritional Components Apparent Disappearance Rates (%)								
DM	77.44 ^{c}	77.48 ^{c}	77.81 ^{b}	78.42 ^{a}	0.151	<0.01	0.587	<0.01
OM	65.28 ^{c}	65.32 ^{bc}	65.66 ^{b}	66.23 ^{a}	0.165	<0.01	0.444	<0.01
CP	73.04 ^{c}	73.67 ^{bc}	73.99 ^{b}	74.93 ^{a}	0.305	<0.01	0.549	<0.01
EE	64.25 ^{c}	64.63 ^{c}	65.66 ^{b}	66.23 ^{a}	0.217	<0.01	0.749	<0.01
NDF	53.71	53.66	53.45	53.81	0.434	0.863	0.709	0.920
ADF	52.02 ^{b}	52.27 ^{b}	52.55 ^{ab}	53.08 ^{a}	0.361	<0.05	0.877	0.062
Total gas production (L/d)								
Day6	7.82 ^{a}	7.74 ^{a}	7.61 ^{b}	7.52 ^{b}	0.066	<0.01	<0.01	<0.05
Day7	7.86 ^{a}	7.79 ^{a}	7.64 ^{b}	7.55 ^{b}	0.068	<0.01	<0.01	<0.05
Day8	7.76 ^{a}	7.69 ^{ab}	7.55 ^{bc}	7.47 ^{c}	0.071	<0.01	<0.01	<0.05

In the table, the superscript "a–c" represents the differences in data between the treatment groups. Different letters indicate differences between the groups ($p < 0.05$), while the same or no letters indicate no significant differences ($p > 0.05$); SEM, standard error of the mean. DM, dry matter; OM, organic matter; CP, crude protein; EE, ether extract; NDF, neutral detergent fibre; ADF, acid detergent fibre.

3.3. The rumen Fermentation Parameters

As shown in Table 4, the pH, NH_3-N, MCP, the content of propionate, isobutyrate, butyrate, valerate, and the acetate/propionate (A/P) were quadratically increased ($p < 0.05$) with an increasing COS level. In addition, the A/P were linearly decreased ($p < 0.05$) with an increasing COS level. The pH, MCP and the content of propionate and butyrate were highest, while the NH_3-N and the content of acetate, isobutyrate, valerate, and the A/P were lowest ($p < 0.05$) in the 0.08% COS group. There were no differences ($p > 0.05$) in the total VFA and the content of isovalerate.

Table 4. Effects of different levels of COS on rumen fermentation parameters of beef cattle.

Items	Dietary COS Level, Feeding Basis				SEM	*p*-Value		
	CON	0.02% COS	0.04% COS	0.08% COS		Main Effect	Linear	Quadratic
pH	6.42 ^{c}	6.43 ^{c}	6.46 ^{b}	6.49 ^{a}	0.007	<0.01	0.858	<0.01
NH_3-N (mg/dL)	7.98 ^{a}	7.93 ^{a}	7.69 ^{b}	7.35 ^{c}	0.090	<0.01	0.739	<0.01
MCP (mg/mL)	0.59 ^{b}	0.59 ^{b}	0.61 ^{ab}	0.62 ^{a}	0.010	0.104	0.733	<0.01
Total VFA (mmol/mL)	112.53	113.09	113.83	113.43	1.249	0.104	0.591	0.557
Acetate (mmol/mL)	68.95 ^{a}	68.15 ^{ab}	67.37 ^{ab}	66.70 ^{b}	0.897	<0.01	0.584	0.07
Propionate (mmol/mL)	25.83 ^{b}	27.13 ^{a}	27.66 ^{a}	27.66 ^{a}	0.524	<0.01	0.059	<0.01
Isobutyrate (mmol/mL)	0.72 ^{a}	0.69 ^{b}	0.66 ^{c}	0.65 ^{c}	0.008	<0.01	0.272	<0.01
Butyrate (mmol/mL)	13.74 ^{b}	13.76 ^{b}	14.91 ^{a}	15.23 ^{a}	0.287	0.635	0.880	<0.01
Isovalerate (mmol/mL)	1.48	1.51	1.47	1.48	0.036	<0.01	0.410	0.467
Valerate (mmol/mL)	1.82 ^{a}	1.85 ^{a}	1.76 ^{b}	1.71 ^{b}	0.027	0.76	0.334	<0.01
Acetate/propionate	2.67 ^{a}	2.51 ^{b}	2.43 ^{b}	2.42 ^{b}	0.044	<0.01	0.042	<0.01

In the table, the superscript "a–c" represents the differences in data between the treatment groups. Different letters indicate differences between the groups ($p < 0.05$), while the same or no letters indicate no significant differences ($p > 0.05$); SEM, standard error of the mean. pH, potential of hydrogen; MCP, microbial protein.

3.4. The Rumen Microbial Community

Following the assessment of nutrient disappearance rates and rumen fermentation indicators, the addition level of 0.08% COS emerged as the most effective among the three treatment groups. Consequently, samples from the 0.08% COS and CON groups underwent analysis to examine the impact of COS on the rumen microbial community of beef cattle. There were no significant differences between the two groups in the α-diversity indices of Chao1, observed_species, PD_whole_tree, and Shannon ($p > 0.05$; Figure 2a). Additionally, the principal coordinate analysis (PCoA) based on the Bray–Curtis distance metric (Figure 2b) revealed that the addition of COS did not induce significant alterations in the composition of rumen bacteria ($p > 0.05$). At the phylum level, the top 20 rumen bacterial communities were evaluated in Figure 2c; predominant communities in both treatment groups included *Bacteroidota*, *Proteobacteria*, and *Firmicutes*. Two differential communities were identified, the relative abundance of the *Latescibacterota* in the COS group was significantly decreased compared to that in the CON group, whereas the relative abundance of the *WPS-2* was significantly increased in the COS group ($p < 0.05$; Figure 2e). Moreover, at the genus level, the top 20 rumen bacterial communities were evaluated in Figure 2d; predominant communities in both treatment groups included *Prevotella*, *Succinivibrionaceae_UCG-002*, and *Rikenellaceae-RC9_gut group*. Seventeen differential communities were discerned. Compared to the CON group, the relative abundance of *Desulfuromonadaceae*, *Endomicrobium*, *Eubacterium-ruminantium_group*, *Eubacterium-Venturiosum_group*, *Halomonas*, *Metanosphaera*, *Probable_genus_10*, *Rheinheimera*, *Ruminococcus_govreauii_group*, *Subgroup_10*, and *WPS-2* exhibited a significant increase ($p < 0.05$), and the relative abundance of *Acetobacter*, *Anaerotruncus*, *Dorea*, *Erysipelotrichaceae_UCG-003*, *Escherichia-Shigella*, and *Subdoligranulum* was decreased ($p < 0.05$) in the COS group (Figure 2f).

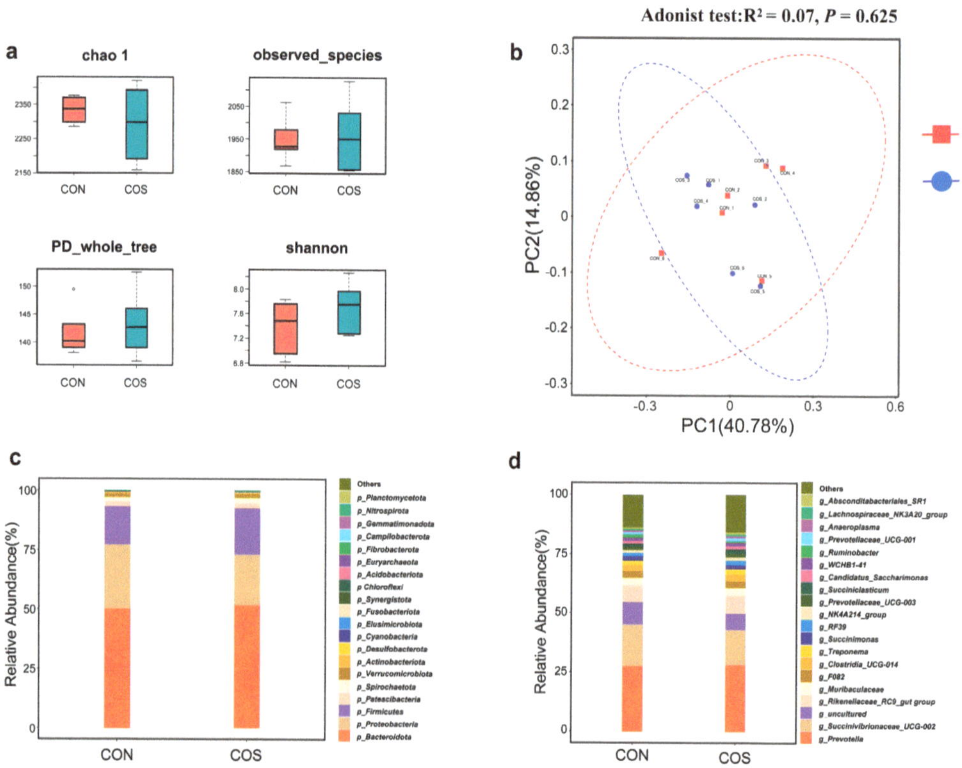

Figure 2. *Cont.*

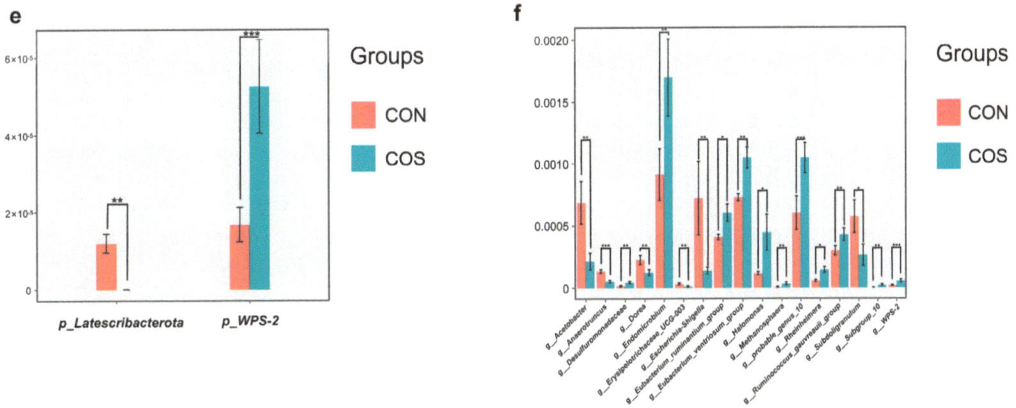

Figure 2. Effects of adding 0.08%COS to the diet on the rumen microbial community of beef cattle (n = 6). (**a**) α-Diversity index, including Chao 1, observed_species, PD_whole_tree, and Shannon. (**b**) β-Diversity index, PCoA analysis based on Bray-Curtis distance. The composition of rumen bacterial communities (top 20) at the phylum level (**c**) and genus level (**d**). The differential microbial communities between the two treatment groups at the phylum level (**e**) and genus level (**f**). And the "*" in (**e**,**f**), respectively, represent: * ($0.05 < p < 0.1$), ** ($0.01 < p < 0.05$), *** ($p < 0.01$).

4. Discussion

Multiple studies have confirmed through in vitro and in vivo experiments that chitosan can be a safe and efficient feed additive [33–36]. The COS in this study is an oligomeric derivative of chitosan, obtained by deacetylation of chitin. It has the same safety profile as chitosan and a stronger biological activity and action than chitosan [37]. However, COS is rarely employed in beef cattle, and there has been no attempt to ascertain whether it can retain its biological activity in the complex digestive system of beef cattle. Therefore, the objective of this study was to determine whether COS can influence the rate of nutrient disappearance and the rumen fermentation process through an in vitro fermentation test.

4.1. Effects of Adding Different Levels of COS to the Diet on Rumen Protozoa Count of Beef Cattle In Vitro

The rumen is an important digestive organ in ruminants, and its stable internal environment, such as pH, microbial community, and VFA content, is a key factor in measuring the growth and health of ruminants [38]. Rumen protozoa are the largest microbial community in the rumen environment, and their quantity and morphology can be observed under a microscope through certain methods. It is essential that the RUSITEC system, which simulates the rumen digestion process, maintains high stability during operation in order to ensure the accuracy of experimental data. Therefore, it is possible to visually evaluate the operation of the system by collecting daily rumen fermentation broth and observing the number of protozoa in it. In this experiment, the number of protozoa in the fermentation broth exhibited a rapid decline between days 1 and 5, followed by a gradual stabilisation from day 6 to day 8 (Figure 1), which is consistent with the description provided by Li [39]. This suggests that samples can be collected when the number of protozoa is stable. As for the reason for the rapid decrease in the number of protozoa between day 1 and day 5, Shen [40] pointed out in his research that due to the need to add substrates twice a day in the continuous in vitro method, although N_2 is introduced, it still cannot avoid contact with the air. The protozoa need an adaptive process, which leads to a rapid decrease in the number. And ultimately, this is an in vitro method, with an experimental period of up to eight days. Only the addition of rumen fluid on the first day, diluted with artificial saliva, can lead to a decrease in the number of protozoa. After the fifth day, the number of protozoa in this device has stabilised, indicating that it has adapted to the new environment. In this study, it was observed that the number of rumen protozoa in the 0.08% COS group was lower than that in the CON group during the sampling period. In previous studies on chitosan, similar results were obtained, namely that chitosan reduced the number of protozoa [33,41]. Previous studies have shown that changes in the number of rumen protozoa are related to eating and drinking behaviour [42,43], protein supply in the diet [44], and abundance of rumen bacterial communities [45]. Interestingly, although studies have shown a close correlation between the number of rumen protozoa and the proportion of nutrients in the rumen, the digestibility of nutrients, and the rumen microbiota [46,47], in this study, the adaptive period COS did not have a significant impact on rumen protozoa. However, during the sampling period, it was found that the number of protozoa with 0.08% COS was significantly lower or had a tendency to be lower than the CON group. It is possible that the addition of COS resulted in the consumption of nutrients in the fermentation substrate being accelerated in the early stages. When the number of rumen protozoa stabilises from the sixth day, the consumption of nutrients required for their growth in the pre-fermentation stage leads to a decrease in the number of protozoa.

4.2. Effects of Adding Different Levels of COS to the Diet on Nutrient Disappearance Rates and Total Gas Production of Rumen Nutrients in Beef Cattle In Vitro

In this study, it was observed that the disappearance rates of DM, OM, CP, EE, and ADF in the solid-state components of the fermentation broth increased to varying degrees after the addition of different proportions of COS, the 0.08% COS exhibited the best effects. In previous studies, Liu [48] observed that the addition of COS did not affect the apparent digestibility of the nutrients. Different results were also obtained in studies related to chitosan. Araújo [49] and Mingoti [50] observed that chitosan had no significant effect on the apparent digestibility of nutrients in beef cattle and cows. In contrast, Wenchelova [51] demonstrated that chitosan could result in a reduction in the apparent digestibility of nutrients in the rumen of sheep. Goiri [52] and Zhang [53] found that chitosan was able to improve rumen digestibility of nutrients in sheep. The different results may be due to the different diets in the tests, the different test animals, and the differences between in vivo and in vitro methods. Furthermore, the microbiological outcomes of this study indicate that COS had the potential to elevate the relative abundance of *Metanosphaera* and *Ruminococcus*. The *Metanosphaera* can promote the utilisation of fermentation substrates in the rumen [54], while the *Ruminococcus* can accelerate cellulose degradation, improve rumen digestion ability, and regulate rumen function [55,56]. In summary, the increased disappearance rate of rumen nutrients such as DM, CP, EE, and ADF can be explained by the increase in relative abundance of *Metanosphaera* and *Ruminococcus* caused by the addition of COS, which promotes the utilisation of nutrients such as DM, CP, EE, and ADF in the rumen. DM is closely related to OM, leading to an increase in the disappearance rate of OM. The increase in nutrient disappearance rate can promote the digestion of feed, improve feed utilisation efficiency, and promote animal growth.

The total gas production can be used to evaluate the efficacy of rumen fermentation and it is also one of the important indicators of equipment stability. Previous studies have indicated that total gas production is related to the degree of degradation of fermentation substrates [57]. In this study, the total gas production decreased with the increase in the addition of COS in the diet. In previous studies, chitosan led to a decrease [58,59] or no change [60] in CH_4 production, and there seems to be no consistent conclusion. Liu [61] proposed that the change in acetate concentration is positively correlated with the abundance of acetate-type *Methanogens*. However, in this study, the concentration of acetate significantly decreased, and the total gas production also decreased. However, no significant change in the abundance of acetate-type methanogenic bacteria was observed in the results of microbiology. On the contrary, we observed an increase in the abundance of the *Metanosphaera* microbiota, which can utilise H_2 to produce CH_4 [52]. Due to the fact that only 2 mol H_2 can generate 1 mol CH_4, it may lead to a decrease in total gas production but an increase in methane production. In addition, an increase in the abundance of *WPS-2* and a decrease in the abundance of *Anaerotruncatus* were observed. *WPS-2* feeds on the intake of H_2, CO, and CO_2 [62–64], and the abundance of *Anaerotruncus* is positively correlated with natural gas production [65]. The comprehensive results of microbiology can explain why the total gas production in this study decreased with the increase in the addition of COS in the diet. Reducing gas production can to some extent alleviate greenhouse gas emissions. However, it should be noted that methane production may increase, and we need to increase the detection of gas components in subsequent experiments.

4.3. *Effects of Adding Different Levels of COS to the Diet on Rumen Fermentation Parameters of Beef Cattle In Vitro*

In this study, adding different levels of COS to the diet had different effects on the concentration and proportion of different VFA in the fermentation broth. Research has shown that the addition of chitosan can change the mode of rumen fermentation in ruminants, shifting the rumen from acetate fermentation to propionate fermentation, reducing the level of acetate in the rumen, increasing the level of propionate and reducing the A/P [50]. In addition, studies have shown that when the mode of rumen fermentation is changed, the total VFA concentration does not change [66,67], which is consistent with the results of our study. Meanwhile, in Liu's previous study, it was mentioned that consistent with the findings in the present study, COS led to an increase in the content of butyrate, but there was no significant difference in the content and proportion of isobutyrate and valerate in their study. The microbiological results confirmed our findings. Among the communities with observed differences, endophytic microorganisms were able to ferment sugars into acetate and butyrate [68], leading to an increase in their abundance and an increase in acetate content. In previous studies, it was found that the main products of the *Subdoligranulum* community were small amounts of acetate and succinic acid [69], and a decrease in their relative abundance would lead to a decrease in acetate production. The relative abundance of *Anaerotruncus* is positively correlated with the concentrations of propionate, butyrate, and total volatile fatty acids [62]. Park [70] found that the microbial community of *Halomonas* mainly produces amylase and can promote the consumption of acetate. As its abundance increases, the consumption of acetate also increases. *Eubacteria* are associated with the production of VFA in the gastrointestinal tract, especially propionate and butyrate, which can convert monosaccharides into butyrate [71,72]. In our study, the increase in the relative abundance of *Endomicrobium*, *Eubacteria ruminantium_group*, and *Eubacteria Venturiosum_group* in the COS group accurately explained the increase in butyrate content, while the decrease in acetate content may be due to the decrease in the relative abundance of *Acetobacter* and *Subdoligranum*, as well as the increase in the abundance of *Halomonas*, which has a greater impact on acetate production than *Endomicrobium*. Under the joint influence of these microorganisms, the content of VFA components has changed. Meanwhile, the increased utilisation of CP, EE, and ADF by the rumen leads to the production of more VFA, and the increase in EE disappearance rate also produces more propionate [73–75], which explains the increase in propionate levels in this study. Overall, the main components of VFA are acetate, propionate, and butyrate, which account for more than 95% of the total VFA. Compared to these, changes in the concentration of other acids have little effect on the total VFA, with a decrease in acetate concentration and an increase in propionate. Under the combined influence of these factors, the reason why the total VFA concentration has not changed can be explained. Regarding NH_3-N and MCP, the addition of COS to the diet significantly reduced the rumen NH_3-N concentration ($p < 0.01$), and the 0.04% COS and 0.08% COS groups significantly increased the MCP content. The changes in NH_3-N concentration are consistent with the results of the Goiri [50] and Zanferari [67] studies, but many studies on changes in MCP have produced inconsistent results. For example, the studies of Seankamthorn [76] and Rey [41] mentioned that the addition of chitosan has no significant effect on MCP. Whereas Gandra's research also found that chitosan as an additive led to a decrease in MCP levels [77]. However, he also suggested that MCP synthesis may be positively correlated with rumen pH [77]. In this study, rumen pH increased significantly under the influence of COS ($p < 0.05$), which may have promoted the synthesis of MCP. In addition, the increase in rumen pH was also supported by the research of Kirwan [78], who pointed out that chitosan or COS contain NH_2 groups, which can provide additional NH_4^+ after degradation in the rumen, thereby increasing rumen pH. COS leads to a decrease in the more acidic acetate and an increase in the less acidic propionate in VFA, which may also be the reason for the increase in pH. In summary, the transition from acetate fermentation to propionate fermentation in the rumen can increase the rumen pH, as confirmed by our results. Therefore, the addition of COS can maintain the rumen pH within a relatively safe range, reduce the risk of rumen acidosis, and lower concentrations of NH_3-N can also promote the synthesis of rumen microorganisms, maintaining a relatively stable rumen environment.

4.4. Effects of Adding COS to the Diet on the Rumen Microbial Community of Beef Cattle In Vitro

In this study, at the phylum level, we observed *Bacteroidota, Proteobacteria,* and *Firmicutes* as communities with relatively high abundance. At the genus level, *Prevotella, Succinivibrionaceae UCG-002,* and *Rikenellaceae RC9_gut group* were communities with relatively high abundance, which is consistent with the dominant communities in rumen fluid in our previous study [30]. This indicates that our bovine rumen fluid fistula and RUSITEC system have good stability, and the data obtained can also be reproduced well in real animal experiments. At the phylum level, we also observed two distinct bacterial communities, *Latescriberota* and *WPS-2*, between the two treatment groups. *Latescoribacterota* has rarely been mentioned in previous studies, and its effects on animals have not yet been elucidated. However, Arcadi detected this community in the seawater and sedimentary rocks of Wurkano Island and pointed out that this community may be related to seawater acidification and has little correlation with the changes in various indicators in this study [79]. Several studies have shown that changes in diet structure or the addition of other substances to the diet can affect the relative abundance of *WPS-2* communities in animal bodies, which is beneficial to the animals. It has been suggested that the survival mode of *WPS-2* may be mainly through phagocytosis of H_2, CO, and CO_2, but the specific physiological function and effects of *WPS-2* on animals are still unclear [62–64]. At the genus level, we found a total of 17 different communities. Previous studies have mentioned that chitosan and its derivatives, due to the presence of R-NH_3^+ ions on the surface, can interact with negative ions on the microbial surface, causing peptidoglycan hydrolysis in the cell wall, leading to cell wall lysis, and have a stronger effect on Gram-positive bacteria than Gram-negative bacteria [80,81]. However, no significant difference was observed between Gram-negative and Gram-positive bacteria in this study. Studies have shown that the addition of chitosan can replace starch-degrading bacteria (*Bacteroidetes*) with fibre-degrading bacteria (*Firmicutes* and *Fibrobacteria*) in the rumen, resulting in a shift in rumen fermentation from the acetate mode to the propionate mode and a decrease in the relative abundance of methanogenic bacteria (*Proteobacteria*) [82]. We also observed similar results to Tong et al. such as an increase in the relative abundance of *Eubacterium_ruminantium_group, Eubacterium_ventriosum_group, Ruminococcus_gauvreauii_group* (*Firmicutes*) and the increase in relative abundance of *Acetobacter, Escherichia-Shigella* (*Proteobacteria*) and *Anaerotruncus* (*Bacteroidota*). The changes in rumen microbiota are closely related to the conditions of the rumen fluid donor animals, the diet and substrate composition, the equipment troubleshooting, and the in vitro method used in this experiment, and the rumen microbiota is also in a dynamic process in the rumen. We also found that the addition of COS led to an increase in the relative abundance of *Endomycobium,* which can reduce free energy wastage and optimise energy utilisation in the diet by generating ATP through substrate phosphorylation [77]. Increasing the relative abundance of *Ruminococcus* can improve rumen digestibility and regulate rumen function [53,54]. It also reduces the relative abundance of the conditionally pathogenic bacteria *Dorea* and *Escherichia Shigella,* which is important for maintaining animal health [83,84]. In summary, in this study, due to the chemical properties of COS, its addition led to changes in some microbial communities, resulting in changes in the disappearance rate of some nutrients and rumen fermentation indicators. At the same time, it regulated rumen function, reduced the abundance of harmful bacteria, and ensured rumen health.

5. Conclusions

In conclusion, different addition ratios of COS increased the disappearance rate of DM, CP, EE, and ADF in beef cattle to varying degrees. In terms of rumen fermentation, COS has been observed to increase the pH, MCP, and the content of propionate and butyrate. Additionally, the gas production, NH_3-N, A/P, the content of acetate, isobutyrate, valerate, and protozoa count were significantly decreased, which gradually transitions rumen fermentation from an acetate mode to a propionate mode. The COS did not alter the diversity of rumen microbiota. However, it can cause changes in the relative abundance of microbial communities, including *Metanosphaera*, *Ruminococcus*, *Endomycobium*, and *Eubacterium*. Although gas production has decreased, there is a risk of an increase in CH_4 content. Moreover, after combining the feed intake of beef cattle and the amount added in this study, the cost of COS used in the experiment was calculated to be about RMB 1.5/cattle/day. However, in the actual production process, with the continuous optimisation of extraction technology and large-scale purchases, the cost will only decrease. COS has good functions in improving nutrient digestion and maintaining rumen health. Therefore, we can consider COS as an additive that can be accepted by farmers. As the specific gas composition was not determined in this study, further work and animal experiments are needed in the future to objectively and comprehensively evaluate whether COS can be used as a safe and efficient additive.

Author Contributions: Conceptualisation, J.H., Q.G. and D.X.; methodology, J.H., W.S., Q.G. and D.X.; software, W.S.; validation, J.H., J.L., Q.G., W.L., M.X. and H.X.; formal analysis, J.H., J.L., W.L., M.X. and H.X.; investigation, J.H., J.L., Q.G., W.L., M.X. and H.X.; resources, J.H., J.L., Q.G., W.L., M.X. and H.X.; data curation, J.H., J.L., Q.G., W.S., W.L., M.X., H.X. and D.X.; writing—original draft, J.H., Q.G. and D.X.; writing—review and editing, J.H. and D.X.; visualisation, J.H., J.L., Q.G. and D.X.; supervision, J.H., W.S. and D.X.; project administration, J.H., W.S. and D.X.; funding acquisition, D.X. All authors have read and agreed to the published version of the manuscript.

Funding: Supported by earmarked Fund for China Agriculture Research System (CARS-37), National Natural Science Foundation of China (31872991).

Institutional Review Board Statement: The animal experiments conducted in this study were approved by the Hunan Agricultural University Biomedical Ethics Research Committee (HUNAU-2023-42, 25 March 2023).

Informed Consent Statement: Not applicable. Because the experimental animals were purchased by the Animal Training Center of Hunan Agricultural University, no further informed consent is required.

Data Availability Statement: The data that support the findings of this study are available on request from the authors.

Acknowledgments: We would like to express our gratitude to Hu Liu and Hui Mi for linguistic assistance during the preparation of this manuscript, and thanks to Li Xiang for his invaluable contribution to the RUSITEC system, which was employed in this study.

Conflicts of Interest: The authors explicitly state that there are no conflicts of interest.

References

1. Wei, M.; Zhou, H.; Ma, J.; Khan, N.; Cao, J.M.; Hu, X.D. Research on Technical Efficiency of Feed Use for Sustainable Beef Cattle Breeding in China: Evidence from 169 Beef Cattle Farmers. *Sustainability* **2022**, *14*, 16430. [CrossRef]
2. Shaw, C.A.; Park, Y.; Gonzalez, M.; Duong, R.A.; Pandey, P.K.; Brooke, C.G.; Hess, M. A Comparison of Three Artificial Rumen Systems for Rumen Microbiome Modeling. *Fermentation* **2023**, *9*, 953. [CrossRef]
3. Huang, Q.; Chen, Y.; Wang, X.; Wei, Y.; Pan, M.; Zhao, G. Effects of Phlorotannins from *Sargassum* on In Vitro Rumen Fermentation, Microbiota and Fatty Acid Profile. *Animals* **2023**, *13*, 2854. [CrossRef]

4. Zhao, W.; Abdelsattar, M.M.; Wang, X.; Zhang, N.; Chai, J. In Vitro Modulation of Rumen Fermentation by Microbiota from the Recombination of Rumen Fluid and Solid Phases. *Microbiol. Spectr.* **2023**, *11*, e0338722. [CrossRef] [PubMed]
5. Zhou, Y.N.; Liu, S.J.; Yang, D.Y.; Zhang, X.W.; Feng, Z.Y.; Cui, Z.H. Effects of alfalfa hay, oat hay, and their mixed feeding on the morphology and digestive enzyme activity of rumen tissue in lactating yak calves. *Chin. J. Anim. Nutr.* **2023**, *35*, 3141–3153. (In Chinese)
6. Soltis, M.P.; Moorey, S.E.; Egert-McLean, A.M.; Voy, B.H.; Shepherd, E.A.; Myer, P.R. Rumen Biogeographical Regions and Microbiome Variation. *Microorganisms* **2023**, *11*, 747. [CrossRef]
7. Li, Y.; Zang, Y.; Zhao, X.; Liu, L.; Qiu, Q.; Ouyang, K.; Qu, M. Dietary Supplementation with Creatine Pyruvate Alters Rumen Microbiota Protein Function in Heat-Stressed Beef Cattle. *Front. Microbiol.* **2021**, *12*, 715088. [CrossRef]
8. Gao, Z.; Liu, B.; La, S.; Li, D.; Zhu, X.; Sun, H.; Ma, S.; Cui, Y.; Shi, Y. Alfalfa Hay Substitution for Wheat Straw Improves Beef Quality Via Rumen Microflora Alteration. *Heliyon* **2023**, *9*, e20803. [CrossRef] [PubMed]
9. Liu, X.; Yang, Z.; Yang, J.; Wang, D.; Niu, J.; Bai, B.; Sun, W.; Ma, S.; Cheng, Y.; Hao, L. A Comparative Study of Growth Performance, Blood Biochemistry, Rumen Fermentation, and Ruminal and Fecal Bacterial Structure between Yaks and Cattle Raised under High Concentrate Feeding Conditions. *Microorganisms* **2023**, *11*, 2399. [CrossRef]
10. Zhang, C.; Yu, Q.; Wang, J.; Yu, Y.; Zhang, Y.; Sun, Y. Effects of Dietary Supplementation with Clostridium butyricum on Growth Performance, Apparent Digestibility, Blood Metabolites, Ruminal Fermentation and Bacterial Communities of Fattening Goats. *Front. Nutr.* **2022**, *9*, 888191. [CrossRef]
11. Muxika, A.; Etxabide, A.; Uranga, J.; Guerrero, P.; Caba, K. Chitosan as a Bioactive Polymer: Processing, Properties and Applications. *Int. J. Biol. Macromol.* **2017**, *105*, 1358–1368. [CrossRef] [PubMed]
12. Lodhi, G.; Kim, Y.S.; Hwang, J.W.; Kim, S.K.; Jeon, Y.J.; Je, J.Y.; Ahn, C.B.; Moon, S.H.; Jeon, B.T.; Park, P.J. Chitooligosaccharide and Its Derivatives: Preparation and Biological Applications. Biomed. *Res. Int.* **2014**, *2014*, 654913. [CrossRef] [PubMed]
13. Rostami, H.; Seidavi, A.; Dadashbeiki, M.; Asadpour, Y.; Simões, J.; Shah, A.A.; Laudadio, V.; Losacco, C.; Perillo, A.; Tufarelli, V. Supplementing Dietary Rosemary (*Rosmarinus officinalis* L.) Powder and Vitamin E in Broiler Chickens: Evaluation of Humoral Immune Response, Lymphoid Organs, and Blood Proteins. *Environ. Sci. Pollut. Res. Int.* **2018**, *25*, 1–7. [CrossRef]
14. Naveed, M.; Phil, L.; Sohail, M.; Hasnat, M.; Baig, M.; Ihsan, A.U.; Shumzaid, M.; Kakar, M.U.; Mehmood Khan, T.; Akabar, M.D.; et al. Chitosan Oligosaccharide (Cos): An Overview. *Int. J. Biol. Macromol.* **2019**, *129*, 827–843. [CrossRef] [PubMed]
15. Jung, S.H.; Lee, Y.H.; Jeong, D.Y.; Jeong, Y.R.; Jung, B.O.; Park, J.K. Matrix-Assisted Laser Desorption/Ionization-Time of Flight Mass Spectrometry Analysis for Molecular Characterization of Chitosan. *Korean Chitin-Chitosan Soc.* **2021**, *26*, 76–82. [CrossRef]
16. Chen, Z.W.; Wang, R.X.; Liu, Q.; Liu, F.; Zhuang, S. The Effects of Oligochitosan as a Substitute for Antibiotics on The Growth Performance, Slaughter Performance, Intestinal Barrier Function, and Muscle Quality of Cherry Valley Meat Ducks. Chin. *J. Anim. Vet. Sci.* **2021**, *52*, 1927–1941. (In Chinese)
17. Huang, B.; Xiao, D.; Tan, B.; Xiao, H.; Wang, J.; Yin, J.; Duan, J.; Huang, R.; Yang, C.; Yin, Y. Chitosan Oligosaccharide Reduces Intestinal Inflammation That Involves Calcium-Sensing Receptor (CaSR) Activation in Lipopolysaccharide (LPS)-Challenged Piglets. *J. Agric. Food Chem.* **2016**, *64*, 245–252. [CrossRef] [PubMed]
18. Lan, R.; Li, S.; Chang, Q.; Zhao, Z. Chitosan Oligosaccharides Protect Sprague Dawley Rats from Cyclic Heat Stress by Attenuation of Oxidative and Inflammation Stress. *Animals* **2019**, *9*, 1074. [CrossRef] [PubMed]
19. Deitmers, J.H.; Gresner, N.; Suedekum, K.H. Opportunities and Limitations of a Standardisation of The Rumen Simulation Technique (Rusitec) for Analyses of Ruminal Nutrient Degradation and Fermentation and on Microbial Community Characteristics. *Anim. Feed. Sci. Technol.* **2022**, *289*, 115325. [CrossRef]
20. Guo, Y.; Xiao, L.; Jin, L.; Yan, S.; Niu, D.; Yang, W. Effect of Commercial Slow-Release Urea Product on In Vitro Rumen Fermentation and Ruminal Microbial Community Using Rusitec Technique. *J. Anim. Sci. Biotechnol.* **2022**, *13*, 56. [CrossRef]
21. NY/T815-2004. Agricultural Industry Standard of the People's Republic of China—Beef Cattle Breeding Standard.
22. Adebayo, A.M.; Zhang, X.M.; Wang, M.; Wang, R.; Wen, J.N.; Hao, L.Z.; He, J.H.; Shen, W.J.; Ma, Z.Y.; Tan, Z.L. Proper Motility Enhances Rumen Fermentation and Microbial Protein Synthesis with Decreased Saturation of Dissolved Gases in Rumen Simulation Technique. *J. Dairy. Sci.* **2022**, *105*, 231–241. [CrossRef]
23. McDougall, E.I. Studies on Ruminant Saliva. 1. The Composition and Output of Sheep's Saliva. *Biochem. J.* **1948**, *43*, 99–109. [CrossRef]
24. AOAC. *Official Methods of Analysis*, 18th ed.; AOAC Int.: Gaithersburgs, MD, USA, 2006.
25. Van Soest, P.J.; Robertson, J.B.; Lewis, B.A. Methods for Dietary Fiber, Neutral Detergent Fiber, and Nonstarch Polysaccharides in Relation to Animal Nutrition. *J. Dairy Sci.* **1991**, *74*, 3583–3597. [CrossRef]
26. Kišidayová, S.; Durkaj, D.; Mihaliková, K.; Váradyová, Z.; Puchalska, J.; Szumacher, S.M.; Cieślak, A.; Gizejewski, Z. Rumen Ciliated Protozoa of the Free-Living European Bison (*Bison bonasus*, Linnaeus). *Front. Microbiol.* **2021**, *12*, 658448. [CrossRef]
27. Chen, J.; Lei, X.J.; Wang, L.; Zhang, Y.L.; Wang, D.D.; Zhao, L.C.; Liu, T.; Yang, Y.T.; Yao, J.H. Effects of Rumen-Protected Leucine on Production Performance and Starch Digestion in the Small Intestine of Lactating Goats. *Anim. Feed. Sci. Technol.* **2022**, *287*, 115270. [CrossRef]
28. Broderick, G.A.; Kang, J.H. Automated Simultaneous Determination of Ammonia and Total Amino Acids in Ruminal Fluid and In Vitro Media. *J. Dairy Sci.* **1980**, *63*, 64–75. [CrossRef]
29. Makkar, H.P.; Sharma, O.P.; Dawra, R.K.; Negi, S.S. Simple Determination of Microbial Protein in Rumen Liquor. *J. Dairy Sci.* **1982**, *65*, 2170–2173. [CrossRef]

30. Gao, Q.; He, J.; Wang, J.; Yan, Y.; Liu, L.; Wang, Z.; Shen, W.; Wan, F. Effects of Dietary D-Lactate Levels on Rumen Fermentation, Microflora and Metabolomics of Beef Cattle. *Front. Microbiol.* **2024**, *15*, 1348729. [CrossRef]
31. Magoč, T.; Salzberg, S.L. Flash: Fast Length Adjustment of Short Reads to Improve Genome Assemblies. *Bioinformatics* **2011**, *27*, 2957–2963. [CrossRef]
32. Chen, S.; Zhou, Y.; Chen, Y.; Gu, J. Fastp: An Ultra-Fast All-In-One Fastq Preprocessor. *Bioinformatics* **2018**, *34*, i884–i890. [CrossRef]
33. El-Zaiat, H.M.; Al-Marzooqi, W.; Al-Kharousi, K. Exploring Rumen Fermentation and Microbial Populations in Dhofari Goats Fed a Chitosan-Added Diet. *Anim. Biotechnol.* **2024**, *35*, 2337748. [CrossRef] [PubMed]
34. Santos, M.V.D.; Goes, R.H.T.B.; Takiya, C.S.; Cabral, L.D.S.; Mombach, M.A.; Oliveira, R.T.; Silva, N.G.D.; Anschau, D.G.; Freitas-Júnior, J.E.; Araújo, M.L.G.M.L.; et al. Effect of Increasing Doses of Chitosan to Grazing Beef Steers on The Relative Population and Transcript Abundance of Archaea and Cellulolytic and Amylolytic Bacterias. *Anim. Biotechnol.* **2023**, *34*, 246–252. [CrossRef] [PubMed]
35. Jiménez-Ocampo, R.; Valencia-Salazar, S.; Pinzón-Díaz, C.E.; Herrera-Torres, E.; Aguilar-Pérez, C.F.; Arango, J.; Ku-Vera, J.C. The Role of Chitosan as a Possible Agent for Enteric Methane Mitigation in Ruminants. *Animals* **2019**, *9*, 942. [CrossRef]
36. Vendramini, T.H.A.; Takiya, C.S.; Silva, T.H.; Zanferari, F.; Rentas, M.F.; Bertoni, J.C.; Consentini, C.E.C.; Gardinal, R.; Acedo, T.S.; Rennó, F.P. Effects of a Blend of Essential Oils, Chitosan or Monensin on Nutrientintake and Digestibility of Lactating Dairy Cows. *Anim. Feed. Sci. Technol.* **2016**, *214*, 12–21. [CrossRef]
37. Wang, Y.; Zhao, K.; Li, L.; Song, X.; He, Y.; Ding, N.; Liu, Z. A Review of The Immune Activity of Chitooligosaccharides. *Food Sci. Technol.* **2023**, *43*, e97822. [CrossRef]
38. Chen, L.; Qiu, Q.; Jiang, Y.; Wang, K.; Lin, Z.; Li, Z.; Bibi, F.; Yang, Y.; Wang, J.; Nie, W.; et al. Large-Scale Ruminant Genome Sequencing Provides Insights into Their Evolution and Distinct Traits. *Science* **2019**, *364*, eaav6202. [CrossRef] [PubMed]
39. Li, Z.; Qiu, H.; Lan, X.; Wang, Z.; Shen, W.; Wan, F.; Xiao, D.; He, J. Appropriate Particle Size of Rice Straw Promoted Rumen Fermentation and Regulated Bacterial Microbiota in a Rumen Simulation Technique System. *Front. Vet. Sci.* **2023**, *10*, 1185191. [CrossRef]
40. Shen, W.J. *Development and Operation Parameter Study of a New Dual Outflow Rumen Simulation System*; Hunan Agricultural University: Changsha, China, 2012. (In Chinese)
41. Rey, J.; Díaz, O.X.; Atxaerandio, R.; Mandaluniz, N.; García-Rodríguez, A.; González-Recio, O.; López-García, A.; Ruiz, R.; Goiri, I. Effect of Chitosan on Ruminal Fermentation and Microbial Communities, Methane Emissions, and Productive Performance of Dairy Cattle. *Animals* **2023**, *13*, 2861. [CrossRef] [PubMed]
42. Wang, X.D.; Luo, Q.J.; Zang, C.J.; Xie, J.X.; Cai, J. Effects of Different Levels of Paraformaldehyde Addition on The Number of Rumen Microorganisms and Digestive Enzyme Activity in Sheep. Chin. *J. Anim. Vet. Sci.* **2019**, *46*, 719–731. (In Chinese)
43. Wang, S.; Jia, Q.Z.; Chen, G.Y. The Effects of Flavonoids from Oxytropis Vulgaris on Rumen Protozoa and Cellulose Degrading Bacteria. *China Feed* **2015**, 11–13+16. (In Chinese)
44. Newbold, C.J.; Fuente, G.; Belanche, A.; Ramos-Morales, E.; McEwan, N.R. The Role of Ciliate Protozoa in the Rumen. *Front. Microbiol.* **2015**, *6*, 01313. [CrossRef]
45. Coleman, G.S.; Sandford, D.C. The Engulfment and Digestion of Mixed Rumen Bacteria and Individual Bacterial Species by Single and Mixed Species of Rumen Ciliate Protozoa Grown In Vivo. *J. Agric. Sci.* **1979**, *92*, 729–742. [CrossRef]
46. Solomon, R.; Wein, T.; Levy, B.; Eshed, S.; Dror, R.; Reiss, V.; Zehavi, T.; Furman, O.; Mizrahi, I.; Jami, E. Protozoa Populations Are Ecosystem Engineers That Shape Prokaryotic Community Structure and Function of The Rumen Microbial Ecosystem. *ISME J.* **2022**, *16*, 1187–1197. [CrossRef] [PubMed]
47. Williams, C.L.; Thomas, B.J.; McEwan, N.R.; Rees-Stevens, P.; Creevey, C.J.; Huws, S.A. Rumen Protozoa Play a Significant Role in Fungal Predation and Plant Carbohydrate Breakdown. Front. *Microbiol.* **2020**, *11*, 521741.
48. Liu, Z.J.; Liang, G.F.; Zhang, T.; Yao, J.H.; Cao, Y.C. The Effects of Malic Acid and Chitosan Oligosaccharides on In Vitro Rumen Fermentation Characteristics and Methane Emissions in Dairy Goats. *J. Anim. Ecol.* **2023**, *44*, 32–37. (In Chinese)
49. Araújo, A.P.C.D.; Venturelli, B.C.; Santos, M.C.B.; Gardinal, R.; Cônsolo, N.R.B.; Calomeni, G.D.; Rennó, F.P. Chitosan Affects Total Nutrient Digestion and Ruminal Fermentation in Nellore Steers. *Anim. Feed Sci. Technol.* **2015**, *8*, 114–118. [CrossRef]
50. Mingoti, R.D.; Freitas, J.E., Jr.; Gandra, J.R.; Gardinal, R.; Calomeni, G.D.; Barletta, R.V.; Rennó, F.P. Dose Response of Chitosan on Nutrient Digestibility, Blood Metabolites and Lactation Performance in Holstein Dairy Cows. *Livest. Sci.* **2016**, *5*, 35–39. [CrossRef]
51. Wencelova, M.; Varadyova, Z.; Mihalikova, K.; Kisidayova, S.; Jalc, D. Evaluating the Effects of Chitosan, Plant Oils, and Different Diets on Rumen Metabolism and Protozoan Population in Sheep. Turk. *J. Vet. Anim. Sci.* **2014**, *38*, 26–33.
52. Goiri, I.; Oregui, L.M.; Garcia-Rodriguez, A. Use of Chitosans to Modulate Ruminal Fermentation of a 50:50 Forage-to-Concentrate Diet in Sheep. *J. Anim. Sci.* **2010**, *88*, 749–755. [CrossRef]
53. Zhang, Q.; Han, Y.; Yang, Y.; Zhou, P.; Shen, X. Effects of The Seleno-Chitosan on Daily Gain, Wool Yield, and Blood Parameter in The Chinese Merino Sheep. *Biol. Trace Elem. Res.* **2022**, *200*, 4704–4711. [CrossRef]
54. Smith, P.E.; Kelly, A.K.; Kenny, D.A.; Waters, S.M. Differences in The Composition of The Rumen Microbiota of Finishing Beef Cattle Divergently Ranked for Residual Methane Emissions. *Front. Microbiol.* **2022**, *13*, 855565. [CrossRef] [PubMed]
55. Kims, H.; Sung, H.G. Effects of Different Fiber Substrates on In Vitro Rumen Fermentation Characteristics and Rumen Microbial Community in Korean Native Goats and Hanwoo Steers. *Fermentation* **2022**, *8*, 611. [CrossRef]

56. Carhuapoma-Delacruz, V.; Auqui-Acharte, G.S.; Valencia-Mamani, N.; Gonzales-Huamán, T.J.; Guillen-Domínguez, H.M.; Esparza, M. Fibrolytic Bacteria Isolated from The Rumen of Alpaca, Sheep and Cattle with Cellulose Biodegrading Capacity. *Rev. Cient.-Fac. De Cienc. Vet.* **2022**, *32*, 1–7.
57. Krieg, J.; Seifried, N.; Steingass, H.; Rodehutscord, M. In Situ and In Vitro Ruminal Starch Degradation of Grains from Different Rye, Triticale and Barley Genotypes. *Animal* **2017**, *11*, 1745–1753. [CrossRef] [PubMed]
58. Thao, N.T.; Phesatcha, K.; Matra, M.; Phesatcha, B.; Wanapat, M. Sources of Rumen Enhancers Including Nitrate, Chitosan Extract and Shrimp Shell Meal Could Modulate Nutrient Degradability and In Vitro Gas Fermentation. *J. Appl. Anim. Res.* **2022**, *50*, 394–399. [CrossRef]
59. Goiri, I.; Garcia, R.A.; Oregui, L.M. Effect of Chitosans on In Vitro Rumen Digestion and Fermentation of Maize Silage. *Anim. Feed Sci. Technol.* **2009**, *148*, 276–287. [CrossRef]
60. Belanche, A.; Pinloche, E.; Preskett, D.; Newbold, C.J. Effects and Mode of Action of Chitosan and Ivy Fruit Saponins on The Microbiome, Fermentation and Methanogenesis in The Rumen Simulation Technique. *FEMS Microbiol. Ecol.* **2016**, *92*, fiv160. [CrossRef] [PubMed]
61. Liu, J.S.; Marison, I.W.; Stockar, U. Microbial Growth by a Net Heat Up-Take: A Calorimetric and Thermodynamic Study on Acetotrophic Methanogenesis by Methanosarcina Barkeri. *Biotechnol. Bioeng.* **2001**, *75*, 170–180. [CrossRef] [PubMed]
62. Tian, X.Z.; Li, J.X.; Luo, Q.Y.; Zhou, D.; Long, Q.M.; Wang, X.; Lu, Q.; Wen, G.L. Effects of Purple Corn Anthocyanin on Blood Biochemical Indexes, Ruminal Fluid Fermentation, and Rumen Microbiota in Goats. *Front. Vet. Sci.* **2021**, *8*, 715710. [CrossRef]
63. Zhu, Y.; Tao, Z.; Chen, X.; Xiao, J.; Zhang, Y.; Wang, Z. Effects of Broussonetia Papyrifera Fermented Feed on Production Performance, Egg Quality, and Caecal Microbiota of Laying Hens During the Late Laying Period. *Ital. J. Anim. Sci.* **2022**, *21*, 659–672. [CrossRef]
64. Ye, Y.; Li, Z.; Wang, P.; Zhu, B.; Zhao, M.; Huang, D.; Ye, Y.; Ding, Z.; Li, L.; Wan, G.; et al. Effects of Probiotic Supplements on Growth Performance and Intestinal Microbiota of Partridge Shank Broiler Chicks. *PeerJ* **2021**, *9*, e12538. [CrossRef] [PubMed]
65. Kong, F.; Liu, Y.; Wang, S.; Zhang, Y.; Wang, W.; Yang, H.; Lu, N.; Li, S. Nutrient Digestibility, Microbial Fermentation, and Response in Bacterial Composition to Methionine Dipeptide: An In Vitro Study. *Biology* **2022**, *11*, 93. [CrossRef] [PubMed]
66. Dias, A.O.C.; Goes, R.; Gandra, J.R.; Gandra, J.R.; Takiya, C.S.; Takiya, C.S.; Jacauna, A.G.; Oliveira, R.T.; Souza, C.J.S.; Vaz, M.S.M. Increasing Doses of Chitosan to Grazing Beef Steers: Nutrient Intake and Digestibility, Ruminal Fermentation, and Nitrogen Utilization. *Anim. Feed Sci. Technol.* **2017**, *225*, 73–80. [CrossRef]
67. Zanferari, F.; Vendramini, T.H.A.; Rentas, M.F.; Gardinal, R.; Calomeni, G.D.; Mesquita, L.G.; Takiya, C.S.; Rennó, F.P. Effects of Chitosan and Whole Raw Soybeans on Ruminal Fermentation and Bacterial Populations, and Milk Fatty Acid Profile in Dairy Cows. *J. Dairy Sci.* **2018**, *101*, 10939–10952. [CrossRef] [PubMed]
68. Méheust, R.; Castelle, C.J.; Matheus-Carnevali, P.B.; Farag, I.F.; He, C.; Chen, L.X.; Amano, Y.; Hug, L.A.; Banfield, J.F. Groundwater Elusimicrobia are Metabolically Diverse Compared to Gut Microbiome Elusimicrobia and Some Have a Novel Nitrogenase Paralog. *ISME J.* **2020**, *14*, 2907–2922. [CrossRef] [PubMed]
69. Cui, S.; Guo, S.; Zhao, Q.; Li, Y.; Ma, Y.; Yu, Y. Alterations of Microbiota and Metabolites in The Feces of Calves with Diarrhea Associated with Rotavirus and Coronavirus Infections. *Front. Microbiol.* **2023**, *14*, 1159637. [CrossRef] [PubMed]
70. Park, J.K.; Jeon, J.M.; Yang, Y.H.; Kim, S.H.; Yoon, J.J. Efficient Polyhydroxybutyrate Production Using Acetate by Engineered Halomonas Sp.Jjy01 Harboring Acetyl-Coa Acetyltransferase. *Int. J. Biol. Macromol.* **2024**, *254 Pt 1*, 127475. [CrossRef]
71. Flaiz, M.; Poehlein, A.; Wilhelm, W.; Mook, A.; Daniel, R.; Dürre, P.; Bengelsdorf, F.R. Refining and Illuminating Acetogenic Eubacterium Strains for Reclassification and Metabolic Engineering. *Microb. Cell Factories* **2024**, *23*, 24. [CrossRef] [PubMed]
72. Karim, M.R.; Iqbal, S.; Mohammad, S.; Morshed, M.N.; Haque, M.A.; Mathiyalagan, R.; Yang, D.C.; Kim, Y.J.; Song, J.H.; Yang, D.U. Butyrate's (a Short-Chain Fatty Acid) Microbial Synthesis, Absorption, and Preventive Roles Against Colorectal and Lung Cancer. *Arch. Microbiol.* **2024**, *206*, 137. [CrossRef]
73. Vinyard, J.R.; Ravelo, A.; Sarmikasoglou, E.; Monteiro, H.F.; Arce-Cordero, J.A.; Johnson, M.L.; Agustinho, B.C.; Lobo, R.R.; Yungmann, M.G.; Winter, A.H.R.; et al. Effects of Exogenous Amylolytic or Fibrolytic Enzymes Inclusion on In Vitro Fermentation of Lactating Dairy Cow Diets in a Dual-Flow Continuous-Culture System. *J. Dairy Sci.* **2023**, *106*, 1002–1012. [CrossRef]
74. Xie, L.; Xing, J.; Qi, X.; Lu, T.; Jin, Y.; Akhtar, M.F.; Li, L.; Liu, G. Effects of Concentrate Feeding Sequence on Growth Performance, Nutrient Digestibility, VFA Production, and Fecal Microbiota of Weaned Donkeys. *Animals* **2023**, *13*, 2893. [CrossRef] [PubMed]
75. Quadros, D.G.; Whitney, T.R.; Weishuhn, C. Influence of Particle Size of Cottonseed Hulls and Bermudagrass Hay on Intake, Sorting Behavior, and Rumen Fluid Parameters of Feedlot Lambs. *Anim. Feed Sci. Technol.* **2022**, *285*, 115218. [CrossRef]
76. Seankamsorn, A.; Cherdthong, A.S.S.; Wanapat, M. Influence of Chitosan Sources on Intake, Digestibility, Rumen Fermentation, and Milk Production in Tropical Lactating Dairy Cows. *Trop. Anim. Health Prod.* **2021**, *53*, 241. [CrossRef]
77. Gandra, J.R.; Takiya, C.S.; Oliveira, E.R.D.; Paiva, P.G.D.; Goes, R.; Gandra, É.R.D.S.; Araki, H.M.C. Nutrient Digestion, Microbial Protein Synthesis, and Blood Metabolites of Jersey Heifers Fed Chitosan and Whole Raw Soybeans. *Rev. Bras. Zootec.* **2016**, *45*, 130–137. [CrossRef]
78. Kirwan, S.F.; Pierce, K.M.; Serra, E.; McDonald, M.; Rajauria, G.; Boland, T.M. Effect of Chitosan Inclusion and Dietary Crude Protein Level on Nutrient Intake and Digestibility, Ruminal Fermentation, and N Excretion in Beef Heifers Offered a Grass Silage Based Diet. *Animals* **2021**, *11*, 771. [CrossRef] [PubMed]
79. Arcadi, E.; Rizzo, C.; Calogero, R.; Sciutteri, V.; Fabiano, F.; Consoli, P.; Andaloro, F.; Romeo, T. Microbial Communities Inhabiting Shallow Hydrothermal Vents as Sentinels of Acidification Processes. *Front. Microbiol.* **2023**, *14*, 1233893. [CrossRef] [PubMed]

80. Kong, M.; Chen, X.G.; Xing, K.; Park, H.J. Antimicrobial Properties of Chitosan and Mode of Action: A State of The Art Review. *Int. J. Food Microbiol.* **2010**, *144*, 51–63. [CrossRef]
81. Shah, A.M.; Qazi, I.H.; Matra, M.; Wanapat, M. Role of Chitin and Chitosan in Ruminant Diets and Their Impact on Digestibility, Microbiota and Performance of Ruminants. *Fermentation* **2022**, *8*, 549. [CrossRef]
82. Tong, J.J.; Zhang, H.; Jia, W.; Yun, L.; Mao, S.Y.; Xiong, B.H.; Jiang, L.S. Effects of Different Molecular Weights of Chitosan on Methane Production and Bacterial Community Structure In Vitro. *J. Integr. Agric.* **2020**, *19*, 1644–1655. [CrossRef]
83. Guo, Y.; Li, Z.; Deng, M.; Li, Y.; Liu, G.; Liu, D.; Liu, Q.; Liu, Q.; Sun, B. Effects of a Multi-Strain Probiotic on Growth, Health, and Fecal Bacterial Flora of Neonatal Dairy Calves. *Anim. Biosci.* **2022**, *35*, 204–216. [CrossRef]
84. Li, Z.; Teng, Y.; Feng, S.; Hu, Z.; Zhao, J.; Ding, H.; Fang, Y.; Liu, H.; Ma, X.; Guo, J.; et al. Microbial Responses and Changes in Metabolic Products in Bovine Uteri Infected with Staphylococcus Aureus. *Int. J. Biol. Macromol.* **2024**, *262 Pt 2*, 130039. [CrossRef]

Disclaimer/Publisher's Note: The statements, opinions and data contained in all publications are solely those of the individual author(s) and contributor(s) and not of MDPI and/or the editor(s). MDPI and/or the editor(s) disclaim responsibility for any injury to people or property resulting from any ideas, methods, instructions or products referred to in the content.

animals

MDPI

Article

Differences in Chemical Composition, Polyphenol Compounds, Antioxidant Activity, and In Vitro Rumen Fermentation among Sorghum Stalks

Xingzhou Tian, Jixiao Qin, Qingyuan Luo, Yiqing Xu, Shuanglong Xie, Rui Chen, Xu Wang and Qi Lu *

Key Laboratory of Animal Genetics, Breeding and Reproduction in the Plateau Mountainous Region, Ministry of Education, College of Animal Science, Guizhou University, Guiyang 550025, China; tianxingzhou@yeah.net (X.T.); zrz2942037436@163.com (J.Q.); natty579@163.com (Q.L.); tacool1001@163.com (Y.X.); xsljty@163.com (S.X.); chenruisc@163.com (R.C.); wangxu10216610@163.com (X.W.)

* Correspondence: qlu@gzu.edu.cn

Simple Summary: Polyphenol compounds are important physiological activity materials that resist corrosion and oxidation, can be maintained in a state of freshness, and can be used safely, among other advantages. In China, the most important cereal crop is sorghum, which is an important raw material for brewing. However, sorghum stalks are usually considered to have no value and are thrown away or burned, resulting in environmental pollution and the possible waste of a usable resource. Notably, the sorghum stalk contains an abundance of polyphenols that exhibit great development potential as unconventional resources. However, studies on sorghum stalks with respect to their polyphenol compounds and their effects on in vitro rumen fermentation in goats have been relatively rare to date. Accordingly, in the present study, six types of sorghum (*Sorghum bicolor* L. Moench) varieties used for brewing traditional Chinese liquor were selected, and we compared their stalks in terms of their chemical composition, polyphenol compounds, antioxidant activity, and effects on in vitro rumen fermentation in goats. Our findings provide new details regarding the use of this unconventional feed for goats, as well as theories for developing this kind of resource for use as ruminant feed.

Abstract: The aim of the study was to examine the differences in the chemical composition, polyphenol compounds, antioxidant activity, and in vitro rumen fermentation among six varieties of sorghum stalks. The results show that maoliangnuo 1 (M1) contained a higher ($p < 0.05$) level of dry matter, and jinzhong 405 (J4) contained a higher ($p < 0.05$) level of crude protein content. The concentrations of neutral detergent fiber, acid detergent fiber, and cellulose were significantly higher ($p < 0.05$) in stalk jinliangnuo (JN). The levels of chlorogenic acid, homoorientin, isovitexin, vitexin, rhoifolin, genistin, quercetin, apigenin, aloe emodin, emodin, and total polyphenols were all significantly ($p < 0.05$) higher in maohongnuo 6 (M6) than in the other stalks. Moreover, stalk M6 contained higher ($p < 0.05$) levels of total antioxidant capacity (TAC), glutathione peroxidase (GPX), catalase (CAT), and 2,2-diphenyl-1-picrylhydrazyl (DPPH) free-radical scavenging capacity. There were significant ($p < 0.05$) positive correlations between total polyphenols and TAC, superoxide dismutase, GPX, CAT, and DPPH free-radical scavenging capacity. The total gas production was significantly ($p < 0.05$) influenced by the sorghum stalk variety and incubation time. Stalk J4 displayed higher values for the ($p < 0.05$) immediately soluble fraction and the potential extent of gas production, while stalk M6 exhibited a significantly lower ($p < 0.05$) insoluble fraction level. Furthermore, stalk M6 exhibited a significantly higher level of ($p < 0.05$) ruminal fluid propionic acid, but its level of butyric acid and its ratio of acetic acid to propionic acid were both significantly lower ($p < 0.05$). Taken together, the results reported in this paper indicate that the chemical composition, polyphenol compounds, antioxidant activity, and in vitro rumen fermentation all vary greatly among different varieties of sorghum stalks.

Keywords: sorghum stalk; polyphenol compound; antioxidant activity; in vitro gas production

Citation: Tian, X.; Qin, J.; Luo, Q.; Xu, Y.; Xie, S.; Chen, R.; Wang, X.; Lu, Q. Differences in Chemical Composition, Polyphenol Compounds, Antioxidant Activity, and In Vitro Rumen Fermentation among Sorghum Stalks. *Animals* **2024**, *14*, 415. https://doi.org/10.3390/ani14030415

Academic Editors: Alejandro Plascencia, Dingfu Xiao and Tiantian Meng

Received: 20 November 2023
Revised: 21 January 2024
Accepted: 26 January 2024
Published: 27 January 2024

Copyright: © 2024 by the authors. Licensee MDPI, Basel, Switzerland. This article is an open access article distributed under the terms and conditions of the Creative Commons Attribution (CC BY) license (https://creativecommons.org/licenses/by/4.0/).

1. Introduction

Polyphenols are important secondary metabolites in plants. They are present in plants of various colors, and exhibit strong antioxidant activities due to their unique molecular structure [1]. Indeed, polyphenol compounds have a special structure, which is characterized by hydroxyl groups on aromatic rings [2]. They mainly act to remove reactive species of chlorine, nitrogen, and oxygen, or to chelate metal ions in the initial and propagation stages of the oxidation process, resulting in increased antioxidant activity [2]. Moreover, polyphenols may promote the transfer of hydrogen atoms from the active hydroxyl group to the free radicals, indicating an antioxidant effect [3]. An increasing number of studies have shown that polyphenols exhibit a variety of biological activities and can positively impact ruminant health. For example, Aderao et al. [4] showed that polyphenol-rich plants could result in lower acetate:propionate ratios and reduced methane production, and thus enhance green livestock production. Similarly, Cattani et al. [5] suggested that polyphenols could induce a shift in the partition of energy, and stimulate microbial growth, in ruminal fluid in vitro.

Today, the full utilization of existing feed resources and the increased development of unconventional feeds both contribute to the sustainable development of the feeding industry [6]. In this regard, it is noteworthy that plants are rich in bioactive compounds which play very important roles in protecting against the effects of free radicals in ruminants [7]. Various studies have demonstrated that crop stalks could be utilized as safe natural antioxidants for animals, because they contain high amounts of natural antioxidants such as polyphenol compounds, and exhibit strong antioxidant activity [8,9]. Thus, the use of stalk polyphenol to improve antioxidant activity in animals may be an effective way to promote sustainable development in agriculture.

As the liquor capital of China, Guizhou province is famed across the world for its traditional Chinese beverages, including Moutai liquor [10]. In China, sorghum is the main raw material used for brewing traditional liquor; the level of production is high, and planting areas are extensive [11]. However, sorghum stalks are usually considered to have no value, and are thrown away or burned, resulting in environmental pollution and the possible waste of a usable resource. Notably, sorghum stalks contain abundant fibers such as hemicelluloses and cellulose, which can stimulate gastrointestinal peristalsis, maintain normal digestive function in ruminants and serve as a potential source of energy for ruminant feeding [12]. For instance, Elseed et al. [13] reported that there was considerable variation among sorghum stalk varieties in terms of their chemical composition and relative chemical proportions. Similarly, Billa et al. [14] showed that sweet sorghum pith and bark fractions also exhibited substantial differences with respect to their composition, so that bark showed higher levels of total cellulose, hemicellulose, and lignin content, compared with pith, while pith was found to be twice as rich in sucrose and glucose, compared with bark. Specifically, sorghum is rich in various polyphenol compounds, and is characterized by high levels of 2,2-diphenyl-1-picrylhydrazyl (DPPH) free-radical scavenging capacity, leading to in vitro antioxidant activity [15]. Wang et al. [16] showed that sheep can effectively utilize nutrients in sorghum stalks, and reported higher nutrient degradability both in vitro and in vivo, indicating that sorghum stalks may offer high nutritive value to ruminants. Indeed, sorghum stalk, as a cheap crop, may be an ideal source of unconventional feed for ruminants. However, previous studies have analyzed total polyphenols using spectrophotometric methods; as a result, the subgroups of polyphenols are not yet fully understood [17,18], and the application of sorghum stalk remains restricted to some extent. Interestingly, polyphenol subgroups in sorghum stalks may be detected by high-performance liquid chromatography-tandem mass spectrometry (HPLC–MS) technology. We hypothesized that sorghum stalks would contain high levels of polyphenol compounds, exhibit high antioxidant activity, and improve ruminal fluid parameters in goats. Therefore, the aim of this study was to compare the chemical composition, polyphenol compounds, antioxidant activity, and in vitro incubation of different varieties of sorghum stalks with the ruminal fluid parameters of goats.

2. Materials and Methods

2.1. Plant Materials

Guizhou province is a major producer of Chinese liquor, and sorghum is the main raw material used for brewing. Because Guizhou province enjoys a subtropical monsoon climate, the local sorghums are particularly suitable for brewing products such as Moutai liquor. Sorghum stalks are a byproduct of the liquor industry, and these may be used as a source of unconventional feed for ruminants. For the present study, six widely available and commonly types of special sorghum (*Sorghum bicolor* L. Moench) varieties used for brewing liquor in Guizhou Province were selected. Accessions of sorghum stalk samples were obtained from the Guizhou University farm (Guiyang, China), and the scientific names of the sorghum varieties were as follows: maohongnuo 6 (M6), hongyingzi (HZ), jinliangnuo (JN), sweet sorghum (SS), maoliangnuo 1 (M1), and jinzhong 405 (J4). The main information descriptors, such as the varieties, origins, and abbreviations of sorghums, are summarized in Table 1. Six types of sorghum varieties were cultivated under the same conditions in a completely randomized design with three duplicates per sorghum variety. After sorghum was harvested at the yellow ripe stage, the sorghum stalk was cut (6–8 cm above the soil surface) using a cutting machine (TU43; Mitsubishi, Tokyo, Japan). After the seeds were harvested, the whole sorghum stalk was removed to the Institute of Animal Nutrition and Feed Laboratory (Guizhou University, China), and all samples were chopped into 2–3 cm lengths. Duplicates of sorghum stalk samples (each stalk was divided into 3 duplicates) were dried at 65 °C in a drying oven, and then smashed by a hammer-plate grinder (QE-200, Zhejiang Yili Industry and Trade Co., Ltd., Jinhua, China); each duplicate sorghum stalk (each sample had 3 duplicates) was then mixed and filtered through 80 mesh to prepare air-dried samples (n = 3); they were stored at 4 °C for further measurement. Pictures of ground sorghum stalk samples are shown in Figure 1.

Table 1. The information for six varieties of sorghums in this study.

Number	Sorghum Varieties	Origin	Abbreviation
I	Maohongnuo 6	China	M6
II	Hongyingzi	China	HZ
III	Jinliangnuo	China	JN
IV	Sweet sorghum	China	SS
V	Maoliangnuo 1	China	M1
VI	Jinzhong 405	China	J4

2.2. Chemical Composition

Approximately 2.0 g of sorghum stalk was weighed, and dry matter (DM) was analyzed by drying in an electric vacuum drying oven at 105 °C for 2 h according to AOAC Method 930.15 [19]. Crude protein (CP) was analyzed by the Kjeldahl method, using automatic Kjeldahl apparatus (K1100, Hanon Technologies, Jinan, China). CP was calculated (Method 988.05) according to the following formula: total nitrogen × 6.25. Gross energy (GE) was analyzed using a bomb calorimeter with an O_2 gas carrier (WGR-WR3, Changsha BENTE Instrument Co., Ltd., Changsha, China). Ash content was analyzed using a muffle furnace at 550 °C for 5 h (Method 942.05). Organic matter (OM) content was calculated as one hundred percent minus ash. Calcium (Ca) and phosphorus (P) were analyzed by the dry ash method (Method 927.02), and the photometric method (Method 965.17), respectively. Neutral detergent fiber (NDF), acid detergent fiber (ADF), and acid detergent lignin (ADL) were analyzed using an automatic fiber analyzer (Fibertherm FT 12, Gerhardt, Germany) according to the methods of Van Soest et al. [20]. All measurements were performed in triplicate. Hemicellulose and cellulose were calculated by trial and error methods, as follows: hemicellulose = NDF − ADF; cellulose = ADF − ADL. Total digestible nutrients (TDN), metabolizable energy (ME), and digestible energy (DE) were calculated according to the following equations, following Van Le et al. [21]: TDN (%) =

82.38 − (0.7515 × ADF); DE (Mcal/kg) = %TDN × 0.01 × 4.4); ME (MJ/kg) = (DE × 0.82) × 4.185.

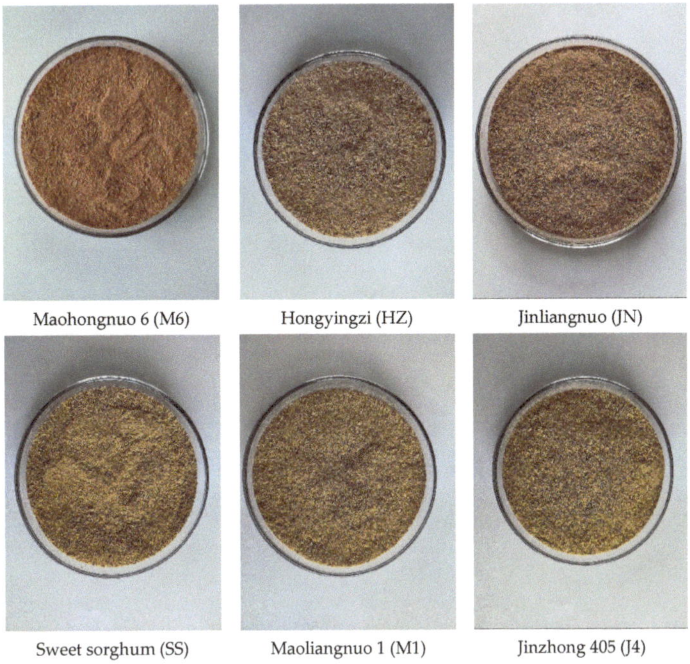

Figure 1. The pictures of different varieties of sorghum stalk grind samples.

2.3. Polyphenol Compounds

Subgroups of polyphenol compounds in sorghum stalks were analyzed according to the method of Wang et al. [22]. Briefly, approximately 0.5 g of sorghum stalk was weighed, added to 4 mL of 1% hydrochloric acid–methanol solution, mixed evenly, and then sonicated for 30 min. In addition, the mixture was centrifuged at 11,000× *g* for 30 min at 4 °C (TG-16G, Hunan Kaida Scientific Instrument Co., Ltd., Changsha, China) to prepare sorghum stalk extract for further analysis. Next, the supernatant was immediately transferred to a 0.22 µm nylon syringe, and the quantitative determination of polyphenol compounds was achieved using an HPLC–MS/MS machine (AGLIENT1260; Agilent Technologies Inc., Santa Clara, CA, USA) and a diode-array detector. The HPLC conditions were as follows: Agilent ZORBAX Eclipse Plus C18 column (3.5 µm × 2.1 mm × 150 mm); mobile phase A: acetonitrile; mobile phase B: 0.1% formic acid; injection temperature 4 °C, column temperature 35 °C; flow rate 0.3 mL/min; and injection volume 3 µL. A standard curve was established, and individual polyphenols were calculated according to the chromatogram peak area.

2.4. Antioxidant Activity

Antioxidant activity parameters for total antioxidant capacity (TAC; kit number A015-1-1), superoxide dismutase (SOD; kit number A001-1-1), glutathione peroxidase (GPX; kit number A005-1-1), catalase (CAT; kit number A007-1-1), and 2,2-diphenyl-1-picrylhydrazyl (DPPH) free-radical scavenging capacity (kit number A153-1-1) of the sorghum stalk extracts (as describe in Section 2.3 above) were analyzed according to the instructions of matched test kits provided by the Nanjing Jiancheng Bioengineering Institute (Nanjing, China).

2.5. In Vitro Gas Production

All experimental animal-care procedures were approved by the Rules of Animal Welfare and Experimental Animal Ethics of Guizhou University (Guiyang, China). Six healthy Qianbei brown goat does (Guizhou indigenous goat breed, Xishui, China) with similar body weights (42.50 ± 0.67 kg) were ruminal fluid donors. The goats were fed a concentrate/roughage ratio of 30:70 with a 13% CP total mixed ration according to the feeding standard of Qianbei brown goats (DB52/T 1377-2018). Before morning feeding, ruminal fluid was collected from animals using an ororuminal probe and the auto-suction pump (AP-9950; Tianjin Autoscience Instrument Co., Ltd., Tianjin, China), and the first 100 mL of ruminal fluid was discarded to avoid saliva. Ruminal fluid was stored in a 39 °C container, and immediately transferred to the in vitro rumen fermentation laboratory of the Institute of Animal Nutrition and Feed, Guizhou University. Next, the ruminal fluid was immediately passed through 4 layers of cheesecloth. Rumen fluid from the six goats was then mixed evenly in equal volume, to prepare an experimental ruminal fluid sample. In vitro gas production was detected using the procedure described by Menke and Steingass [23], as follows: (1) preparation of syringes: 0.50 g of each substrate was weighed by an analytical balance (with results correct to four decimal places), and removed to a 100 mL glass gas-tight syringe (Changzhou Mingyue Medical Equipment Co., Ltd., Changzhou, China). The syringe was incubated in a 39 °C water bath (SYG-2–8; Tianjin Taist Instrument Co., Ltd., Tianjin, China) after the plunger was greased with Vaseline. (2) Preparation of artificial saliva: buffer solution (35 g $NaHCO_3$, 4 g NH_4HCO_3, and 1 L distilled water), macromineral solution (5.7 g Na_2HPO_4, 6.2 g KH_2PO_4, 0.6 g $MgSO_4$, and 1 L distilled water), micromineral solution (13.2 g $CaCl_2{\cdot}2H_2O$, 10.0 g $MnCl_2{\cdot}4H_2O$, 1.0 g $CoCl_2{\cdot}6H_2O$, 0.8 g $FeCl_2{\cdot}6H_2O$, and 1 L distilled water), and resazurin aqueous solution (100 mg/100 mL) were added to a round flat-bottomed flask. (3) Detection of gas production: the ratio of artificial saliva:ruminal fluid was 2:1, and the solution was mixed well. Then, 30 mL of mixed solution was injected into a syringe, and the total gas production (GP) was read and calculated at 3, 6, 9, 12, 24, 48, 72, and 96 h. Each duplicate of sorghum stalk was incubated in two syringes, giving a total of six syringes per sample of in vitro gas production (n = 6). The GP was calculated using the method of Ørskov and McDonald [24], according to the following formula: $y = a + b \times (1 - e^{-ct})$, where y is the gas production volume at t h, a is the immediately soluble fraction, b is the insoluble fraction, c is the rate constant, t is the incubation time, and a + b is the potential extent for the GP. The organic matter digestibility (OMD), metabolizable energy (ME), and effective degradability (ED) were calculated using the following equations [25]: OMD (%) = 0.986 × GP (24 h) + 0.0606 × CP + 11.03; ME (MJ/kg) = −0.20 + 0.1410 × OMD; and ED (%) = a + bc/(k + c), where k = 0.031 h.

In addition, about 70% of the roughage nutrition disappeared within 24 h of in vitro gas production, indicating that it might ultimately all be degradable with this diet of ruminants [24]. For the purposes of the present study, then, rumen fermentation parameters were detected at 24 h incubation time. The syringe fermentation was stopped, and the pH value was detected immediately using a portable pH meter (pH 818, Guangdong, China). In addition, a fermentation:HCL (6 mol/L) ratio of 4:1 was prepared and mixed for further analysis of ammonia nitrogen (NH_3-N) and volatile fatty acid (VFA). The ruminal fluid of NH_3-N was detected using the steam distillation method described by Bremner and Keeney [26]. VFAs were detected using a Thermo TRACE 1310-ISQ gas chromatography–mass spectrum (GC–MS) machine (Thermo, Waltham, MA, USA). The GC conditions were as follows: chromatographic column: an Agilent HP-INNOWAX capillary column (30 m × 0.25 mm × 0.25 μm); injection temperature: 250 °C; ion power temperature: 230 °C; transmission line temperature: 250 °C; quadrupole temperature: 150 °C. Temperature programming was conducted as follows: the reaction temperature was initially 90 °C; it was then increased to 120 °C at 10 °C/min, then to 150 °C at 5 °C/min, and, finally, to 250 °C at 25 °C/min; this latter temperature was held for 2 min. Helium was used as the carrier gas, the rate was 1.0 mL/min; the injection volume was 1 μL, and the split ratio was

10:1. The MS conditions were as follows: electron impact ionization; ion energy of 70 eV; and use of the single ion monitoring scanning method. Individual VFAs consisted of acetic acid, propionic acid, and butyric acid, their sum was the total VFA (TVFA) concentration, and the ratio of acetic acid content to propionic acid content was the acetic: propionic ratio of the ruminal fluid.

2.6. Statistical Analysis

The sorghum stalk duplicate was set as the experimental unit for chemical composition, polyphenol compounds, and antioxidant activity parameters (n = 3); the syringe was set as the experimental unit for gas production kinetics, ruminal fluid fermentation, and GP parameters (n = 6). All data were analyzed using Statistical Analysis System software (Version 9.1.3; SAS Institute Inc., Cary, NC, USA). Data on sorghum-stalk chemical composition, polyphenol compounds, antioxidant activity, gas production kinetics, and ruminal fluid fermentation parameters were detected using one-way ANOVA: $Y_{ij} = \mu + \tau_i + \varepsilon_{ij}$, where Y_{ij} means observation, μ means overall mean, τ_i means effect of the treatment, and ε_{ij} means random error with a mean of 0 and variance σ^2. Data on GP were obtained using two-factorial ANOVA: $Y_{ijn} = \mu + S_i + T_j + (S^*T)_{ij} + \varepsilon_{ijn}$, where Y_{ijn} means observation, μ means overall mean, S_i means effect of sorghum stalk, T_j means incubation time, $(S^*T)_{ij}$ means effect of interaction between sorghum stalk and incubation time, and ε_{ijn} means random error with mean 0 and variance σ^2 [27]. The Pearson correlation coefficient (r) was used to determine the relationship between the total polyphenol content and antioxidant activity parameters in sorghum stalk extracts. Differences were considered to be statistically significant at the level of $p < 0.05$.

3. Results

3.1. Chemical Composition

No significant differences ($p > 0.05$) in GE were observed among the six varieties of sorghum stalks (Table 2). M1 contained a higher ($p < 0.05$) level of DM, and J4 contained a higher ($p < 0.05$) level of CP, relative to the other samples. In addition, M1 and J4 contained lower ($p < 0.05$) levels of OM and higher ($p < 0.05$) levels of ash, respectively. Significantly ($p < 0.05$) higher concentrations of NDF, ADF, and cellulose were observed in stalk JN, along with lower levels of TDN, DE, and ME. Stalk M6 had significantly ($p < 0.05$) higher levels of ADL, hemicellulose, Ca, and P.

Table 2. Comparison of chemical composition of six varieties of sorghum stalks.

Items	Varieties						SEM	*p*-Value
	M6	HZ	JN	SS	M1	J4		
DM, %	28.64 ^{b}	28.32 ^{c}	28.69 ^{b}	19.11 ^{e}	29.20 ^{a}	25.29 ^{d}	0.0361	<0.0001
CP, % of DM	3.78 ^{e}	5.58 ^{c}	5.09 ^{d}	5.56 ^{c}	5.99 ^{b}	7.15 ^{a}	0.0607	<0.0001
GE, MJ/kg of DM	16.83	17.33	17.95	17.60	17.73	17.82	0.3495	0.3513
OM, % of DM	94.02 ^{c}	94.04 ^{c}	94.72 ^{b}	95.51 ^{a}	93.61 ^{d}	93.53 ^{d}	0.0382	<0.0001
Ash, % of DM	5.98 ^{b}	5.96 ^{b}	5.28 ^{c}	4.49 ^{d}	6.39 ^{a}	6.47 ^{a}	0.0386	<0.0001
NDF, % of DM	71.62 ^{b}	68.14 ^{c}	73.21 ^{a}	50.87 ^{f}	60.50 ^{e}	62.15 ^{d}	0.4013	<0.0001
ADF, % of DM	45.65 ^{b}	42.01 ^{c}	49.41 ^{a}	31.35 ^{f}	36.32 ^{e}	37.64 ^{d}	0.2024	<0.0001
ADL, % of DM	8.08 ^{a}	6.31 ^{c}	6.96 ^{b}	3.95 ^{f}	4.96 ^{e}	5.41 ^{d}	0.0838	<0.0001
Hemicellulose, % of DM	25.97 ^{a}	26.14 ^{a}	23.80 ^{b}	19.52 ^{c}	24.18 ^{b}	24.51 ^{b}	0.3087	<0.0001
Cellulose, % of DM	37.57 ^{b}	35.69 ^{c}	42.45 ^{a}	27.40 ^{f}	31.36 ^{e}	32.23 ^{d}	0.1833	<0.0001
Ca, % of DM	1.16 ^{a}	0.58 ^{b}	0.53 ^{b}	0.57 ^{b}	0.67 ^{b}	0.67 ^{b}	0.1142	0.0485
P, % of DM	0.20 ^{a}	0.13 ^{bc}	0.18 ^{ab}	0.13 ^{bc}	0.07 ^{d}	0.09 ^{dc}	0.0139	0.0014
TDN, % of DM	48.08 ^{e}	50.81 ^{d}	45.25 ^{f}	58.82 ^{a}	55.09 ^{b}	54.09 ^{c}	0.1524	<0.0001
DE, Mcal/kg	2.12 ^{e}	2.24 ^{d}	1.99 ^{f}	2.59 ^{a}	2.42 ^{b}	2.38 ^{c}	0.0067	<0.0001
ME, MJ/kg	7.26 ^{e}	7.67 ^{d}	6.83 ^{f}	8.88 ^{a}	8.32 ^{b}	8.17 ^{c}	0.0230	<0.0001

Values with the same superscript letters in the same line are of no significant difference ($p > 0.05$); those with different letters are of significant difference ($p < 0.05$). M6, maohongnuo 6; HZ, hongyingzi; JN, jinliangnuo; SS, sweet sorghum; M1, maoliangnuo 1; J4, jinzhong 405; DM, dry matter; CP, crude protein; GE, gross energy; OM, organic matter; NDF, neutral detergent fiber; ADF, acid detergent fiber; ADL, acid detergent lignin; Ca, calcium; P, phosphorus; TDN, total digestible nutrient; ME, metabolizable energy; DE, digestible energy.

3.2. Polyphenol Compounds

Protocatechualdehyde, epicatechin, vitexin-2-o-rhamnosid, rutin, isorhamnetin-3-o-neohespeidoside, hyperoside, hesperidin, resveratrol, psoralen, bergapten, asiatic acid, rhein, galangin, chrysophanol, and physcion were not detected among the six varieties of sorghum stalks (Table 3). Genistein was detected in M6, but not in any of the other five stalks. Among the samples, there were no significant differences ($p > 0.05$) in levels of gallic acid, nobiletin, or ursolic acid. Levels of chlorogenic acid, homoorientin, isovitexin, vitexin, rhoifolin, genistin, quercetin, apigenin, aloe emodin, emodin, and total polyphenols were significantly ($p < 0.05$) higher in M6 than in the other stalks. The HZ contained significantly higher ($p < 0.05$) levels of caffeic acid, kaempferol 3-rutinoside, and aurantio-obtusin; and lower ($p < 0.05$) levels of protocatechuic acid. Compared to the other stalks, JN contained a higher ($p < 0.05$) level of neohesperidin, naringin, and naringenin. M1 contained a higher ($p < 0.05$) level of tiliroside; however, its level of total polyphenols was lower ($p < 0.05$) than those of the other five stalks. The catechin and kaempferol contents in M1 were significantly ($p < 0.05$) higher than those in the other stalks. The level of umbelliferone in J4 was greater ($p < 0.05$) than that of the other sorghum stalks.

Table 3. Comparison of polyphenol compounds of six varieties of sorghum stalks.

Items, ng/g	Varieties						SEM	*p*-Value
	M6	HZ	JN	SS	M1	J4		
Gallic acid	37.04	30.56	33.00	36.19	32.68	36.31	1.9178	0.3213
Protocatechuic acid	356.25 ^{cd}	318.64 ^{d}	317.67 ^{d}	450.13 ^{ab}	398.15 ^{bc}	477.08 ^{a}	19.0688	0.0028
Protocatechualdehyde	nd	nd	nd	nd	nd	nd	-	-
Chlorogenic acid	157.43 ^{a}	95.95 ^{b}	45.22 ^{d}	54.87 ^{cd}	60.22 ^{cd}	74.77 ^{bc}	5.9469	<0.0001
Catechin	18.01 ^{c}	31.65 ^{b}	17.55 ^{c}	21.52 ^{bc}	46.42 ^{a}	26.38 ^{bc}	2.9806	0.0032
Caffeic acid	270.64 ^{bc}	341.14 ^{a}	164.13 ^{d}	260.42 ^{c}	239.26 ^{c}	309.66 ^{ab}	10.6066	0.0001
Epicatechin	nd	nd	nd	nd	nd	nd	-	-
Homoorientin	354.18 ^{a}	292.93 ^{b}	148.67 ^{d}	185.45 ^{d}	313.60 ^{ab}	237.44 ^{c}	13.0793	<0.0001
Vitexin-2-o-rhamnosid	nd	nd	nd	nd	nd	nd	-	-
Rutin	nd	nd	nd	nd	nd	nd	-	-
Neohesperidin	12.52 ^{b}	13.73 ^{b}	36.33 ^{a}	10.03 ^{b}	14.86 ^{b}	11.39 ^{b}	2.2072	0.0009
Isorhamnetin-3-o-neohespeidoside	nd	nd	nd	nd	nd	nd	-	-
Isovitexin	1072.89 ^{a}	192.14 ^{d}	397.67 ^{b}	151.45 ^{d}	172.96 ^{d}	234.49 ^{c}	10.7116	<0.0001
Vitexin	1089.96 ^{a}	196.60 ^{d}	418.07 ^{b}	177.68 ^{d}	185.70 ^{d}	274.38 ^{c}	11.2289	<0.0001
Hyperoside	nd	nd	nd	nd	nd	nd	-	-
Kaempferol 3-rutinoside	23.30 ^{c}	78.70 ^{a}	20.75 ^{c}	59.55 ^{ab}	32.30 ^{c}	56.55 ^{b}	5.5120	0.0011
Rhoifolin	347.25 ^{a}	96.14 ^{bc}	78.77 ^{c}	95.20 ^{bc}	80.97 ^{c}	129.83 ^{b}	11.1883	<0.0001
Naringin	107.36 ^{ab}	74.02 ^{bc}	127.33 ^{a}	47.53 ^{c}	46.81 ^{c}	52.29 ^{c}	9.4232	0.0028
Hesperidin	nd	nd	nd	nd	nd	nd	-	-
Genistin	6588.96 ^{a}	1136.61 ^{b}	941.66 ^{bc}	837.33 ^{c}	1033.64 ^{bc}	940.99 ^{bc}	55.2716	<0.0001
Umbelliferone	23.62 ^{bc}	38.75 ^{ab}	18.04 ^{c}	18.45 ^{c}	30.25 ^{bc}	47.39 ^{a}	4.0702	0.0086
Tiliroside	17.59 ^{c}	43.15 ^{ab}	23.56 ^{bc}	55.99 ^{a}	26.29 ^{bc}	37.62 ^{abc}	5.0550	0.0133
Resveratrol	nd	nd	nd	nd	nd	nd	-	-
Quercetin	8.23 ^{a}	6.15 ^{b}	5.05 ^{b}	4.67 ^{b}	4.41 ^{b}	8.81 ^{a}	0.4364	0.0006
Apigenin	20462.63 ^{a}	2476.39 ^{b}	2239.26 ^{b}	1008.49 ^{c}	1481.75 ^{c}	2693.96 ^{b}	166.8909	<0.0001
Naringenin	397.06 ^{b}	335.71 ^{c}	473.37 ^{a}	163.80 ^{e}	255.09 ^{d}	245.85 ^{d}	14.3457	<0.0001
Genistein	76.88	nd	nd	nd	nd	nd	-	-
Kaempferol	54.87 ^{c}	40.93 ^{c}	36.01 ^{c}	410.56 ^{b}	563.29 ^{a}	419.53 ^{b}	22.2789	<0.0001
Psoralen	nd	nd	nd	nd	nd	nd	-	-
Aurantio-obtusin	17.98 ^{b}	43.56 ^{a}	25.37 ^{b}	12.49 ^{b}	20.68 ^{b}	16.23 ^{b}	4.1026	0.0215
Bergapten	nd	nd	nd	nd	nd	nd	-	-
Asiatic acid	nd	nd	nd	nd	nd	nd	-	-
Rhein	nd	nd	nd	nd	nd	nd	-	-
Aloe emodin	320.27 ^{a}	87.92 ^{b}	71.81 ^{bc}	32.28 ^{c}	53.84 ^{bc}	58.92 ^{bc}	11.7452	<0.0001
Nobiletin	5.43	6.74	8.12	6.04	6.97	6.45	0.8341	0.5113
Galangin	nd	nd	nd	nd	nd	nd	-	-
Emodin	24.95 ^{a}	14.33 ^{c}	15.95 ^{b}	nd	nd	nd	0.3292	<0.0001
Chrysophanol	nd	nd	nd	nd	nd	nd	-	-
Physcion	nd	nd	nd	nd	nd	nd	-	-
Ursolic acid	2.44	3.16	2.72	2.37	2.36	2.31	0.2598	0.4200
Total polyphenols	31847.74 ^{a}	5995.59 ^{b}	5666.09 ^{b}	4102.51 ^{d}	5102.50 ^{c}	6398.64 ^{b}	215.0871	<0.0001

Values with the same superscript letters in the same line are of no significant difference ($p > 0.05$); those with different letters are of significant difference ($p < 0.05$). M6, maohongnuo 6; HZ, hongyingzi; JN, jinliangnuo; SS, sweet sorghum; M1, maoliangnuo 1; J4, jinzhong 405. nd, not detected.

3.3. Antioxidant Activity Parameters

As shown in Table 4, stalk M6 showed higher ($p < 0.05$) levels of TAC, GPX, CAT, and DPPH free-radical scavenging capacity relative to the other five stalks. Similarly, compared to stalk JN, SS, and M1, stalk M6 contained a significantly higher level of SOD ($p < 0.05$).

Table 4. Comparison of antioxidant activity parameters of six varieties of sorghum stalks.

Items	Varieties						SEM	p-Value
	M6	HZ	JN	SS	M1	J4		
TAC, U/mL	13.22 ^{a}	7.42 ^{b}	7.08 ^{b}	6.59 ^{b}	6.88 ^{b}	7.42 ^{b}	0.4245	<0.0001
SOD, U/mL	168.05 ^{a}	159.06 ^{ab}	153.75 ^{b}	123.48 ^{d}	140.92 ^{c}	162.06 ^{ab}	3.9206	<0.0001
GPX, U/mL	92.32 ^{a}	80.27 ^{b}	82.92 ^{b}	68.61 ^{d}	74.88 ^{c}	81.89 ^{b}	1.6207	<0.0001
CAT, U/mL	35.41 ^{a}	27.44 ^{bc}	26.74 ^{bc}	24.34 ^{c}	26.61 ^{bc}	30.38 ^{b}	1.2326	0.0006
DPPH scavenging capacity, %	85.65 ^{a}	75.10 ^{b}	73.04 ^{bc}	62.97 ^{d}	68.27 ^{c}	73.48 ^{bc}	1.6999	<0.0001

Values with the same superscript letters in the same line are of no significant difference ($p > 0.05$); those with different letters are of significant difference ($p < 0.05$). M6, maohongnuo 6; HZ, hongyingzi; JN, jinliangnuo; SS, sweet sorghum; M1, maoliangnuo 1; J4, jinzhong 405; TAC, total antioxidant capacity; SOD, superoxide dismutase; GPX, glutathione peroxidase; CAT, catalase; DPPH, 2,2-diphenyl-1-picrylhydrazyl free-radical scavenging capacity.

3.4. Pearson Correlation Coefficients

There were significant ($p < 0.05$) positive correlations between total polyphenols and TAC, SOD, GPX, CAT, and DPPH free-radical scavenging capacity in sorghum stalk extracts (Table 5).

Table 5. Pearson correlation coefficients between total polyphenols and antioxidant activity parameters.

Items		TAC	SOD	GPX	CAT	DPPH Scavenging Capacity
Total polyphenols	r	0.968	0.530	0.750	0.814	0.805
	p	<0.0001	0.0237	0.0003	<0.0001	<0.0001

TAC, total antioxidant capacity; SOD, superoxide dismutase; GPX, glutathione peroxidase; CAT, catalase; DPPH, 2,2-diphenyl-1-picrylhydrazyl free-radical scavenging capacity.

3.5. Total Gas Production

The GP was significantly ($p < 0.05$) influenced by both sorghum stalk varieties and incubation times (Table 6). However, there was no correlation ($p = 0.0785$) between sorghum stalk variety and incubation time with respect to GP.

Table 6. Comparison of total gas production of six varieties of sorghum stalks.

Varieties, mL	Incubation Time, h									SEM	p-Value		
	3	6	9	12	24	36	48	72	96		Stalk	Time	S × T
M6	7.33	12.67	17.00	22.33	41.67	52.33	60.50	67.67	70.00	3.6796	<0.0001	<0.0001	0.0785
HZ	12.17	20.00	25.00	29.50	50.83	62.50	70.17	76.83	79.83				
JN	9.17	16.17	20.83	25.83	46.50	57.83	68.50	79.83	83.50				
SS	9.17	34.33	41.50	47.33	59.83	75.83	81.17	87.33	88.67				
M1	7.67	22.17	27.50	33.33	53.67	66.83	71.00	75.33	80.33				
J4	7.33	20.33	25.83	32.00	51.83	67.83	79.33	86.50	89.67				

Stalk, effect of sorghum stalk; time, effect of incubation time; S × T, effect of sorghum stalk and incubation time interactions. M6, maohongnuo 6; HZ, hongyingzi; JN, jinliangnuo; SS, sweet sorghum; M1, maoliangnuo 1; J4, jinzhong 405.

3.6. Gas Production Kinetics

The stalk J4 displayed higher ($p < 0.05$) a and a + b values, relative to the other sorghum stalks (Table 7). Compared to the other five sorghum stalks, stalk M6 had a significantly lower ($p < 0.05$) b level, and stalk SS had a significantly higher ($p < 0.05$) c level. Additionally, stalk SS showed significantly higher ($p < 0.05$) levels of OMD and ME, and stalks SS and J4 had ED values that were significantly higher ($p < 0.05$) than those of M6 and JN.

Table 7. Comparison of in vitro gas production kinetics of six varieties of sorghum stalks.

Items	Varieties						SEM	*p*-Value
	M6	HZ	JN	SS	M1	J4		
a, mL	3.97 ^{c}	5.08 ^{b}	2.91 ^{c}	3.75 ^{c}	3.62 ^{c}	6.25 ^{a}	0.3201	<0.0001
b, mL	73.24 ^{b}	86.18 ^{a}	88.83 ^{a}	89.05 ^{a}	85.25 ^{a}	97.16 ^{a}	3.8328	0.0157
a + b, mL	77.20 ^{c}	91.27 ^{ab}	91.74 ^{ab}	92.80 ^{ab}	88.88 ^{bc}	103.41 ^{a}	3.8099	0.0074
c, % h	0.035 ^{b}	0.036 ^{b}	0.029 ^{b}	0.057 ^{a}	0.042 ^{ab}	0.036 ^{b}	0.0049	0.0114
OMD, %	52.34 ^{c}	61.49 ^{abc}	57.19 ^{bc}	70.36 ^{a}	64.31 ^{ab}	62.57 ^{abc}	3.5647	0.0263
ME, MJ/kg	7.18 ^{c}	8.47 ^{abc}	7.86 ^{bc}	9.72 ^{a}	8.87 ^{ab}	8.62 ^{abc}	0.5026	0.0263
ED, %	42.58 ^{c}	50.70 ^{abc}	45.75 ^{bc}	59.26 ^{a}	52.01 ^{ab}	58.10 ^{a}	2.7503	0.0019

Values with the same superscript letters in the same line are of no significant difference ($p > 0.05$); those with different letters are of significant difference ($p < 0.05$). a, immediately soluble fraction; b, insoluble fraction; a + b, potential extent of gas production; c, rate constant; OMD, organic matter digestibility; ME, metabolizable energy; ED, effective degradability. M6, maohongnuo 6; HZ, hongyingzi; JN, jinliangnuo; SS, sweet sorghum; M1, maoliangnuo 1; J4, jinzhong 405.

3.7. Ruminal Fluid Fermentation Parameters

No significant differences ($p > 0.05$) were detected in the ruminal fluid with respect to pH, NH_3-N, acetic acid, butyric acid, or TVFA concentrations (Table 8). In contrast, stalk M6 contained significantly higher ($p < 0.05$) levels of ruminal-fluid propionic acid; however, the ratio of acetic acid to propionic acid in its ruminal fluid was significantly lower ($p < 0.05$), compared with the other five varieties of sorghum stalks.

Table 8. Comparison of rumen fermentation parameters of six varieties of sorghum stalks at 24 h.

Items	Varieties						SEM	*p*-Value
	M6	HZ	JN	SS	M1	J4		
pH	6.42	6.38	6.36	6.22	6.23	6.28	0.0873	0.3296
NH_3-N, mg/dL	5.95	5.90	6.10	6.94	6.92	6.98	0.3158	0.0649
Acetic acid, mmol/L	62.55	66.35	66.67	66.26	66.22	64.49	1.4283	0.3421
Propionic acid, mmol/L	21.87 ^{a}	17.12 ^{b}	17.41 ^{b}	16.86 ^{b}	16.77 ^{b}	18.61 ^{b}	0.6622	0.0011
Butyric acid, mmol/L	4.42	5.57	5.32	5.55	5.75	5.55	0.4155	0.3176
TVFA, mmol/L	88.18	89.04	89.74	88.67	88.73	88.64	1.5106	0.9853
Acetic acid to propionic acid ratio	2.87 ^{b}	3.88 ^{a}	3.84 ^{a}	3.94 ^{a}	3.97 ^{a}	3.47 ^{a}	0.1586	0.0025

Values with the same superscript letters in the same line are of no significant difference ($p > 0.05$); those with different letters are of significant difference ($p < 0.05$). M6, maohongnuo 6; HZ, hongyingzi; JN, jinliangnuo; SS, sweet sorghum; M1, maoliangnuo 1; J4, jinzhong 405; NH_3-N, ammonia nitrogen; TVFA, total volatile fatty acid.

4. Discussion

The level of fiber is among the most important factors in feed that can affect the health and performance of ruminants [28]. Crude fibers are the main components of plant cell walls; these include cellulose, hemicellulose, and ADL, amongst others. In the context of the present study, it is of interest that the cell wall is associated with polyphenol compounds (such as p-coumaric and ferulic acids), and that p-Coumaric acid is the predominant p-hydroxycinnamic associated with the cell wall in sweet sorghum [14]. Hence, stalk M6 had high levels of total polyphenols, and thus contained lower levels of CP and higher levels of ADL. Elseed et al. [13] studied the chemical composition of sorghum stalk and found, depending on the varieties, that the proportions of CP, NDF, ADF, and hemicellulose in whole sorghum stalks ranged from 3.2–7.4%, 60.6–78.0%, 45.0–60.0%, and 4.2–18.5%, respectively. These findings are consistent with our results. In addition, Manea et al. [17]

analyzed five varieties of sorghum stalks and found that their CP content was less than 2%; compared with this, our results indicated levels of CP which were two–three times higher. Agbagla-Dohnani et al. [29] showed that rice stalks exhibited great variability in chemical composition, such as DM, silica, ash, CP, NDF, ADF, and ADL. In addition, Firdous and Gilani [30] reported that plant maturity had a much greater impact on the chemical composition of whole sorghum plants, as well as leaf and stem fractions. Differences in chemical composition observed in other studies likely reflect differences between the parts of the plants used, and their maturity, in addition to soil, weather, and environmental characteristics. The data obtained in the present study correspond with the findings of Billa et al. [14], who indicated that different parts (bark and pith) of sweet sorghum showed variation in chemical composition.

Another previous study demonstrated that sorghum stalks contain many natural polyphenol compounds that show high antioxidant activity, and that their extracts can change bacterial morphology and internal structure, strongly inhibiting the growth of foodborne pathogens [31]. Similarly, Chen et al. [32] found that sweet sorghum stalk extract contains abundant levels of p-hydroxybenzoic acid, as well as caffeic, gentisic, chlorogenic, coumaric, and gallic acids, with potential antimicrobial effects and strong antioxidant properties. In addition, the authors of [33] reported that sorghum exhibits red, white, yellow, and brown colors, and these contain different totals of polyphenols. In the present study, the six varieties of sorghum stalks had different colors, resulting in different polyphenol compounds. We found that sorghum M6 was rich in 25 polyphenol compounds; sorghum HZ and JN were rich in 24 polyphenol compounds; and sorghums SS, M1, and J4 were rich in 23 polyphenol compounds. Sorghum stalk color is thus related to polyphenol content, and polyphenol content has also been shown to exhibit a strong positive correlation with DPPH free-radical scavenging capacity [15]. Using HPLC-ESI–MS/MS technology, Luo et al. [34] found that extracts of red sorghum bran contain abundant polyphenol compounds such as taxifolin, taxifolin hexoside, procyanidins, and epicatechin. In addition, Kalisz et al. [35] showed that red *Malinowy* rhubarb stalks exhibited an intense pink-red color, and also contained a high polyphenol content. Similarly, in the present study, we found that stalk M6 was red in color (Figure 1) and contained a higher level of total polyphenol compounds, suggesting that M6 may exhibit strong potential antioxidant and antimicrobial effects (e.g., against *Staphylococcus aureus* or *Escherichia coli*) [36].

Free radicals are byproducts of metabolism which influence the homeostasis between the generation and scavenging of radicals in vivo; this homeostasis mainly depends on the antioxidant system [37]. Moreover, free radicals are highly reactive molecules that bind and destroy body cells, and are the main cause of diseases and ageing [38]. Polyphenols are secondary metabolites characterized by one or more hydroxyl groups that bind to one or more aromatic rings; they are powerful antioxidants that complement and enhance the functions of antioxidant enzymes to protect against oxidative stress [39]. Additionally, polyphenols can achieve a high level of antioxidant activity in vitro by scavenging free radicals or limiting their formation [40]. Rodríguez-Muela et al. [41] showed that the inclusion of polyphenol-rich plants in the diet of lambs could improve plasma antioxidant activity parameters. In the present study, we found that stalk M6 contained a higher level of total polyphenols, and its extract showed higher antioxidant activity parameters, including TAC, SOD, GPX, CAT, and DPPH free-radical scavenging capacity. This could have occurred because (1) polyphenol compounds can improve antioxidant activity, to protect against oxidative stress in ruminants; (2) polyphenol compounds are oxidized by free radicals, resulting in the formation of less reactive and more stable molecules [42]; or (3) polyphenol compounds can chelate metal ions and form stable complexes with metal ions of the transition group and are thus involved in the antioxidant protection of the cell [7]. In line with our findings, Shih et al. [43] found that Moringa extract contained higher levels of polyphenol compounds, showed stronger hydrogen peroxide scavenging activity, and exhibited higher levels of SOD activity in vitro. However, chemical composition parameters varied widely among the six varieties of sorghum stalk, so that it could not be determined if antioxidant activity difference was responsible for polyphenols or chemical composition. This point needs to be addressed in future studies.

Polyphenol compounds are excellent natural antioxidants and free-radical scavengers that can eliminate free radicals in the bodies of animals [44]. Moreover, polyphenol compounds may also interact with iron and exhibit a stronger ability to scavenge free radicals in vitro [45]. Shen et al. [46] showed that polyphenol compounds exhibit scavenging activity with respect to strong superoxide radicals, hydroxyl radicals, as well as ferric reducing antioxidant power and moderate metal ion-chelating activity. Thus, significant positive correlations were observed between total polyphenols and TAC, SOD, GPX, CAT, and DPPH free-radical scavenging capacity. Our results are also in line with those reported by Kiselova et al. [47], who showed that the antioxidant capacity of plant extracts largely resulted from their polyphenol compounds, and that antioxidant activity and polyphenol content showed a stronger positive correlation. In short, our findings provide evidence that polyphenol compounds display higher levels of in vitro antioxidant activity. It is therefore reasonable to suggest that dietary supplementation with polyphenol-rich sorghum stalks might inhibit inflammatory reactions and improve the activities of antioxidants in goats; however, this needs to be validated by future in vivo experiments in ruminants.

In vitro gas production can reflect the type and degree of fermentation, and thereby serve as an important indicator for the nutritional value of feed [48]. Vasta et al. [49] showed that polyphenol compounds could inhibit ruminant methane production by inhibiting fibrolytic bacteria, decreasing fiber digestibility and H_2 production, and reducing the protozoa population of the ruminal fluid. In the present study, M6 contained the lowest level of GP during the entire incubation period, probably because it contained higher levels of polyphenol compounds such as apigenin, genistin, vitexin, isovitexin, and aloe emodin, as well as higher total polyphenols; however, this assumption needs further validation. Nevertheless, our results are consistent those of with Lu et al. [18], who showed that high-polyphenol feed could inhibit methane production, and result in a low level of GP, in in vitro ruminal fluid incubation.

In the current study, J4 showed the highest a value, perhaps because it contained the highest level of CP. Recalling the suggestion by Tovar-Gomez et al. [50] that low levels of ADF and cellulose in crop stalks result in higher levels of b and c, we also note that, in the present study, stalk SS had higher c values. These differences might result from different levels of ADF and cellulose in sorghum stalks. In addition, Elseed et al. [13] showed that sorghum stalks exhibited lower degradability and lower potential feeding value when they contained a higher cell-wall content. In the current study, stalk M6 had low a, b, a + b and c values, perhaps due to its ADL content being higher than that of other sorghum stalks. Moreover, active substances, such as polyphenols in plants, may protect CP from degradation in the ruminal fluid of goats [51]. Thus, in the present study, SS had higher levels of OMD, ME, and ED, possibly because of its high chemical composition for TDN, DE, and ME values.

Polyphenol compounds could regulate ruminal-fluid ciliate and protozoal Gram-positive fibrolytic bacteria, leading to a reduction in VFA production, specifically for the decreased production of acetic acid [49]. Thus, polyphenols can influence gastrointestinal tract function and improve health in ruminants [52]. In addition, Odongo et al. [51] have shown that polyphenols can directly inhibit the growth of methanogens and hydrogen-producing microbes and decrease protozoal numbers, thereby decreasing ruminal-fluid CH_4 production in ruminants. Interestingly, 3-(4-hydroxypheny1) propionic acid in ruminal fluid may result from the chemical reduction of dietary phenolic monomers by ruminal microorganisms [53]. Specifically, polyphenols could produce a positive fermentation pattern with a better ratio of acetic acid to propionic acid [54]. In the present study, we found that stalk M6 increased ruminal-fluid propionic acid and decreased the ratio of acetic acid to propionic acid. The possible reasons may be that (1) polyphenols are involved in propionic acid in ruminal fluid, possibly as the result of the interaction of polyphenols and the nonpolyphenolic polymer lignin [4]; (2) polyphenols destroy the integrity of the bacterial cell membrane structure, resulting in the exudation of intracellular molecules and an increase in the electrolyte content in the cell culture medium [31]; and (3) polyphenols could increase microbial protein flow from the rumen and increase the efficiency of substrate utilization [54]. In short, because stalk M6 contained an abundance of polyphenols, this suggests that the propionic acid ratio of the ruminal fluid was increased, and that methane emissions were effectively suppressed. These findings are similar to those reported by Wang et al. [55], who showed that the addition of polyphenol-rich *Castanea mollissima* Blume could significantly increase the propionic acid content and reduce the acetic acid/propionic acid ratio in in vitro rumen fermentation. However, further experimental data from in vivo feeding trials with ruminants are required to validate these findings.

5. Conclusions

The results of the present study indicate that the chemical composition, polyphenol compounds, antioxidant activity, and in vitro rumen fermentation differed greatly among six varieties of sorghum stalks. Specifically, stalk M6 showed higher levels of total polyphenols, antioxidant activity parameters, and propionic acid, with lower levels of GP and a lower ratio of acetic acid to propionic acid. Further research is needed involving in vivo feeding trials with ruminants to analyze how polyphenol compounds from sorghum stalks affect microorganisms and antioxidant potential in ruminal fluid.

Author Contributions: X.T.: writing—original draft, data curation, methodology, project administration; J.Q.: preparation of in vitro gas production; Q.L. (Qingyuan Luo): detection of chemical composition, preparation of samples; Y.X., S.X., R.C. and X.W.: resources, investigation, laboratory testing; Q.L. (Qi Lu): data analysis, writing—original draft, writing—review and editing, software. All authors have read and agreed to the published version of the manuscript.

Funding: This study was supported by the National Natural Science Foundation of China (32260849; 32302779), the Key Science and Technology Project of Guizhou Province (Qiankehe Support [222] Key 033), the Science and Technology Project of Guizhou Province (Qiankehe foundation-ZK [2021] General 164), the Cultivating Project of Guizhou University (2019-33), and the Basic Research Project of Guizhou University (2023-16).

Institutional Review Board Statement: This study was approved by the Guizhou University, Subcommittee of Experimental Animal Ethics (Ethic Approval Date: 30 August 2023; Ethic Approval Code: EAE-GZU-2023-E027).

Informed Consent Statement: Not applicable.

Data Availability Statement: Data are contained within the article.

Acknowledgments: The authors would like to thank MDPI for their help in improving the English of the manuscript.

Conflicts of Interest: The authors declare no conflicts of interest.

References

1. Shen, N.; Wang, T.; Gan, Q.; Liu, S.; Wang, L.; Jin, B. Plant flavonoids: Classification, distribution, biosynthesis, and antioxidant activity. *Food Chem.* **2022**, *383*, 132531. [CrossRef]
2. Campos, M.R.S. Bioactive compounds: Health benefits and potential applications. In *Phenolic Compounds: Structure, Classification, and Antioxidant Power*; Vuolo, M.M., Lima, V.S., Junior, M.R.M., Eds.; Elsevier Inc.: Amsterdam, The Netherlands, 2019; pp. 33–50.
3. Bešlo, D.; Golubić, N.; Rastija, V.; Agić, D.; Karnaš, M.; Šubarić, D.; Lučić, B. Antioxidant activity, metabolism, and bioavailability of polyphenols in the diet of animals. *Antioxidants* **2023**, *12*, 1141. [CrossRef]
4. Aderao, G.N.; Sahoo, A.; Bhatt, R.S.; Kumawat, P.K.; Soni, L. In vitro rumen fermentation kinetics, metabolite production, methane and substrate degradability of polyphenol rich plant leaves and their component complete feed blocks. *J. Anim. Sci. Technol.* **2018**, *60*, 26. [CrossRef]
5. Cattani, M.; Tagliapietra, F.; Bailoni, L.; Schiavon, S. Synthetic and natural polyphenols with antioxidant properties stimulate rumen microbial growth in vitro. *Anim. Prod. Sci.* **2012**, *52*, 44–50. [CrossRef]
6. Uushona, T.; Chikwanha, O.C.; Katiyatiya, C.L.F.; Tayengwa, T.; Strydom, P.E.; Mapiye, C. Ruminant meat production and quality enhancement, nematode suppression and greenhouse gas emission mitigation: A sustainable paradigm for valorisation of acacia leaves. *Anim. Feed Sci. Tech.* **2022**, *284*, 115187. [CrossRef]

7. Bešlo, D.; Došlic, G.; Agic, D.; Rastija, V.; Šperanda, M.; Gantner, V.; Lucic, B. Polyphenols in ruminant nutrition and their effects on reproduction. *Antioxidants* **2022**, *11*, 970. [CrossRef] [PubMed]
8. Ramos, M.; Laveriano, E.; Sebastián, L.S.; Perez, M.; Jiménez, A.; Lamuela-Raventos, R.M.; Garrigós, M.C.; Vallverdú-Queralt, A. Rice straw as a valuable source of cellulose and polyphenols: Applications in the food industry. *Trends Food Sci. Tech.* **2023**, *131*, 14–27. [CrossRef]
9. Elzaawely, A.A.; Maswada, H.F.; El-Sayed, M.E.A.; Ahmed, M.E. Phenolic compounds and antioxidant activity of rice straw extract. *Int. Lett. Nat. Sci.* **2017**, *64*, 1–9. [CrossRef]
10. Li, J.; Ouyang, Z.; Liu, P.; Zhao, X.; Wu, R.; Zhang, C.; Lin, C.; Li, Y.; Guo, X. Distribution and characteristics of microplastics in the basin of Chishui River in Renhuai, China. *Sci. Total Environ.* **2021**, *773*, 145591. [CrossRef]
11. Kangama, C.O.; Xu, R.M. Introduction of sorghum (*Sorghum bicolor* (L.) Moench) into China. *Afr. J. Biotechnol.* **2005**, *4*, 575–579.
12. Cao, C.; Yang, Z.; Han, L.; Jiang, X.; Ji, G. Study on in situ analysis of cellulose, hemicelluloses and lignin distribution linked to tissue structure of crop stalk internodal transverse section based on FTIR microspectroscopic imaging. *Cellulose* **2015**, *22*, 139–149. [CrossRef]
13. Elseed, A.M.A.F.; Eldaim, N.I.N.; Amasaib, E.O. Chemical composition and in situ dry matter degradability of stover fractions of five sorghum varieties. *J. Appl. Sci. Res.* **2007**, *3*, 1141–1145.
14. Billa, E.; Koullas, D.P.; Monties, B.; Koukios, E.G. Structure and composition of sweet sorghum stalk components. *Ind. Crop. Prod.* **1997**, *6*, 297–302. [CrossRef]
15. Choi, S.C.; Kim, J.M.; Lee, Y.G.; Kim, C. Antioxidant activity and contents of total phenolic compounds and anthocyanins according to grain colour in several varieties of *Sorghum bicolor* (L.) moench. *Cereal Res. Commun.* **2019**, *47*, 228–238. [CrossRef]
16. Wang, J.; Zhang, Z.; Liu, H.; Xu, J.; Liu, T.; Wang, C.; Zheng, C. Evaluation of gas production, fermentation parameters, and nutrient degradability in different proportions of sorghum straw and ammoniated wheat straw. *Fermentation* **2022**, *8*, 415. [CrossRef]
17. Manea, V.; Tanase, A.; Casarica, A.; Albulescu, R.; Radulescu, G.; Campeanu, G.; Israel-Roming, F.; Stoina, G. Study of the chemical composition of sweet sorghum stalks depleted in carbohydrates with applications in obtaining bioethanol. *Analele Ştiinţifice AleUniv. Alexandru Ioan Cuza Secţiunea Genet. Şi Biol. Mol.* **2010**, *9*, 87–96.
18. Lu, Q.; Luo, Q.; Li, J.; Wang, X.; Ban, C.; Qin, J.; Tian, Y.; Tian, X.Z.; Chen, X. Evaluation of the chemical composition, bioactive substance, gas production, and rumen fermentation parameters of four types of distiller's grains. *Molecules* **2022**, *27*, 613. [CrossRef]
19. AOAC. *Official Methods of Analysis*, 18th ed.; Association of Official Analtical Chemists: Arlington, MA, USA, 2005.
20. Van Soest, P.V.; Robertson, J.B.; Lewis, B.A. Methods for dietary fiber, neutral detergent fiber, and nonstarch polysaccharides in relation to animal nutrition. *J. Dairy Sci.* **1991**, *74*, 3583–3597. [CrossRef]
21. Van Le, H.; Nguyen, D.V.; Vu Nguyen, Q.; Malau-Aduli, B.S.; Nichols, P.D.; Malau-Aduli, A.E.O. Fatty acid profiles of muscle, liver, heart and kidney of Australian prime lambs fed different polyunsaturated fatty acids enriched pellets in a feedlot system. *Sci. Rep.* **2019**, *9*, 1238. [CrossRef]
22. Wang, X.; Li, J.X.; Zhou, D.; Qin, J.X.; Xu, Y.Q.; Lu, Q.; Tian, X.Z. Effects of *Rosa roxburghii* tratt seed on the growth performance, meat quality, and sensory evaluation characteristics in growing rabbits. *Meat Sci.* **2024**, *208*, 109394. [CrossRef]
23. Menke, K.H.; Steingass, H. Estimation of the energetic feed value obtained from chemical analysis and in vitro gas production using rumen fluid. *Anim. Res. Dev.* **1988**, *28*, 7–55.
24. Ørskov, E.R.; McDonald, I. The estimation of protein degradability in the rumen from incubation measurements weighted according to rate of passage. *J. Agr. Sci.* **1979**, *92*, 499–503. [CrossRef]
25. Tian, X.Z.; Paengkoum, P.; Paengkoum, S.; Thongpe, S.; Ban, C. Comparison of forage yield, silage fermentative quality, anthocyanin stability, antioxidant activity, and in vitro rumen fermentation of anthocyanin-rich purple corn (*Zea mays* L.) stover and sticky corn stover. *J. Integr. Agr.* **2018**, *17*, 2082–2095. [CrossRef]
26. Bremner, J.M.; Keeney, D.R. Steam distillation methods for determination of ammonium, nitrate and nitrite. *Anal. Chim. Acta* **1965**, *32*, 485–495. [CrossRef]
27. Kaps, M.; Lamberson, W.R. *Biostatistics for Animal Science*; CABI Publishing: Cambridge, MA, USA, 2004.
28. Waghorn, G.C.; Clark, D.A. Feeding value of pastures for ruminants. *N. Zeal. Vet. J.* **2004**, *52*, 320–331. [CrossRef] [PubMed]
29. Agbagla-Dohnani, A.; Nozière, P.; Clément, G.; Doreau, M. In sacco degradability, chemical and morphological composition of 15 varieties of European rice straw. *Anim. Feed Sci. Tech.* **2001**, *94*, 15–27. [CrossRef]
30. Firdous, R.; Gilani, A.H. Changes in chemical composition of sorghum as influenced by growth stage and cultivar. *Asian Austral. J. Anim.* **2001**, *14*, 935–940. [CrossRef]
31. Chen, H.; Xu, Y.; Chen, H.; Liu, H.; Yu, Q.; Han, L. Isolation and identification of polyphenols from fresh sweet sorghum stems and their antibacterial mechanism against foodborne pathogens. *Front. Bioeng. Biotech.* **2022**, *9*, 770726. [CrossRef] [PubMed]
32. Chen, H.; Tian, X.; Yu, Q.; Hu, W.; Chen, J.; Zhou, L. Sweet sorghum stalks extract has antimicrobial activity. *Ind. Crops Prod.* **2021**, *170*, 113746. [CrossRef]
33. Rhodes, D.H.; Stephen, K. Sorghum (*Sorghum bicolor* (L.) Moench) genotypes with contrasting polyphenol compositions differentially modulate inflammatory cytokines in mouse macrophages. *J. Chem.* **2016**, *2016*, 9640869. [CrossRef]

34. Luo, X.; Cui, J.; Zhang, H.; Duan, Y.; Zhang, D.; Cai, M.; Chen, G. Ultrasound assisted extraction of polyphenolic compounds from red sorghum (*Sorghum bicolor* L.) bran and their biological activities and polyphenolic compositions. *Ind. Crops Prod.* **2018**, *112*, 296–304. [CrossRef]
35. Kalisz, S.; Oszmiański, K.; Kolniak-Ostek, J.; Grobelna, A.; Kieliszek, M.; Cendrowski, A. Effect of a variety of polyphenols compounds and antioxidant properties of rhubarb (*Rheum rhabarbarum*). *LWT—Food Sci. Technol.* **2020**, *118*, 108775. [CrossRef]
36. Martelli, G.; Giacomini, D. Antibacterial and antioxidant activities for natural and synthetic dual-active compounds. *Eur. J. Med. Chem.* **2008**, *158*, 91–105. [CrossRef] [PubMed]
37. Warraich, U.E.A.; Hussain, F.; Kayani, H.U.R. Aging-oxidative stress, antioxidants and computational modeling. *Heliyon* **2020**, *6*, e04107. [CrossRef] [PubMed]
38. Andrés, C.M.C.; de la Lastra, J.M.P.; Juan, C.A.; Plou, F.J.; Pérez-Lebeña, E. From reactive species to disease development: Effect of oxidants and antioxidants on the cellular biomarkers. *J. Biochem. Mol. Toxic.* **2023**, *37*, e23455. [CrossRef]
39. Stagos, D. Antioxidant activity of polyphenolic plant extracts. *Antioxidants* **2020**, *9*, 19. [CrossRef] [PubMed]
40. Chiva-Blanch, G.; Visioli, F. Polyphenols and health: Moving beyond antioxidants. *J. Berry Res.* **2012**, *2*, 63–71. [CrossRef]
41. Rodríguez-Muela, C.; Rodríguez, H.E.; Arzola, C.; Díaz-Plascencia, D.; Ramírez-Godínez, J.A.; Flores-Mariñelarena, A.; Mancillas-Flores, P.F.; Corral, G. Antioxidant activity in plasma and rumen papillae development in lambs fed fermented apple pomace. *J. Anim. Sci.* **2015**, *93*, 2357–2362. [CrossRef]
42. Mathew, S.; Abraham, T.E.; Zakaria, Z.A. Reactivity of phenolic compounds towards free radicals under in vitro conditions. *J. Food Sci. Tech.* **2015**, *52*, 5790–5798. [CrossRef]
43. Shih, M.C.; Chang, C.M.; Kang, S.M.; Tsai, M.L. Effect of different parts (leaf, stem and stalk) and seasons (summer and winter) on the chemical compositions and antioxidant activity of Moringa oleifera. *Int. J. Mol. Sci.* **2011**, *12*, 6077–6088. [CrossRef]
44. Avila-Nava, A.; Medina-Vera, I.; Toledo-Alvarado, H.; Corona, L.; Márquez-Mota, C.C. Supplementation with antioxidants and phenolic compounds in ruminant feeding and its effect on dairy products: A systematic review. *J. Dairy Res.* **2023**, *90*, 216–226. [CrossRef]
45. Perron, N.R.; Brumaghim, J.L. A review of the antioxidant mechanisms of polyphenol compounds related to iron binding. *Cell Biochem. Biophys.* **2009**, *53*, 75–100. [CrossRef]
46. Shen, Y.; Zhang, H.; Cheng, L.; Wang, L.; Qian, H.; Qi, X. In Vitro and in vivo antioxidant activity of polyphenols extracted from black highland barley. *Food Chem.* **2016**, *194*, 1003–1012. [CrossRef] [PubMed]
47. Kiselova, Y.; Ivanova, D.; Chervenkov, T.; Gerova, D.; Yankova, T. Correlation between the in vitro antioxidant activity and polyphenol content of aqueous extracts from Bulgarian herbs. *Phytother. Res.* **2010**, *20*, 961–965. [CrossRef]
48. Sallam, S.M.A.; Nasser, M.E.A.; El-Waziry, A.M.; Bueno, I.C.S.; Abdalla, A.L. Use of an in vitro rumen gas production technique to evaluate some ruminant feedstuffs. *J. Appl. Sci. Res.* **2007**, *3*, 34–41.
49. Vasta, V.; Daghio, M.; Cappucci, A.; Buccioni, A.; Serra, A.; Viti, C.; Mele, M. Invited review: Plant polyphenols and rumen microbiota responsible for fatty acid biohydrogenation, fiber digestion, and methane emission: Experimental evidence and methodological approaches. *J. Dairy Sci.* **2019**, *102*, 3781–3804. [CrossRef]
50. Tovar-Gomez, M.R.; Emile, J.C.; MichaletDoreau, B.; Barriere, Y. In situ degradation kinetics of maize hybrid stalks. *Anim. Feed Sci. Technol.* **1997**, *68*, 77–88. [CrossRef]
51. Odongo, N.E.; Garcia, M.; Viljoen, G.J. *Sustainable Improvement of Animal Production and Health*; Food and Agriculture Organization of the United Nations: Rome, Italy, 2010; pp. 151–166.
52. Theodorou, M.K.; Kingston-Smith, A.H.; Winters, A.L.; Lee, M.R.F.; Minchin, F.R.; Morris, P.; MacRae, J. Polyphenols and their influence on gut function and health in ruminants: A review. *Environ. Chem. Lett.* **2006**, *4*, 121–126. [CrossRef]
53. Cremin, J.D.; Drackley, J.K.; Grum, D.E.; Hansen, L.R.; Fahey, G.C. Effects of reduced phenolic acids on metabolism of propionate and palmitate in bovine liver tissue in vitro. *J. Dairy Sci.* **1994**, *77*, 3608–3617. [CrossRef]
54. Parmar, P.; Bhatt, S.; Dhyani, S.; Jain, A. Phytochemical studies of the secondary metabolites of *Ziziphus mauritania* Lam. Leaves. *Int. J. Curr. Pharm. Res.* **2012**, *4*, 153–155.
55. Wang, Y.; Yu, S.; Li, Y.; Zhang, S.; Qi, X.; Guo, K.; Guo, Y.; Fortina, R. Pilot study of the effects of polyphenols from chestnut involucre on methane production, volatile fatty acids, and ammonia concentration during in vitro rumen fermentation. *Animals* **2021**, *11*, 108. [CrossRef] [PubMed]

Disclaimer/Publisher's Note: The statements, opinions and data contained in all publications are solely those of the individual author(s) and contributor(s) and not of MDPI and/or the editor(s). MDPI and/or the editor(s) disclaim responsibility for any injury to people or property resulting from any ideas, methods, instructions or products referred to in the content.

animals

MDPI

Article

Protective Effects of Niacin on Rumen Epithelial Cell Barrier Integrity in Heat-Stressed Beef Cattle

Bicheng Zou †, Fan Long †, Fuguang Xue, Chuanbin Chen, Xian Zhang, Mingren Qu and Lanjiao Xu *

Jiangxi Province Key Laboratory of Animal Nutrition, Engineering Research Center of Feed Development, Jiangxi Agricultural University, Nanchang 330045, China; zoubicheng1997@163.com (B.Z.); longfan9807@163.com (F.L.); xuefuguang1024@jxau.edu.cn (F.X.); chenchuanb8297@163.com (C.C.); zhangxian517@163.com (X.Z.); qumingren@jxau.edu.cn (M.Q.)

* Correspondence: xulanjiao@jxau.edu.cn; Tel.: +86-139-70885364

† These authors contributed equally to this work.

Simple Summary: Heat stress can easily cause a series of physiological hazards in beef cattle, affecting the rumen barrier function and, ultimately, beef cattle performance. As a B vitamin with anti-inflammatory and anti-heat stress effects, it is essential to investigate whether niacin has a protective and ameliorative effect on the rumen barrier function in heat-stressed beef cattle. In this study, the plasma levels of IL-1β, IL-2, IL-6, and TNF-α decreased, IL-4 increased, and LPS decreased in beef cattle supplemented with niacin. These indicators are closely related to inflammation and intestinal permeability. The expression of the tight junction proteins mRNA ZO-1 and Occludin, which are closely associated with rumen barrier function, was enhanced. Histological observations showed no damage to the rumen epithelium. The results of this study suggest that niacin supplementation in heat-stressed beef cattle can reduce inflammation, improve intestinal permeability, and protect the rumen epithelial barrier function.

Abstract: The present study investigates the theoretical basis for maintaining normal physiological functions in heat-stressed beef cattle by exploring the effects of niacin supplementation on the permeability of the rumen epithelial cell barrier. Herein, 12 Jinjiang bulls with an average weight of approximately 400 ± 20.0 kg were randomly divided into three groups, thermoneutral (TN), heat-stressed (HS), and heat-stressed niacin-supplemented (HN) groups, with 4 bulls in each group. The experiment spanned 70 days, and the plasma concentrations of D-lactic acid, diamine oxidase (DAO), lipopolysaccharides (LPSs), and inflammatory cytokines were analyzed. Additionally, we assessed the gene expression of tight junction proteins to understand the effect of niacin supplementation on heat-stressed beef cattle. Our results revealed that heat stress significantly increased the D-lactic acid and LPS levels in beef cattle plasma on days 30 and 45 of the experiment ($p < 0.05$). Moreover, it led to a significant rise in DAO levels on day 30 ($p < 0.05$). Niacin supplementation significantly reduced the LPS levels on day 30 ($p < 0.05$). Heat stress significantly elevated the plasma concentrations of inflammatory cytokines interleukin-1β (IL-1β), IL-2, IL-6, and tumor necrosis factor-α (TNF-α) ($p < 0.05$), while reducing the IL-4 concentration ($p < 0.05$). However, niacin supplementation effectively mitigated the concentrations of these inflammatory factors by reducing IL-1β, IL-2, IL-6, and TNF-α concentrations and increasing IL-4 concentrations. The mRNA expressions of tight junction proteins zonula occluden-1 (ZO-1), claudin-1, claudin-4, and claudin-7 were significantly downregulated ($p < 0.05$) in the HS group compared to those in the TN group, and those of ZO-1 and occludin were significantly upregulated ($p < 0.05$) in the HN group compared to those in the HS group. Notably, no significant differences were observed in ruminal papillae length and width among the studied groups ($p > 0.05$). Our findings indicate that heat stress adversely impacted the tight junction structure of the rumen epithelium, leading to a significant reduction in the expression of tight junction protein mRNA. Consequently, heat stress impaired the rumen mucosal barrier function, resulting in increased intestinal permeability. The mechanism underlying this effect may be associated with the decreased expression of tight junction protein genes in the rumen epithelial cells. However, niacin supplementation mitigated the detrimental effects of heat stress on intestinal

Citation: Zou, B.; Long, F.; Xue, F.; Chen, C.; Zhang, X.; Qu, M.; Xu, L. Protective Effects of Niacin on Rumen Epithelial Cell Barrier Integrity in Heat-Stressed Beef Cattle. *Animals* **2024**, *14*, 313. https://doi.org/10.3390/ani14020313

Academic Editors: Dingfu Xiao and Tiantian Meng

Received: 24 November 2023
Revised: 28 December 2023
Accepted: 5 January 2024
Published: 19 January 2024

Copyright: © 2024 by the authors. Licensee MDPI, Basel, Switzerland. This article is an open access article distributed under the terms and conditions of the Creative Commons Attribution (CC BY) license (https://creativecommons.org/licenses/by/4.0/).

permeability in beef cattle and increased the expression of tight junction protein genes in the rumen epithelium, thereby effectively protecting the rumen barrier in heat-stressed beef cattle. These results highlight the potential of nicotinic acid as a protective agent against the negative impacts of heat stress on intestinal integrity in beef cattle.

Keywords: niacin; heat stress; Jinjiang bulls; rumen epithelial barrier; tight junction

1. Introduction

Heat stress in beef cattle is known to induce oxidative stress and inflammation, leading to the impairment of the rumen epithelial barrier and an increased risk of health issues such as rumen microbial dysregulation and rumen acidosis [1]. As a consequence, the overall growth performance of the cattle may be reduced [2]. As the rumen plays a vital role in feed digestion and ruminant nutrition, maintaining the integrity of the rumen epithelial barrier function is crucial for efficient digestion. The rumen epithelium serves as a protective barrier against the luminal contents, engaging in essential functions such as nutrient absorption, pH regulation, immune defense, and overall barrier protection [3].

Heat stress disrupts the metabolism of rumen papillae and alters the rumen fermentation of the beef cattle, which subsequently affects the epithelial barrier function [4]. Micromorphological observations have suggested that heat stress exacerbates the exfoliation of the stratum corneum, leading to the destruction of the rumen epithelium's physical barrier [4]. Therefore, ensuring the homeostasis of ruminal epithelial barrier function is crucial for the health of ruminants. The junctions between rumen epithelial cells, particularly tight junctions (TJs), are essential for maintaining the integrity of the rumen epithelium. These junctions, from the top to the basement membrane, include TJs, gap junctions, adhesion junctions, and desmosome junctions [5,6].

Niacin, a water-soluble vitamin feed additive, is widely used to reduce the incidence of ketosis during early lactation and alleviate heat stress in dairy cows [7,8]. Our previous studies have demonstrated that niacin supplementation improves the performance and rumen fermentation function of heat-stressed beef cattle, which lead to alterations in the rumen microbiota structure [9]. The rumen epithelium holds a unique position for host–microbe interactions, which profoundly impact the net nutrient utilization of the host [10,11]. There exists a direct association between rumen bacteria and the host's capacity to efficiently absorb nutrients and maintain the integrity of the epithelial barrier in the rumen. Correlation analyses further indicate that the rumen microbiota might regulate the immunological barrier function and rumen fermentation metabolism genes, consequently affecting genes linked to rumen epithelial transcription [12,13]. These findings suggest that niacin-induced changes in the rumen epithelial barrier function might be related to its impact on the rumen microbiota.

Furthermore, previous studies conducted by Digby et al. and Gambhir et al. have suggested that niacin-induced activation of GPR109A leads to the inhibition of inflammatory cytokine and chemokine expression in epithelial cells [14,15]. It was reported that niacin had anti-inflammatory functions, potentially enhancing the function of the rumen barrier [16]. Given these findings, the present study aims to investigate the effect of niacin supplementation on the permeability of the rumen epithelial cell barrier in heat-stressed beef cattle. Our hypothesis posits that heat stress may trigger tissue inflammation and disrupt the barrier function of rumen epithelial cells, while niacin supplementation could potentially protect the rumen barrier function. To elucidate these effects, we conducted an examination of the micromorphology and gene expression patterns of rumen epithelial cells in heat-stressed beef cattle, focusing on the role of niacin supplementation.

2. Materials and Methods

2.1. Animal Ethics

The experimental procedures and protocols employed in this study were subjected to rigorous review and approval by the Committee for the Care and Use of Experimental Animals at Jiangxi Agricultural University (JXAULL-20210066).

2.2. Diet, Experiment Design, and Animals

The experimental trial was conducted in Gao'an, Jiangxi Province, during the summer of 2020 (July to September). Twelve Jinjiang bulls, with an average age of 24 ± 2 months and a weight of 400 ± 20.0 kg, were selected for this study. The rearing trial spanned 70 days, including a 10-day acclimatization period followed by a 60-day experimental period. The animals were randomly assigned to three treatment groups: thermoneutral (TN), heat-stressed (HS), and heat-stressed niacin-supplemented (HN) groups. In the TN group, the beef cattle were transported to a climate-controlled chamber and maintained under thermoneutral conditions, with a temperature of 26 ± 2 °C and humidity of 30 ± 2.7%. They were fed a basal diet during the experiment and were in a state of non-heat stress [mean temperature–humidity index (THI) = 71.10 ± 0.62%]. Beef cattle in the HS and HN groups were housed in the same high-temperature, high-humidity barn during the summer months of July through September, and both experienced heat stress, with an average daily temperature of 31.55 ± 2.61 °C, an average daily humidity of 66.66 ± 8.66%, and THI > 72.The HS group was fed a basal diet. And in the HN group, 800 mg/kg of niacin (DM basis) was added to the concentrate, and during the preparation process, niacin was first added to the premix and mixed well and then mixed step by step to form the concentrate (Jiangxi Brother Medicine Co., Ltd., Nanchang, China). All animals were fed twice daily at 6:00 and 16:00. They had constant access to freshwater through automatic waterers. The basal diets and nutritional group compositions are detailed in Table 1.

Table 1. Composition and nutrient levels of the basal diet (air dry basis).

Items	Content (%)
Ingredients,%	
Brewers grains	20.00
Rice straw	20.00
Corn	46.80
Soybean meal	9.00
Premix [a]	2.40
$NaHCO_3$	1.20
NaCl	0.60
Total	100.00
Nutrient composition,%	
DM	95.67
CP	14.44
EE	4.03
Ash	8.31
NDF	26.99
ADF	11.62
Ca	0.69
P	0.36

[a] One kilogram of premix provided the following: vitamin A 150,000 IU, vitamin D3 20,000 IU, vitamin E 3000 IU, Fe 3200 mg, Mn 1500 mg, Zn 2000 mg, Cu 650 mg, I 35 mg, Se10 mg, Co 10 mg, Ca 130 g, P 30 g.

2.3. Temperature–Humidity Index Measurement

During the trial, measurements of relative humidity and temperature inside the barns of cattle in the HS and HN groups were taken at 07:30, 13:30, and 19:30 each day.

Subsequently, the average daily temperature, average relative humidity, and average daily THI were calculated. The THI was calculated using the following formula:

$$THI = (1.8 \times T + 32) - [(0.55 - 0.0055 \times RH) \times (1.8T - 26.8)],$$

where T represents the dry-bulb temperature (°C), and RH denotes the relative humidity (%) [17].

2.4. Sample Collection and Analysis

Blood samples were collected on the specified sampling days, that is, days 1, 15, 30, 45, and 60, and beef cattle fasted for 12 h before sampling. The samples were obtained via the jugular vein and collected into 5 mL vacuum tubes containing sodium heparin. Subsequently, they were subjected to centrifugation at 3500× *g* for 15 min at 4 °C to separate the plasma. The separated plasma was then preserved at −20 °C for subsequent analysis.

2.5. Intestinal Permeability Assay

To assess intestinal permeability, the levels of D-lactic acid, diamine oxidase (DAO), and lipopolysaccharide (LPS) in the plasma were measured. For this purpose, commercially available enzyme-linked immunosorbent assay (ELISA) kits from Shanghai Enzyme Link Biotechnology Co., Ltd. (Shanghai, China) were utilized.

2.6. Inflammatory Cytokines Assays

The concentrations of key inflammatory cytokines interleukin-1 (IL-1), IL-2, IL-4, IL-6, and tumor necrosis factor-α (TNF-α) in the plasma were evaluated using commercially available radioimmunoassay kits (the Beijing Huaying Institute of Biotechnology, Beijing, China).

2.7. Histologic Analysis

On day 60 of the trial, following a 12 h fast, all 12 Jinjiang cattle were slaughtered using the captive bolt knockout method, adhering to national Standard Operating Procedures (GB/T 19477-2004, China [18]). Subsequently, ruminal tissue samples were taken after washing with phosphate-buffered saline. Rumen tissue samples with a total thickness of 1~3 cm were removed from the ventral sac and preserved in a 4% paraformaldehyde solution for histological examination. Additionally, 10 g of rumen epithelium was obtained from the ventral sac site of the rumen, snap-frozen in liquid nitrogen, and stored at −80 °C for subsequent RNA extraction.

The tissue samples were sectioned into 1 × 1 cm pieces and placed in a 4% paraformaldehyde solution for histomorphological analysis. The morphology of the rumen epithelium was examined using hematoxylin and eosin (H&E) staining. After paraffin embedding, the samples were cut into 5 μm sections on the longitudinal plane and mounted on slides. The slides were stained with H&E using conventional methods. The sections were examined under a DM3000 microscope (Leica, Wetzlar, Germany) with three replicates per sample. The height and width of the papillae were measured using the Image-Pro program. Six papillae were randomly selected for each sample, and the average value was calculated.

2.8. Quantitative Real-Time Polymerase Chain Reaction

Total RNA was extracted from rumen tissue samples using the TransZol UP Plus RNA Kit (TransGen Biotech, Beijing, China) following the manufacturer's instructions. The integrity and purity of RNA were detected by gel electrophoresis. And the purity and concentration of RNA samples were determined with a NanoDrop ND-2000 1 spectrophotometer (Thermo Fisher Scientific Inc., Waltham, MA, USA) to ensure that the ratio of A260/A280 nm ratio was between 1.8 and 2.0. Total RNA (1 μg) was then reverse-transcribed into cDNA in a final volume of 20 μL using the cDNA Synthesis SuperMix kit as per the manufacturer's instructions. Primers used for the quantitative real-time polymerase chain reaction (PCR) were provided by Shanghai Jierui Bioengineering Co., Ltd.

(Shanghai, China) and are listed in Table 2. β-actin was used as an internal reference. The quantitative real-time PCR was performed using the PerfectStartTM Green qPCR SuperMix kit, following the manufacturer's directions, using a CFX Connect Real-Time PCR Detection System (Bio-Rad, Hercules, CA, USA). The following methodology was used for the PCR amplification: 42 cycles at 94 °C for 30 s, 5 s, and 62 °C for 30 s of annealing. The data from the quantitative PCR were analyzed using the $2^{-\Delta\Delta Ct}$ method.

Table 2. Information of target genes and primers.

Genes	Accession No.	Primer Sequences (5′ to 3′ Direction)
ZO-1	XM_024982002.1	Forward: TCTCGAAGATAGCCCTGCTG Reverse: AGGTCAAGCAGGAAGAGGAC
Claudin-1	NM_001001854.2	Forward: CTTCATCCTGGCGTTTCTGG Reverse: AGACTTTGCACTGGATCTGC
Claudin-4	XM_027527588.1	Forward: GCAACGACAAGCCCTACTC Reverse: AGCTCAGTCCAGGGAGAAAC
Claudin-7	XM_027519738.1	Forward: ATTCTGAAGGCGGAAATGGC Reverse: GGGCGGTGATGATATTGTCG
Occludin	NM_001082433.2	Forward: AGATGCACGTTCGACCAATG Reverse: ATTACTCCGGGAGGAGAGGT
β-actin	NM_173979.3	Forward: CCCTGGAGAAGAGCTACGAG Reverse: CAGGAAGGAAGGCTGGAAGA

2.9. Statistical Analyses

Statistical analyses were performed using SPSS 23.0 software (IBM Corp, Armonk, NY, USA). Firstly, all experiment data were tested for normal distribution, after which independent samples *t*-tests were conducted to compare the TN and HS groups as well as the HS and HN groups. A significance level of $p < 0.05$ was considered as indicating a significant difference. The various indicators of intestinal mucosal permeability (D-lactate, DAO, and LPS), inflammatory cytokines, TJ protein gene expression, and rumen epithelial tissue were analyzed using these statistical tests.

3. Results

3.1. Subsection

3.1.1. Temperature-Humidity Index

At present, THI is the commonly used index to evaluate the degree of heat stress in livestock and poultry; when THI > 72, the animal is in a state of heat stress, of which, 72 < THI < 78 is mild heat stress, 78 < THI < 88 is moderate heat stress, and THI > 88 is severe heat stress. As depicted in Figure 1, over the entire duration of the experiment, the average daily temperature of the barn housing the test beef cattle from the HS and HN groups was 31.55 °C, with an average daily humidity of 66.66%. The average daily THI recorded was 82.74, encompassing the highest and lowest THI values of 89.89 and 73.88, respectively. Notably, 54 days fell within the THI range of 78–88, indicating that the test beef cattle in both groups experienced thermal stress conditions.

3.1.2. Intestinal Permeability

As shown in Table 3, compared to those in the TN group, the D-lactate levels in beef cattle within the HS group were significantly higher on days 30 and 45 ($p < 0.05$). Concurrently, on day 30, the levels of DAO in beef cattle from the HS group were significantly elevated in comparison to those in the TN group ($p < 0.05$). No significant differences in D-lactate and DAO levels were observed between the HS and HN groups. Moreover, the LPS levels were significantly elevated in the HS group than those in the TN group on days 30 and 45 ($p < 0.05$). Notably, the LPS levels were significantly lower in the HN group than those in the HS group on day 30 ($p < 0.05$).

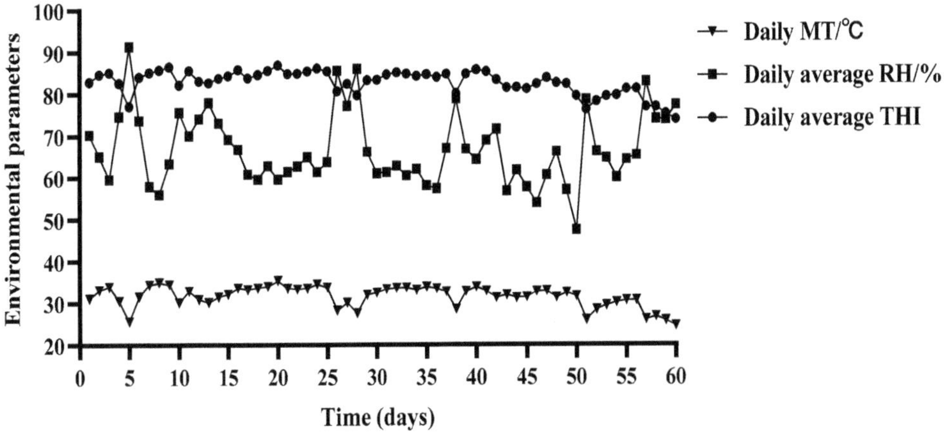

Figure 1. Average daily temperature, relative humidity, and THI of the barn in the HS and HN groups.

Table 3. Effects of niacin on plasma D-lactic acid, DAO, and LPS levels in heat-stressed beef cattle.

Item		TN	HS	HN	*p*-Value	
					TN vs. HS	HS vs. HN
D-lactic acid (umol/mL)	Day 1	15.77 ± 1.88	16.68 ± 2.08	15.41 ± 2.19	0.756	0.906
	Day 15	13.61 ± 2.33	13.66 ± 2.36	12.66 ± 2.07	0.990	0.760
	Day 30	18.69 ± 1.19	23.85 ± 0.92	22.09 ± 0.86	0.014	0.213
	Day 45	16.93 ± 0.69	22.13 ± 1.00	19.67 ± 0.67	0.005	0.086
	Day 60	16.8 ± 1.67	18.03 ± 1.03	16.14 ± 0.20	0.340	0.124
DAO (ng/mL)	Day 1	5.83 ± 1.28	6.71 ± 0.84	6.62 ± 0.95	0.586	0.945
	Day 15	5.93 ± 1.28	6.39 ± 0.3	6.04 ± 0.9	0.617	0.535
	Day 30	7.86 ± 0.29	10.4 ± 0.76	9.33 ± 0.63	0.020	0.260
	Day 45	8.38 ± 0.62	10.13 ± 0.73	9.32 ± 0.42	0.117	0.788
	Day 60	7.45 ± 1.39	8.31 ± 0.79	8.2 ± 0.84	0.609	0.936
LPS (EU/L)	Day 1	12.14 ± 1.90	15.62 ± 1.79	12.78 ± 1.25	0.231	0.241
	Day 15	14.97 ± 1.21	18.85 ± 1.46	16.91 ± 1.58	0.087	0.402
	Day 30	15.44 ± 2.22	24.91 ± 2.22	15.83 ± 0.71	0.013	0.020
	Day 45	13.73 ± 1.50	27.31 ± 2.17	22.76 ± 2.10	0.002	0.125
	Day 60	13.51 ± 1.88	18.67 ± 2.01	17.04 ± 2.02	0.110	0.589

Note: The results are presented as the mean values and standard error. TN = thermoneutral treatment group; HS = heat stress treatment group; HN = heat stress supplemented with niacin treatment group.

3.1.3. Plasma Cytokine Profile

As shown in Table 4, compared with the TN group, plasma IL-1β, IL-2, IL-6, and TNF-α levels in the HS group were significantly increased on days 1, 15, 30, 45, and 60 ($p < 0.05$), and plasma IL-4 levels in the HS group were significantly decreased on days 1, 15, 30, and 45 ($p < 0.05$). Furthermore, compared with the HS group, IL-1 β levels in the HN group were significantly decreased on days 1 and 15 ($p < 0.\ 05$), IL-2 levels were significantly decreased on days 1, 15, 30, and 60 ($p < 0.\ 05$), IL-4 levels were significantly increased on days 15, 30, and 45 ($p < 0.05$), IL-6 levels were significantly decreased on days 1 and 45 ($p < 0.05$), and TNF-α levels were significantly decreased on days 15, 30, 45, and 60 ($p < 0.05$). Collectively, the impact of heat stress resulted in elevated plasma levels of IL-1β, IL-2, IL-6, and TNF-α, along with decreased levels of IL-4. Notably, niacin supplementation led to decreased plasma levels of IL-1β, IL-2, IL-6, and TNF-α and increased levels of IL-4 in heat-stressed beef cattle.

Table 4. Effects of niacin on plasma inflammatory cytokines in heat-stressed beef cattle.

Item (pg/mL)		TN	HS	HN	*p*-Value	
					TN vs. HS	HS vs. HN
IL-1β	Day 1	16.40 ± 1.52	27.46 ± 0.87	23.63 ± 0.83	0.001	0.019
	Day 15	18.17 ± 1.22	27.84 ± 1.80	21.86 ± 1.28	0.004	0.035
	Day 30	16.48 ± 1.26	26.06 ± 1.83	21.19 ± 1.54	0.005	0.087
	Day 45	17.13 ± 2.12	27.42 ± 1.80	21.66 ± 1.63	0.01	0.056
	Day 60	16.32 ± 1.52	26.76 ± 1.84	21.83 ± 1.13	0.005	0.063
IL-2	Day 1	176.57 ± 19.55	313.21 ± 9.53	269.14 ± 8.02	0.001	0.012
	Day 15	189.65 ± 22.12	314.54 ± 21.81	239.82 ± 5.53	0.007	0.016
	Day 30	183.95 ± 17.15	298.66 ± 22.10	231.74 ± 14.15	0.006	0.043
	Day 45	188.20 ± 25.47	314.70 ± 20.99	247.27 ± 23.21	0.009	0.075
	Day 60	206.69 ± 31.41	314.63 ± 18.15	256.18 ± 13.96	0.025	0.043
IL-4	Day 1	10.76 ± 0.87	6.25 ± 0.22	6.68 ± 0.39	0.002	0.369
	Day 15	10.19 ± 0.96	6.21 ± 0.45	8.31 ± 0.64	0.01	0.036
	Day 30	10.35 ± 1.13	6.68 ± 0.52	8.79 ± 0.28	0.026	0.012
	Day 45	10.54 ± 1.39	6.31 ± 0.43	8.33 ± 0.51	0.027	0.022
	Day 60	10.21 ± 1.54	6.48 ± 0.43	7.68 ± 0.39	0.059	0.085
IL-6	Day 1	105.27 ± 7.19	163.80 ± 5.03	139.81 ± 3.94	0.001	0.009
	Day 15	110.06 ± 4.69	166.25 ± 10.49	135.93 ± 6.89	0.003	0.052
	Day 30	111.78 ± 8.55	155.99 ± 10.43	128.08 ± 7.74	0.017	0.075
	Day 45	109.33 ± 9.64	163.76 ± 10.53	127.14 ± 6.73	0.009	0.026
	Day 60	113.20 ± 14.36	159.93 ± 10.88	136.11 ± 6.03	0.041	0.104
TNF-α	Day 1	44.20 ± 3.94	71.76 ± 3.33	64.84 ± 1.82	0.002	0.118
	Day 15	49.37 ± 3.80	75.35 ± 5.07	59.19 ± 2.88	0.006	0.032
	Day 30	43.45 ± 2.80	70.34 ± 5.08	55.25 ± 2.84	0.004	0.041
	Day 45	46.35 ± 5.34	74.13 ± 5.02	56.06 ± 3.45	0.009	0.025
	Day 60	49.94 ± 6.73	74.82 ± 4.25	60.68 ± 3.04	0.020	0.035

Note: The results are presented as the mean values and standard error. TN = thermoneutral treatment group; HS = heat stress treatment group; HN = heat stress supplemented with niacin treatment group.

3.1.4. The Expression and Histological Analysis of Tight Junction Protein-Related Genes in Rumen Epithelium

The relative mRNA expressions of TJ proteins zonula occluden-1 (ZO-1), claudin-1, claudin-4, and claudin-7 were significantly lower in the HS group than in the TN group ($p < 0.05$). However, no significant differences were observed in the relative expression of the occludin gene between the TN and HS groups (Figure 2). The relative expression of the TJ protein ZO-1 and occludin genes was significantly higher in rumen epithelial cells of the HN group cattle than that in those of the HS group ($p < 0.05$). Notably, no significant difference was observed in the relative expression of claudin-1, claudin-4, and claudin-7 genes between the HS and HN groups (Figure 3). In terms of histological observations, the TN group cattle displayed an intact corneum layer (Figure 4A,B). In contrast, the corneum layer of the HS group cattle exhibited signs of damage and shedding (Figure 4C,D). The HN group cattle displayed a minor lesion compared to the cuticle (Figure 4E,F). Notably, as shown in Table 5, no significant differences were observed in ruminal papillae length and width among the treatment groups.

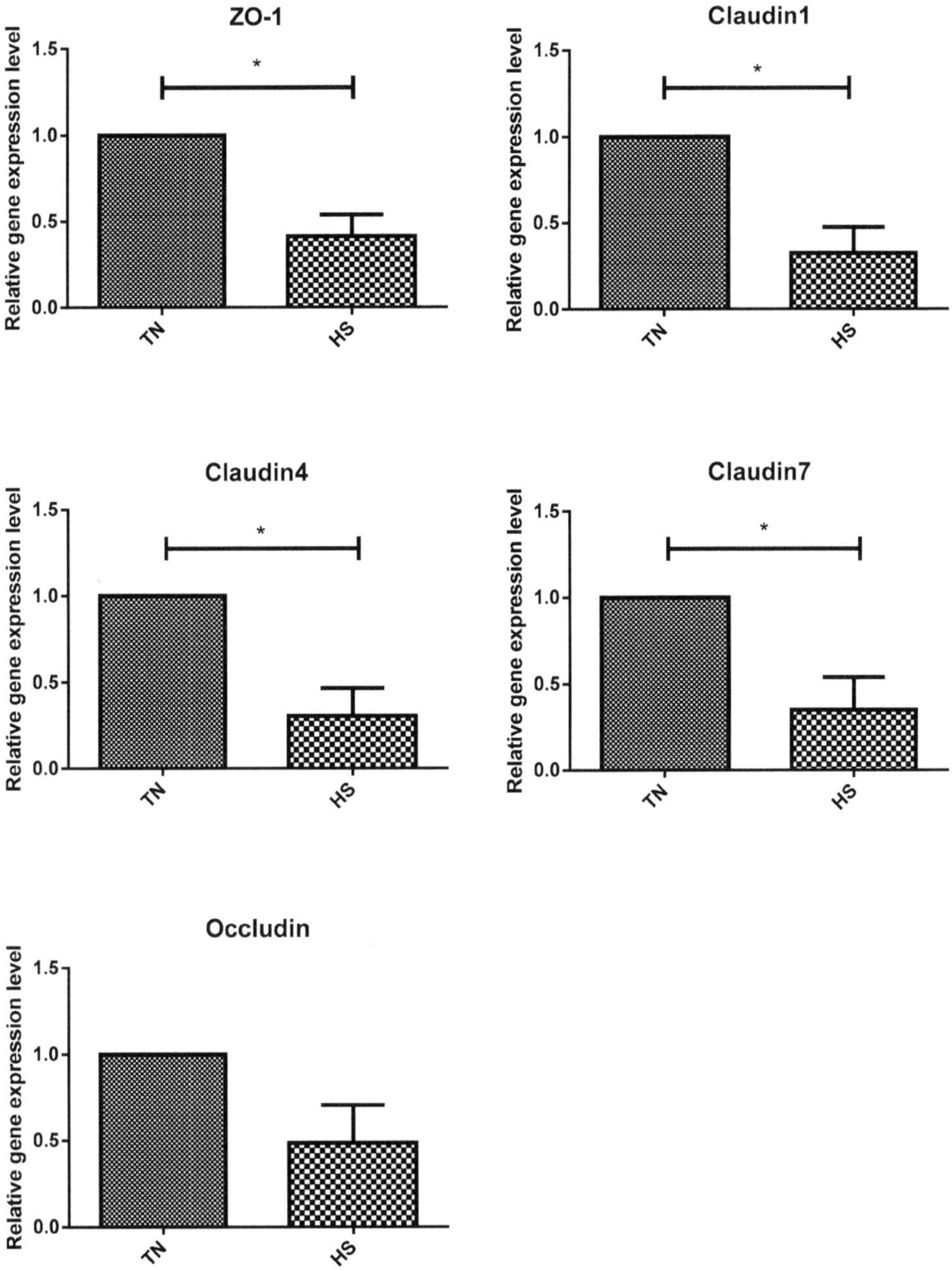

Figure 2. The effect of heat stress on the expression of tight junction protein gene in beef cattle. TN = thermoneutral group. HS = heat stress group. An "*" indicates a significant difference ($p < 0.05$).

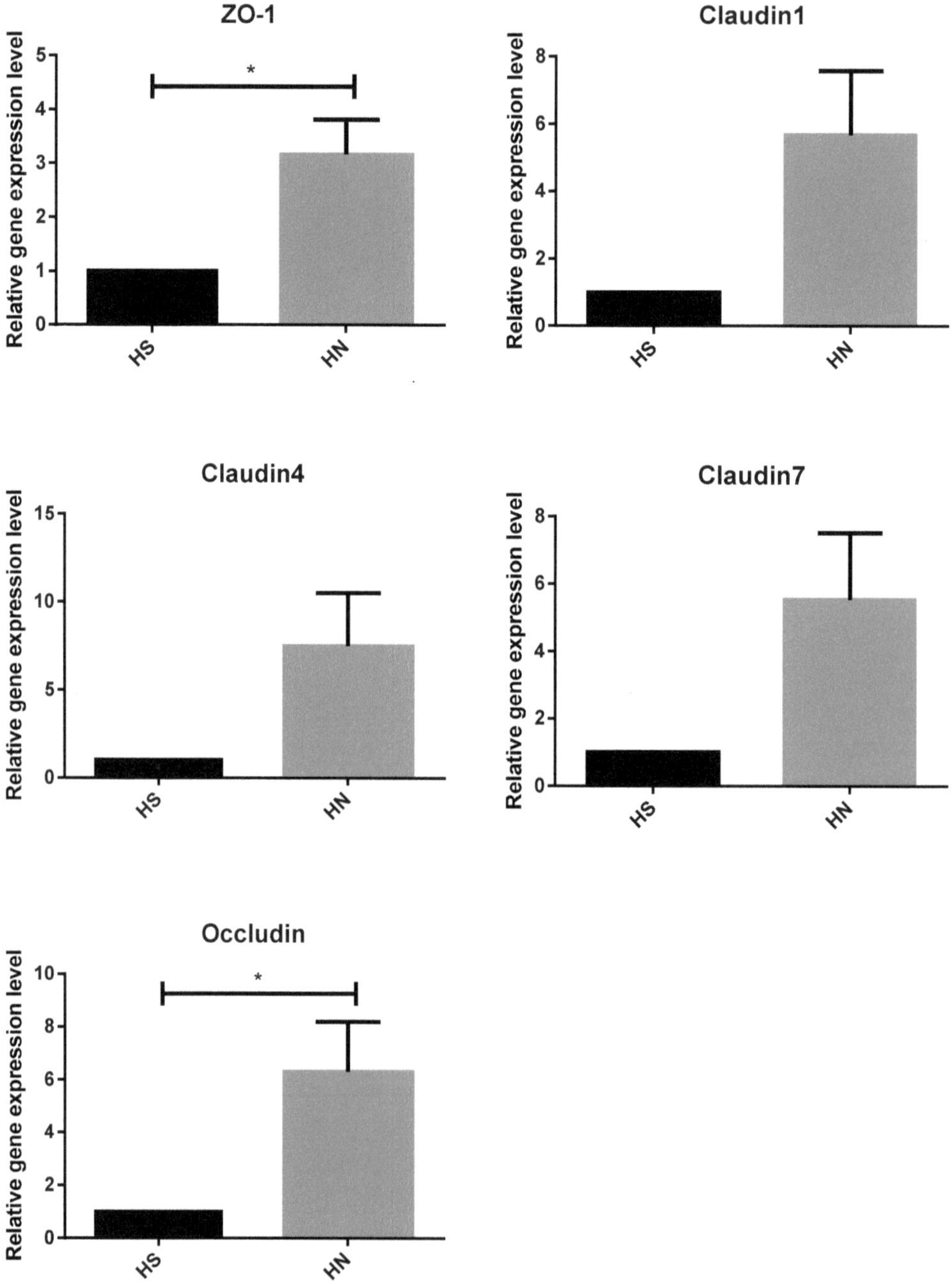

Figure 3. Effects of niacin on the expression of tight junction protein gene in heat-stressed beef cattle. HS = heat stress group. HN = heat stress supplemented with niacin treatment group. An "*" indicates a significant difference.

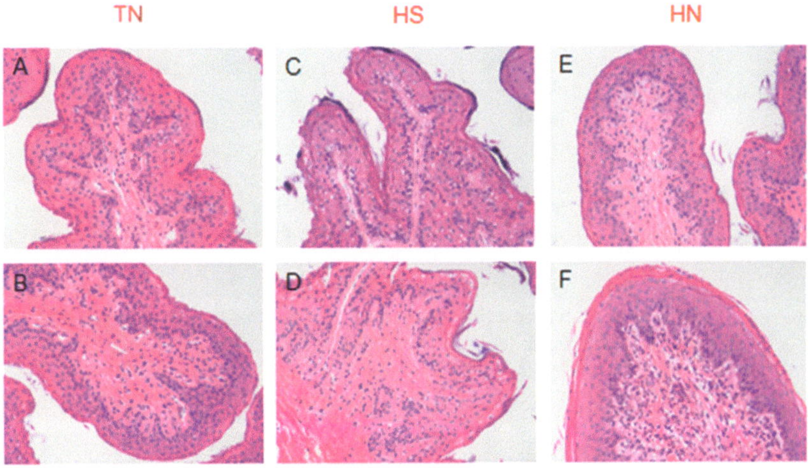

Figure 4. Representative histology sections of the rumen papillae of beef cattle from the thermoneutral group (TN), heat stress group (HS), and heat stress supplemented with niacin group (HN). Representative micrographs of H&E staining are shown (4 × 100 multiple microscope). (**A**,**B**) belong to the TN group. (**C**,**D**) belong to the HS group. (**E**,**F**) belong to the HN group.

Table 5. Effects of niacin on length and width of rumen papillae in heat-stressed beef cattle.

Item	TN	HS	HN	*p*-Value	
				TN vs. HS	HS vs. HN
Rumen papillae length (μm)	1599.07 ± 118.56	1482.88 ± 166.96	1561.53 ± 210.01	0.591	0.779
Rumen papillae width (μm)	405.53 ± 61.33	559.32 ± 98.65	392.96 ± 42.67	0.234	0.173

Note: The result is presented as the mean values and standard error. TN = thermoneutral treatment group; HS = heat stress treatment group; HN = heat stress supplemented with niacin treatment group.

4. Discussion

4.1. Effects of Niacin on Intestinal Permeability and Inflammatory Cytokines in Heat-Stressed Beef Cattle

The preservation of the integrity of the ruminal epithelial barrier function holds significant importance in upholding ruminant gut immunity and overall body health. The rumen serves as a semipermeable paracellular diffusion barrier that plays a multifaceted role in various biological functions, encompassing the modulation of nutrient absorption and the mitigation of immunoreactions [19]. Maintaining the rumen homeostasis is crucial, as disruptions such as heightened release of bacterial LPSs can compromise the functionality of the ruminal epithelial barrier. The absorptive function of the rumen must be highly selective to prevent the simultaneous influx of microorganisms and toxins from the rumen into the bloodstream [20]. Research indicates a direct correlation between the host's inflammatory response and the presence of specific aberrant metabolites in the rumen, particularly LPS [21]. The presence of LPS in the rumen could potentially augment permeability of the rumen wall, thereby impacting the structural integrity of the rumen epithelium and potentially impairing its barrier function [22]. LPS and the substantial release of inflammatory factors can affect the TJs of rumen epithelial cells, impair the barrier function of rumen epithelial cells, and result in elevated permeability of the rumen mucosa [23].

Elevated LPS concentrations within the systemic circulation can serve as a reliable marker, indicating the potential severity of damage to the integrity of the intestinal mucosal barrier [24]. Our findings reveal that LPS plasma levels were significantly higher in beef cattle from the HS group than in those from the TN group. Notably, a significant difference was observed on days 30 and 45 of the experimental period, respectively. These findings indicate a progressive increase in LPS levels with an increase in the duration of heat stress, contributing to heightened intestinal permeability and a discernible impairment of the rumen barrier function. The supplementation of niacin introduced a mitigating effect on LPS levels, suggesting a potential safeguarding role of niacin against increased intestinal permeability in heat-stressed beef cattle.

Heat stress elicits an enhanced expression of pro-inflammatory cytokines and a concomitant reduction in anti-inflammatory cytokines in broilers [25]. Existing research underscores the intrinsic relationship between ruminal epithelial inflammation and the consequential disruption of barrier structure and function [26]. Feng et al. revealed that niacin supplementation effectively downregulates the expression of TNF-α, IL-8, γ-interferon (IFN-γ), and IL-1β [27]. The increase in the concentration of LPS within the systemic circulation serves as a potential indicator signifying the development of a severe injury to the integrity of the intestinal mucosal barrier. Herein, heat stress induced a significant increase in the levels of specific inflammatory cytokines IL-1β, IL-2, IL-6, and TNF-α in the bloodstream of beef cattle, and the release of these pro-inflammatory factors was associated with LPS-induced activation of NF-κB and MAPK pathways, indicating that heat stress not only causes an increase in the intestinal permeability but also induces a systemic inflammatory response. However, niacin supplementation caused a significant decrease in the levels of these indicators. This aligns with the findings of Feng et al. [27], confirming niacin's potential to ease inflammation in heat-stressed cattle. Niacin likely relieves the inflammatory damage caused by heat stress in beef cattle by reducing the levels of pro-inflammatory factors IL-1β, IL-2, IL-6, and TNF-α, while boosting the level of the anti-inflammatory factor IL-4.

4.2. Effects of Niacin on Gene Expression and Tissue Structure of Tight Junction Protein in Heat-Stressed Beef Cattle

It has been well-established in scientific literature that disruptions in rumen metabolism can trigger the release of bacterial LPSs or other harmful substances, thereby compromising the integrity of rumen epithelial cells and the intercellular TJs. Consequently, there is a subsequent reduction in the absorption rate of nutrients across the epithelial lining, an elevation in permeability, impairment of barrier functionality, and an increased translocation of bacterial toxins into the bloodstream. These interconnected events can trigger systemic pathological responses, ultimately causing a significant impact on animal productivity. Animals subjected to heat stress exhibit compromised intestinal health, accompanied by a malfunction of the intestinal barrier [28–30], which is consistent with the findings of the present study. Our study substantiates that heat stress may increase intestinal permeability, exerting a detrimental effect on the intestinal health of beef cattle. This assertion is supported by our assessments of both the intestinal permeability index such as D-lactic acid, DAO, and LPS, and the levels of plasma inflammatory cytokines.

The findings of the present study unveiled a substantial reduction in the mRNA relative expression levels of TJ proteins ZO-1, claudin-1, claudin-4, and claudin-7 in rumen epithelial cells under the influence of heat stress. The significance of direct cellular interactions mediated by TJs is paramount in establishing selective barriers. These TJs are constituted by a complex interplay of integral transmembrane proteins, notably claudins and occludins, along with the adapter proteins ZO-1, ZO-2, and ZO-3 [31–33]. The expression of occludin has been correlated with a spectrum of junctional functions, revealing that augmenting occludin expression enhances barrier function and transmembrane resistance in mammalian epithelial cells [34,35].

Moreover, claudins play an indispensable role in upholding ion permeability and preserving the barrier function at the intercellular junctions [36,37]. Previous research has demonstrated that the overexpression of claudin-1 leads to the enhancement of the barrier function [38]. Notably, ZO-1 is a crucial TJ component that engages in critical interactions with both occludins and claudins, underscoring its pivotal role [39]. Therefore, it is plausible to hypothesize that the effect of heat stress on the rumen epithelium could have incited a responsive mechanism, potentially downregulating the expression of closely interrelated genes within the cellular framework and thereby disrupting the rumen barrier's functionality. Herein, the significant upregulation in the expression of ZO-1 and occludin genes after niacin supplementation suggests a reinforcing effect of niacin on the protective function of the rumen barrier. These positive effects could potentially stem from the enhancement of rumen epithelium metabolism after niacin supplementation.

Our findings from the micromorphological analysis that heat stress aggravates cuticle shedding and contributes to the disruption of the physical barrier of the rumen epithelium are consistent with those of a previous study [4]. Heat stress caused detrimental alterations to the tissue structure of the rumen epithelium and particularly resulted in cuticle degradation. Such disruptions could facilitate the entry of pathogenic bacteria and toxic substances into the body through the compromised rumen wall, consequently impinging on physiological processes. This phenomenon is likely underpinned by the downregulation of ZO-1, claudin-1, and occludin expression in the rumen epithelium. However, niacin supplementation demonstrated a mitigating effect on this condition, which could be attributed to its potential to attenuate the inflammatory response and upregulate TJ protein gene expression.

5. Conclusions

In conclusion, this study underscores that heat stress increases intestinal permeability, disrupts rumen epithelial cell integrity and barrier function, and influences the relative expression of TJ genes. Niacin supplementation emerges as a potential remedy to counteract these heat-induced effects. By attenuating the damage to rumen epithelium caused by heat stress and augmenting the relative expression of TJ protein-related genes, niacin supplementation maintains the integrity of the rumen epithelial barrier function. This intervention holds promise for improving the health and productive performances of livestock under conditions of heat stress.

Author Contributions: L.X. and M.Q. designed the overall study. B.Z., F.L., C.C., X.Z., F.X. and M.Q. performed the experiments. The manuscript was written by B.Z. and L.X. The article's submission was reviewed and approved by all of the authors. All authors have read and agreed to the published version of the manuscript.

Funding: This work was supported by the National Technical System of Beef Cattle Yak Industry (CARS-37), Natural Science Foundation of Jiangxi Province (20192BAB204020), National Natural Science Foundation of China (32260852), and National Natural Science Foundation of China (32060760).

Institutional Review Board Statement: The experimental procedures and protocols used in this study were reviewed and approved by the Committee for the Care and Use of Experimental Animals at Jiangxi Agricultural University (JXAULL-20210066) (June 2020).

Informed Consent Statement: Informed consent was obtained from all subjects involved in the study.

Data Availability Statement: The data presented in this study are available in [Protective Effects of Niacin on Rumen Epithelial Cell Barrier Integrity in Heat-Stressed Beef Cattle].

Acknowledgments: The authors appreciate all of the help from their colleagues and collaborators.

Conflicts of Interest: The authors declare no conflicts of interest.

References

1. Kaldur, T.; Kals, J.; Ööpik, V.; Zilmer, M.; Zilmer, K.; Eha, J.; Unt, E. Effects of heat acclimation on changes in oxidative stress and inflammation caused by endurance capacity test in the heat. *Oxidative Med. Cell. Longev.* **2014**, *2014*, 107137. [CrossRef] [PubMed]
2. St-Pierre, N.R.; Cobanov, B.; Schnitkey, G. Economic losses from heat stress by US livestock industries. *J. Dairy Sci.* **2003**, *86* (Suppl. 1), E52–E77. [CrossRef]
3. McCann, J.C.; Luan, S.; Cardoso, F.C.; Derakhshani, H.; Khafipour, E.; Loor, J.J. Induction of Subacute Ruminal Acidosis Affects the Ruminal Microbiome and Epithelium. *Front. Microbiol.* **2016**, *7*, 701. [CrossRef]
4. Guo, Z.; Gao, S.; Ding, J.; He, J.; Ma, L.; Bu, D. Effects of Heat Stress on the Ruminal Epithelial Barrier of Dairy Cows Revealed by Micromorphological Observation and Transcriptomic Analysis. *Front. Genet.* **2022**, *12*, 768209. [CrossRef] [PubMed]

5. Turner, J.R. Intestinal mucosal barrier function in health and disease. *Nat. Rev. Immunol.* **2009**, *9*, 799–809. [CrossRef]
6. Marchiando, A.M.; Graham, W.V.; Turner, J.R. Epithelial barriers in homeostasis and disease. *Annu. Rev. Pathol.* **2010**, *5*, 119–144. [CrossRef] [PubMed]
7. Havlin, J.M.; Robinson, P.H.; Garrett, J.E. Effects on post-fresh period milk production and fertility as a result of prior niacin supplementation of dairy cows during their fresh period. *Livest. Sci.* **2018**, *214*, 73–78. [CrossRef]
8. Wrinkle, S.R.; Robinson, P.H.; Garrett, J.E. Niacin delivery to the intestinal absorptive site impacts heat stress and productivity responses of high producing dairy cows during hot conditions. *Anim. Feed. Sci. Technol.* **2012**, *175*, 33–47. [CrossRef]
9. Zou, B.; Long, F.; Xue, F.; Qu, M.; Chen, C.; Zhang, X.; Xu, L. Alleviation effects of niacin supplementation on beef cattle subjected to heat stress: A metagenomic insight. *Front. Microbiol.* **2022**, *13*, 975346. [CrossRef]
10. Holmes, E.; Kinross, J.; Gibson, G.R.; Burcelin, R.; Jia, W.; Pettersson, S.; Nicholson, J.K. Therapeutic modulation of microbiota-host metabolic interactions. *Sci. Transl. Transl. Med.* **2012**, *4*, 137rv6. [CrossRef]
11. Malmuthuge, N.; Guan, L.L. Understanding host-microbial interactions in rumen: Searching the best opportunity for microbiota manipulation. *J. Anim. Sci. Biotechnol.* **2017**, *8*, 8. [CrossRef] [PubMed]
12. Sommer, F.; Nookaew, I.; Sommer, N.; Fogelstrand, P.; Bäckhed, F. Site-specific programming of the host epithelial transcriptome by the gut microbiota. *Genome Biol.* **2015**, *16*, 62. [CrossRef] [PubMed]
13. Malmuthuge, N.; Liang, G.; Guan, L.L. Regulation of rumen development in neonatal ruminants through microbial metagenomes and host transcriptomes. *Genome Biol.* **2019**, *20*, 172. [CrossRef] [PubMed]
14. Digby, J.E.; Martinez, F.; Jefferson, A.; Ruparelia, N.; Chai, J.; Wamil, M.; Greaves, D.R.; Choudhury, R.P. Anti-inflammatory effects of nicotinic acid in human monocytes are mediated by GPR109A dependent mechanisms. *Arterioscler. Thromb. Vasc. Biol.* **2012**, *32*, 669–676. [CrossRef] [PubMed]
15. Gambhir, D.; Ananth, S.; Veeranan-Karmegam, R.; Elangovan, S.; Hester, S.; Jennings, E.; Offermanns, S.; Nussbaum, J.J.; Smith, S.B.; Thangaraju, M.; et al. GPR109A as an anti-inflammatory receptor in retinal pigment epithelial cells and its relevance to diabetic retinopathy. *Investig. Ophthalmol. Vis. Sci.* **2012**, *53*, 2208–2217. [CrossRef] [PubMed]
16. Guo, W.; Liu, J.; Li, W.; Ma, H.; Gong, Q.; Kan, X.; Cao, Y.; Wang, J.; Fu, S. Niacin alleviates dairy cow mastitis by regulating the GPR109A/AMPK/NRF2 signaling pathway. *Int. J. Mol. Sci.* **2020**, *21*, 3321. [CrossRef]
17. Dikmen, S.; Mateescu, R.G.; Elzo, M.A.; Hansen, P.J. Determination of the optimum contribution of Brahman genetics in an Angus-Brahman multibreed herd for regulation of body temperature during hot weather. *J. Anim. Sci.* **2018**, *96*, 2175–2183. [CrossRef]
18. *GB/T 19477-2004*; Operating Procedures of Cattle Slaughtering. China National Standards: Beijing, China, 2004.
19. Zhang, K.; Meng, M.; Gao, L.; Tu, Y.; Bai, Y. Sodium Butyrate Improves High-Concentrate-Diet-Induced Impairment of Ruminal Epithelium Barrier Function in Goats. *J. Agric. Food Chem.* **2018**, *66*, 8729–8736. [CrossRef]
20. Aschenbach, J.R.; Zebeli, Q.; Patra, A.K.; Greco, G.; Amasheh, S.; Penner, G.B. Symposium review: The importance of the ruminal epithelial barrier for a healthy and productive cow. *J. Dairy Sci.* **2019**, *102*, 1866–1882. [CrossRef]
21. Mani, V.; Weber, T.E.; Baumgard, L.H.; Gabler, N.K. Growth and Development Symposium: Endotoxin, inflammation, and intestinal function in livestock. *J. Anim. Sci.* **2012**, *90*, 1452–1465. [CrossRef]
22. Khafipour, E.; Krause, D.O.; Plaizier, J.C. A grain-based subacute ruminal acidosis challenge causes translocation of lipopolysaccharide and triggers inflammation. *J. Dairy Sci.* **2009**, *92*, 1060–1070. [CrossRef] [PubMed]
23. Wang, K.; Lei, Q.; Ma, H.; Jiang, M.; Yang, T.; Ma, Q.; Datsomor, O.; Zhan, K.; Zhao, G. Phloretin Protects Bovine Rumen Epithelial Cells from LPS-Induced Injury. *Toxins* **2022**, *14*, 337. [CrossRef] [PubMed]
24. Singleton, K.D.; Wischmeyer, P.E. Oral glutamine enhances heat shock protein expression and improves survival following hyperthermia. *Shock* **2006**, *25*, 295–299. [CrossRef]
25. Gadde, U.D.; Oh, S.; Lee, Y.; Davis, E.; Zimmerman, N.; Rehberger, T.; Lillehoj, H.S. Dietary Bacillus subtilis-based direct-fed microbials alleviate LPS-induced intestinal immunological stress and improve intestinal barrier gene expression in commercial broiler chickens. *Res. Vet. Sci.* **2017**, *114*, 236–243. [CrossRef]
26. Klevenhusen, F.; Hollmann, M.; Podstatzky-Lichtenstein, L.; Krametter-Frötscher, R.; Aschenbach, J.R.; Zebeli, Q. Feeding barley grain-rich diets altered electrophysiological properties and permeability of the ruminal wall in a goat model. *J. Dairy Sci.* **2013**, *96*, 2293–2302. [CrossRef] [PubMed]
27. Feng, L.; Li, S.Q.; Jiang, W.D.; Liu, Y.; Jiang, J.; Wu, P.; Zhao, J.; Kuang, S.Y.; Tang, L.; Tang, W.N.; et al. Deficiency of dietary niacin impaired intestinal mucosal immune function via regulating intestinal NF-κB, Nrf2 and MLCK signaling pathways in young grass carp (*Ctenopharyngodon idella*). *Fish Shellfish Immunol.* **2016**, *49*, 177–193. [CrossRef] [PubMed]
28. Hall, D.M.; Buettner, G.R.; Oberley, L.W.; Xu, L.; Matthes, R.D.; Gisolfi, C.V. Mechanisms of circulatory and intestinal barrier dysfunction during whole body hyperthermia. *Am. J. Physiol. Heart Circ. Physiol.* **2001**, *280*, H509–H521. [CrossRef]
29. Koch, F.; Thom, U.; Albrecht, E.; Weikard, R.; Nolte, W.; Kuhla, B.; Kuehn, C. Heat stress directly impairs gut integrity and recruits distinct immune cell populations into the bovine intestine. *Proc. Natl. Acad. Sci. USA* **2019**, *116*, 10333–10338. [CrossRef]
30. Pearce, S.C.; Sanz-Fernandez, M.V.; Hollis, J.H.; Baumgard, L.H.; Gabler, N.K. Short-term exposure to heat stress attenuates appetite and intestinal integrity in growing pigs. *J. Anim. Sci.* **2014**, *92*, 5444–5454. [CrossRef]
31. Gassler, N.; Rohr, C.; Schneider, A.; Kartenbeck, J.; Bach, A.; Obermüller, N.; Otto, H.F.; Autschbach, F. Inflammatory bowel disease is associated with changes of enterocytic junctions. *Am. J. Physiol. Gastrointest. Liver Physiol.* **2001**, *281*, G216–G228. [CrossRef]

32. Heller, F.; Florian, P.; Bojarski, C.; Richter, J.; Christ, M.; Hillenbrand, B.; Mankertz, J.; Gitter, A.H.; Bürgel, N.; Fromm, M.; et al. Interleukin-13 is the key effector Th2 cytokine in ulcerative colitis that affects epithelial tight junctions, apoptosis, and cell restitution. *Gastroenterology* **2005**, *129*, 550–564. [CrossRef] [PubMed]
33. Poritz, L.S.; Harris, L.R., 3rd; Kelly, A.A.; Koltun, W.A. Increase in the tight junction protein claudin-1 in intestinal inflammation. *Dig. Dis. Sci.* **2011**, *56*, 2802–2809. [CrossRef]
34. McCarthy, K.M.; Skare, I.B.; Stankewich, M.C.; Furuse, M.; Tsukita, S.; Rogers, R.A.; Lynch, R.D.; Schneeberger, E.E. Occludin is a functional component of the tight junction. *J. Cell Sci.* **1996**, *109 Pt 9*, 2287–2298. [CrossRef]
35. Balda, M.S.; Matter, K. Tight junctions and the regulation of gene expression. *Biochim. Biophys. Acta* **2009**, *1788*, 761–767. [CrossRef] [PubMed]
36. Furuse, M.; Furuse, K.; Sasaki, H.; Tsukita, S. Conversion of zonulae occludentes from tight to leaky strand type by introducing claudin-2 into Madin-Darby canine kidney I cells. *J. Cell Biol.* **2001**, *153*, 263–272. [CrossRef] [PubMed]
37. Tsukita, S.; Furuse, M.; Itoh, M. Multifunctional strands in tight junctions. *Nat. Rev. Mol. Cell Biol.* **2001**, *2*, 285–293. [CrossRef]
38. Inai, T.; Kobayashi, J.; Shibata, Y. Claudin-1 contributes to the epithelial barrier function in MDCK cells. *Eur. J. Cell Biol.* **1999**, *78*, 849–855. [CrossRef]
39. Bazzoni, G.; Martinez-Estrada, O.M.; Orsenigo, F.; Cordenonsi, M.; Citi, S.; Dejana, E. Interaction of junctional adhesion molecule with the tight junction components ZO-1, cingulin, and occludin. *J. Biol. Chem.* **2000**, *275*, 20520–20526. [CrossRef]

Disclaimer/Publisher's Note: The statements, opinions and data contained in all publications are solely those of the individual author(s) and contributor(s) and not of MDPI and/or the editor(s). MDPI and/or the editor(s) disclaim responsibility for any injury to people or property resulting from any ideas, methods, instructions or products referred to in the content.

animals

MDPI

Article

Effects of High-Concentrate-Induced SARA on Antioxidant Capacity, Immune Levels and Rumen Microbiota and Function in Goats

Siqin Fan, Mengli Zheng, Ao Ren, Hongxiang Mao, Donglei Long and Lingyuan Yang *

College of Animal Science and Technology, Hunan Agricultural University, Changsha 410125, China; fanjinyi0616@hunau.edu.cn (S.F.); zml199302@sina.com (M.Z.); 18711326324@163.com (A.R.); maohongxiang@cn.wilmar-intl.com (H.M.); ldl15073147518@foxmail.com (D.L.)
* Correspondence: lyyang@hunau.edu.cn

Simple Summary: Our study elucidated the specific alterations in rumen microbiota associated with subacute ruminal acidosis (SARA) triggered by a high-concentrate diet. It demonstrated that SARA can lead to a decrease in microbial richness and diversity in the rumen of goats. SARA could potentially hinder the growth of rumen bacterial communities. Within the context of SARA, the antioxidant capacity of goats diminishes, while inflammatory cytokines increase, potentially raising the risk of health issues in growing goats. Understanding this is crucial for predicting changes in goat microbiota and improving the efficiency and well-being of ruminants through nutritional approaches such as dietary interventions.

Abstract: This study aims to explore the antioxidant, immune, and enzyme metabolism aspects in goats experiencing subacute ruminal acidosis (SARA). Furthermore, we seek to elucidate the relationship between the symbiotic microbiota of goats and their metabolic function. Sixteen goats were equally divided into two groups and fed a normal-concentrate diet (NC, 55% concentrate) or a high-concentrate diet (HC, 90% concentrate) for five weeks. We found that the HC diet reduced the total antioxidant capacity (T-AOC) ($p = 0.022$) and increased interleukin-1β (IL-1β) ($p = 0.015$), interleukin-4 (IL-4) ($p = 0.008$) and interleukin-6 (IL-6) ($p = 0.002$) concentration of goats. Simultaneously, the HC diet significantly increased the concentrations of alkaline phosphatase (ALP) and amylase (AMY) in the blood and rumen fluid of goats ($p < 0.05$). Microbial analysis in the rumen of goats revealed that the HC diet decreased bacterial richness and diversity, as evidenced by the changed observed species, Chao 1, PD whole tree and Shannon when compared to the NC diet ($p < 0.01$). The proportion of Proteobacteria increased while that of Spirochaetes and Fibrobacteres significantly decreased with the HC diet ($p < 0.05$). The Christensenellaceae_R-7_group and Ruminococcaceae_UCG-010 in rumen was notably decreased when a diet was switched from 55% concentrate diet to 90% concentrate diet ($p < 0.05$). Additionally, microbial functional potentials deduced that the HC diet significantly increased the abundance of the citrate cycle (TCA cycle) (ko00020) associated with carbohydrate metabolism ($p = 0.028$). Furthermore, the HC diet significantly increased the glutathione metabolism (ko00480) associated with the metabolism of other amino acids ($p = 0.008$). Our findings suggested that SARA reduced the total antioxidant capacity and increased levels of inflammatory factors in goats, as well as decreased rumen bacterial species and abundance.

Keywords: goat; high concentrate; SARA; immunity; rumen; microbiota

Citation: Fan, S.; Zheng, M.; Ren, A.; Mao, H.; Long, D.; Yang, L. Effects of High-Concentrate-Induced SARA on Antioxidant Capacity, Immune Levels and Rumen Microbiota and Function in Goats. *Animals* **2024**, *14*, 263. https://doi.org/10.3390/ani14020263

Academic Editors: Adriana Bonanno and Gonzalo Hervás

Received: 10 December 2023
Revised: 9 January 2024
Accepted: 12 January 2024
Published: 15 January 2024

Copyright: © 2024 by the authors. Licensee MDPI, Basel, Switzerland. This article is an open access article distributed under the terms and conditions of the Creative Commons Attribution (CC BY) license (https://creativecommons.org/licenses/by/4.0/).

1. Introduction

In recent years, in pursuit of enhanced production efficiency, intensive farming has witnessed the widespread adoption of high-concentrate diets, primarily reliant on cereal starch as the primary energy source. Unfortunately, this extensive use has led to an upsurge in the prevalence of metabolic diseases. SARA is a prevalent gastrointestinal ailment in the field of ruminant farming. It is caused by excessive feeding of highly fermentable carbohydrates and insufficient dietary crude fiber [1]. A distinctive feature of SARA is the prolonged period of low rumen pH, with the rate of absorption and metabolism of rumen VFA serving as a pivotal determinant of rumen pH [2]. During the initial stages of SARA, typical symptoms are often absent, making diagnosis challenging. The primary diagnostic criterion for SARA is a rumen fluid pH persisting between 5.2 and 5.6 for over 3 h [3]. Consequences of SARA encompass reduced feed intake, diminished fiber digestibility, lower milk fat content, heightened occurrence of diarrhea, hoof inflammation, liver abscesses, elevated bacterial endotoxin levels, and increased acute phase protein during inflammation [4]. Owing to the scarcity of high-quality roughage resources in China, often characterized by subpar quality, producers find themselves compelled to heavily rely on starch-rich grain concentrates to fulfill the nutritional demands of animals, all in the pursuit of elevated production performance. Consequently, animals suffer from inadequate consumption of high-quality fiber, leading to the onset of acidosis, particularly prevalent in high-yield dairy cows and intensively fattened cattle and sheep. The emergence of SARA has emerged as a significant impediment to the production of cattle and sheep in China.

Research has demonstrated that in response to acid stress, microorganisms possess the ability to mitigate cell damage by modulating the dynamic equilibrium of intracellular pH (pHi), inducing the expression of stress proteins, and regulating cell membrane morphology and function [5]. Nevertheless, if stress surpasses the self-regulating capacity of microorganisms, intracellular protons and acid ions accumulate gradually, leading to a rapid decline in pHi. This, in turn, causes damage to cell membranes and internal macromolecular structures, disrupting normal cell growth and metabolism, and, in severe cases, resulting in microbial mortality [6–8]. The alterations in physical and chemical properties, reduced proliferation, and potential mortality of rumen microorganisms under acid stress constitute significant factors influencing rumen fermentation and shifts in microbial flora. Mertens [9] defined peNDF as: neutral detergent fiber (NDF) in the diet that can promote rumen liquid and solid phase stratification and affect the chewing of ruminants. Previous studies have shown that increasing the peNDF content of the diet increased the average rumen pH and is an effective means of preventing SARA [10,11].

As sequencing technology continues to advance, we have the opportunity to delve deeper into the influence of dietary changes on the gut microbiome of animals [12–15]. Maintaining relatively high hay feeding and a stable bacterial composition is imperative for the well-being of ruminants. SARA, on the other hand, leads to significant alterations in both rumen and intestinal microbiota [16–18]. Currently, research into the gut microbiota of both animals and humans is advancing, unveiling ever-expanding functions of the gut microbiota. Simultaneously, there is a continuous emergence of developments in gut microbiota preparations [19,20]. This experiment aims to construct a subacute rumen acidosis model in goats and study the changes in antioxidant capacity, immune levels, and microbial community structure and composition under sub healthy rumen conditions, enriching the survival and reproduction patterns of rumen bacteria under diseases, and providing a basis for the development of gut microbiota preparations for ruminants.

2. Materials and Methods

2.1. Animals and Treatment

This study was approved by the Animal Care and Use Committee of Hunan Agricultural University (HAU201408). To conduct this study, we chose sixteen Liuyang black goats at random. These goats were all six months old and had an average weight of 15.3 ± 1.67 kg. The goats were subsequently separated into two categories and given either a regular-concentrate meal (NC, with a ratio of 55% concentrate to 45% forage) or a concentrated diet (HC, with a ratio of 90% concentrate to 10% forage). Before the diets were formulated, the paddy straw was chopped to approximately 2 cm in length. The dry matter intake in the two groups was 572 g/d (NC) and 602 g/d (HC), respectively. Table 1 provided a comprehensive breakdown of the diets' composition and nutritional value. The experimental period spanned 35 days, including a 7-day diet adaptation phase. The diets were evenly distributed to the goats at approximately 08:00 and 18:00 h, respectively. Each goat was housed in its individual enclosure.

Table 1. Ingredients and nutrient levels of the experimental diets (air-dried basis).

Item	NC [1]	HC [2]
	Ingredients composition (%)	
Forage		
Rice straw	45.0	10.0
Concentrate		
Rice with shell	33.2	54.3
Soybean meal	9.60	15.7
Wheat bran	6.00	9.80
Fat powder	3.20	5.20
Calcium carbonate	0.50	0.80
Calcium bicarbonate	1.10	1.80
Sodium chloride	0.60	1.00
Premix [3]	1.00	1.40
	Nutrient levels [4], % of DM	
Crude protein	13.5	17.6
Crude ash	9.34	9.12
Crude fat	4.18	6.01
Neutral detergent fiber	49.8	38.4
Acid detergent fiber	36.5	9.51
Starch	26.57	38.57

[1] NC: normal-concentrate diet. [2] HC: high-concentrate diet. [3] Premix composition per kg diet: 68 mg $FeSO_4 \cdot H_2O$, 44 mg $CuSO_4 \cdot 5H_2O$, 411 μg $CoCl_2 \cdot 6H_2O$, 1.70 mg KIO_3, 211 mg $MnSO_4 \cdot H_2O$, 126 mg $ZnSO_4 \cdot H_2O$, 56 μg Na_2SeO_3, 462 mg $MgSO_4 \cdot 7H_2O$, 737 IU vitamin A, 8.29 mg vitamin E, 5.1 g carrier zeolite powder. [4] Nutrient levels were measured values.

2.2. *Sample Collection*

Upon completion of the 35-day feeding period, plasma samples were collected into tubes using the methodology described in a prior study [21], and subsequently stored at −20 °C for later analysis. Rumen epithelial samples were extracted and preserved in 10% neutral formalin to facilitate histomorphology examinations. Six animals were randomly chosen from each group and humanely euthanized. Rumen fluid was collected, promptly frozen at −80 °C, and later utilized for genomic DNA isolation and the assessment of enzyme metabolism indicators. The pH values in rumen samples of two groups in SARA model were the same before feeding in the morning (0 h), but it was declined in the HC group at 3 h or 6 h after feeding. During the time period from 3 h after the feeding until the sampling at 6 h, the average ruminal pH was below 5.7 in the HC group and remained significantly lower in NC goats [22]. Rumen segments from each goat were gathered to observe the morphology of the rumen epithelium.

2.3. *Measurement of Enzyme Metabolism, Antioxidant and Immune Indicators*

The T-AOC, NO, IFN-γ, HIS, IL-1β, IL-2, IL-4, and IL-6 in the blood were detected using a corresponding enzyme-linked immunosorbent assay (ELISA) kit (Jiangsu Yutong Biological Technology Co., Ltd. Yancheng, China). The blood and rumen fluid metabolites including ALT (alanine aminotransferase), AST (aspartate aminotransferase), ALP (alkaline phosphatase), and AMY (amylase) were detected with the commercial assay kits (Roche Diagnostics (Shanghai) Ltd., Shanghai, China; and determined using an Automatic Biochemistry analyzer (Cobas c 311, Roche, Shanghai, China), and strictly following the manufacturer's instructions, respectively.

2.4. Histological Examination

Morphological observation involved the collection of rumen epithelial segments from every goat. The rumen epithelial tissues were dehydrated, embedded, and subsequently sliced into approximately 5 mm-thick sections using a microtome. Afterwards, they were dyed with hematoxylin and eosin. Images of the rumen epithelium's structure were taken using a light microscope (Pannoramic DESK, P-MIDI, P250, 3DHISTECH, Budapest, Hungary) that had a computer-assisted morphometric system.

2.5. DNA Extraction and PCR Amplification

DNA extraction from rumen fluid was carried out using the E.Z.N.A.® soil DNA Kit (Omega Bio-Tek, Norcross, GA, USA) and verified through 1% agarose gel electrophoresis [23]. The V3–V4 hypervariable regions of the bacterial 16S-rDNA gene were subsequently amplified using universal primers 338F (5′-ACTCCTACGGGAGGCAGCAG-3′) and 806R (5′-GGACTACHVGGGTWTCTAAT-3′). PCR reactions were conducted in a total volume of 20 μL, comprising 1 × FastPfu Buffer, 250 μM dNTPs, 0.1 μM of each primer, 1 U FastPfu Polymerase (Beijing TransGen Biotech, Beijing, China), and 10 ng of template DNA. The PCR protocol consisted of an initial denaturation at 95 °C for 2 min, followed by 30 cycles of denaturation at 95 °C for 30 s, annealing at 55 °C for 30 s, elongation at 72 °C for 30 s, and a final extension at 72 °C for 5 min. Subsequently, the amplification products were confirmed using 2% agarose gel electrophoresis. Purified amplicons were equimolarly pooled and subjected to paired-end sequencing (2 × 300) on an Illumina MiSeq platform (Allwegene, Beijing, China) following standard protocols [24].

2.6. Bacterial Data Processing and Function Predication

The FASTQ files were processed using QIIME (version 1.17) by demultiplexing and quality filtering. The following criteria were applied: (i) Reads with a length of 300 bp were truncated at positions where the average quality score dropped below 20 within a 10 bp sliding window. Reads shorter than 50 bp after truncation were discarded. (ii) Stringent processing involved precise barcode matching, allowing for a maximum of two nucleotide mismatches in primer matching, and excluding reads containing ambiguous characters. (iii) Assembly of sequences was performed only for those with overlaps longer than 10 bp in their overlapping regions. The unassembled reads were eliminated. OTUs were clustered using UPARSE (version 7.1) with a 97% similarity cutoff, and chimeric sequences were identified and removed using UCHIME (https://www.drive5.com/uchime/uchime_download.html, accessed on 8 January 2024). A rarefaction analysis was performed using Mothur v.1.21.1 to evaluate diversity indices including observed_species, PD_whole_tree, and Shannon. Primer 6 software (Primer-E Ltd., London, UK) was used for conducting hierarchical clustering analysis, and Canoco 4.5 was employed for performing Principal Component Analysis (PCA). The bioinformatics tool PICRUSt was used to predict the functional potentials of metagenomes by reconstructing unobserved states through a phylogenetic investigation of communities, relying on data from the 16S rRNA gene [25]. The OTU table was utilized in PICRUSt to predict functional genes, making use of the Kyoto Encyclopedia of Genes and Genomes (KEGG) database.

2.7. Statistical Analysis

All of the experimental data in the tables and figures are presented as the mean ± SD and the differences between groups were assessed by the independent-samples t test using SPSS (IBM SPSS 21.0, Chicago, IL, USA). The data of 16S rRNA sequencing were analyzed on the allwegene platform. Statistical significance was set at $p < 0.05$ and tendencies at $0.05 \leq p \leq 0.10$.

3. Results

3.1. Antioxidant and Immune Indicators of Growing Goats Fed Different Diets

As shown in Figure 1A, the T-AOC was decreased dramatically in plasma by feeding the HC diet to goats compared to the NC diet ($p = 0.022$). Compared with the NC group, the data also showed that the NO level of goats was improved notably by the HC diet ($p = 0.011$) (Figure 1B). And diets supplemented with the HC diet has no significant effect on the IFN-γ (Figure 1C), there were no statistical differences between the groups ($p = 0.185$). The HC diet raised the HIS concentration of goats remarkably ($p = 0.003$) (Figure 1D). Results for plasma IL-1β was significantly boosted for the HC group compared to the NC group ($p = 0.015$) (Figure 1E). The level of IL-4 ($p = 0.008$) and IL-6 ($p = 0.002$) in plasma increased significantly when dietary concentrate ratio increased from 55% to 90% (Figure 1G,H). Compared with the NC group, plasma concentration of IL-2 in the HC group was not significantly different compared to the NC group ($p = 0.110$) (Figure 1F).

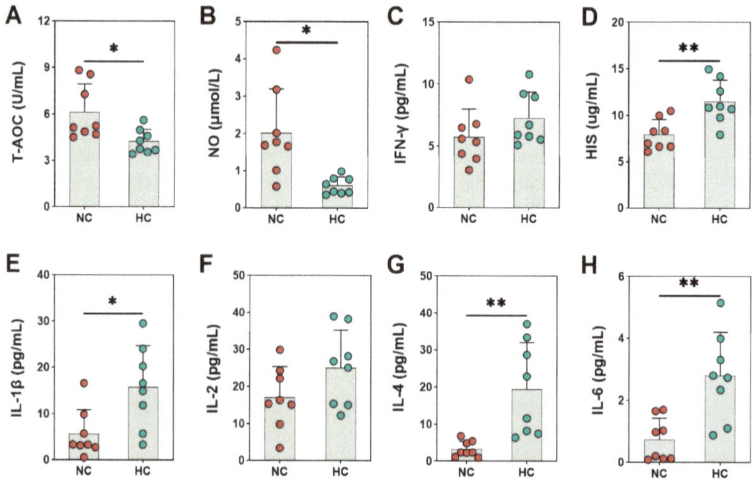

Figure 1. The antioxidant and immune indicators of growing goats fed normal-concentrate diet (NC) and high-concentrate diet (HC). (**A**) T-AOC, total antioxidant capacity; (**B**) NO, Nitric Oxide; (**C**) IFN-γ, interferon-γ; (**D**) HIS, Histamine; (**E**) IL-1β, interleukin-1β; (**F**) IL-2, interleukin-2; (**G**) IL-4, interleukin-4; (**H**) IL-6, interleukin-6. (*) $0.01 < p < 0.05$; (**) $p < 0.01$.

3.2. The Enzyme Metabolism Indicators of Growing Goats Fed Different Diets

The results shown in Figure 2A demonstrated that the ALT concentration in plasma was increased by the HC diet ($p = 0.025$). And the AST concentration in plasma was also increased by the HC diet compared to the NC diet ($p = 0.051$) (Figure 2B). The ALP concentration in plasma was notably increased when a diet administered to steers was switched from 55% concentrate diet to 90% concentrate diet ($p < 0.001$) (Figure 2C). Results for plasma AMY concentration were significantly boosted for the HC group compared to the NC group ($p = 0.015$) (Figure 2D), where ALT ($p < 0.001$) and AST ($p = 0.025$) concentrations in rumen were increased in this experimental model by increasing the grain percentage of the diet to induce SARA (Figure 2E,F). And the HC diet elevated significantly the ALP ($p < 0.001$) and AMY ($p < 0.001$) in rumen compared to the NC diet fed to growing goats (Figure 2G,H). The rumen epithelium from the NC group showed an intact structure while the rumen epithelium of the HC group showed an indication of sloughing in the stratum corneum (Figure 2I–L).

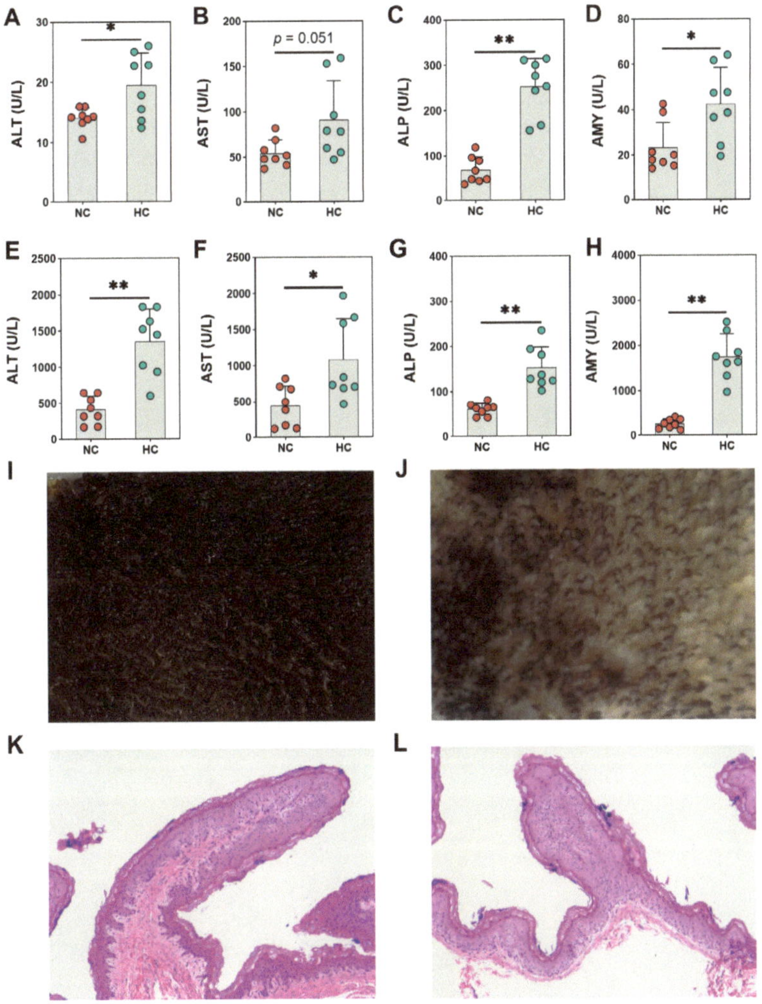

Figure 2. The enzyme metabolism indicators of growing goats fed normal-concentrate diet (NC) and high-concentrate diet (HC). (**A–D**) The enzyme metabolism indicators in plasma. (**E–H**) The enzyme metabolism indicators in rumen. The representative original images (**I**,**J**) and light micrographs (**K**,**L**) of ruminal epithelium in goats fed the NC and HC diets are shown. (*) $0.01 < p < 0.05$; (**) $p < 0.01$.

3.3. Rumen Microflora of Growing Goats Fed Different Diets

The increase in species as the sample size increases was described using species accumulation curves (Figure 3A). The findings indicated the frequency at which novel OTUs (novel species) emerge through ongoing sampling. In a specific interval, when the sample size grows, a steep rise in the curve suggests the presence of numerous species in the community. Conversely, a flattening curve indicates that the species in this habitat do not substantially increase as the sample size increases. Our results indicated that sampling is sufficient for microbiological data analysis. Based on the analysis of hierarchical clustering, the branch length indicates the distance between samples, and samples with greater similarity tend to cluster closer together from Figure 3B. Based on the evidence, it can be verified that the rumen samples from goats that were given regular and high-concentrate diets were grouped distinctly. The PCoA revealed distinct separation between rumen samples in the NC group and those in the HC group, as demonstrated in Figure 3C.

The flower plot represents a sample with each petal, with the core number in the middle representing the number of OTUs common to all samples, and the number on the petal representing the unique number of OTUs for that sample. it can be proven that the HC diet reduces the number of OTUs in the rumen compared to the NC diet fed to growing goats (Figure 3D). The observed species was markedly lower in the HC group than in the NC group ($p < 0.05$) (Figure 3E). In the rumen fluid, the Chao 1, PD whole tree, and the Shannon index notably decreased ($p < 0.01$ for Chao 1 and PD whole tree, $p < 0.05$ for the Shannon index) when the goats' diet was changed from 55% concentrate to 90% concentrate (Figure 3F–H). Hence, the HC diet declined the rumen bacteria richness and diversity.

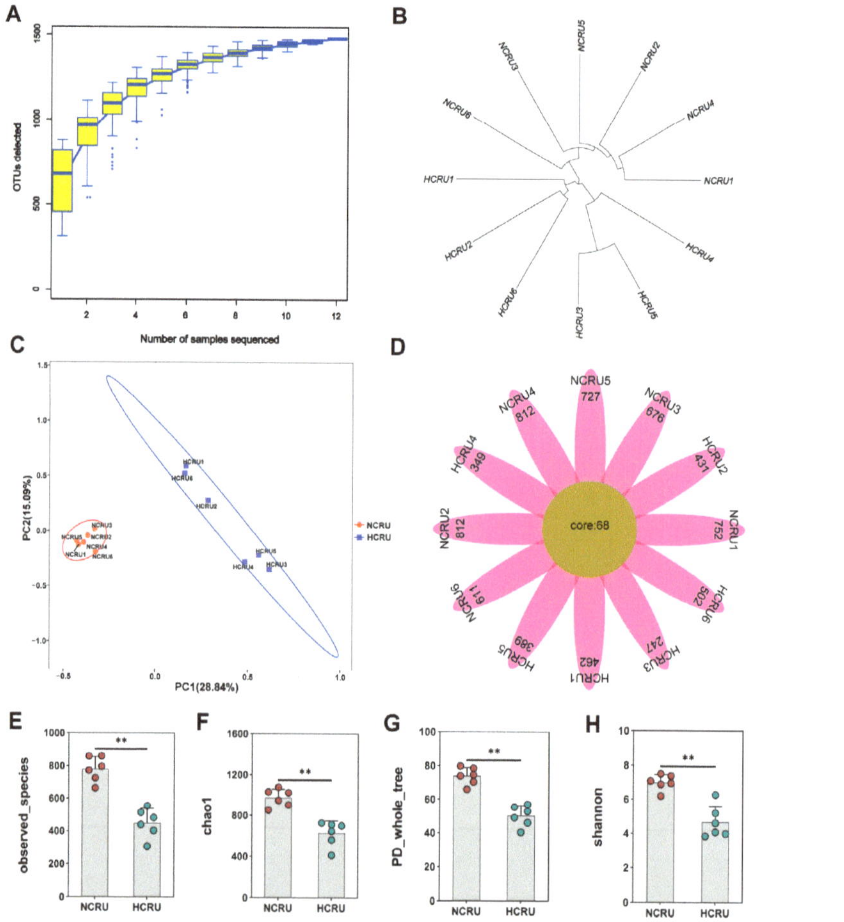

Figure 3. Rumen microflora of growing goats fed normal-concentrate diet (NC) and high-concentrate diet (HC). (**A**) Species accumulation curves. (**B**) Hierarchical clustering (NCRU1–NCRU6 and HCRU1–HCRU6 are rumen samples of goats fed with 55% or 90% concentrate, separately). (**C**) Principal component analysis (PCA) of rumen bacterial community. (**D**) Flower plot. (**E–H**) Alpha diversity indices of rumen bacterial community. (**) $p < 0.01$.

3.4. Rumen Bacterial Community Structure of Growing Goats Fed Different Diets

Bacteroidetes, Proteobacteria and Firmicutes were dominant phyla in the rumen of goats, accounting for more than 87% of the total rumen bacterial community (Figure 4A). Bacteroidetes and Proteobacteria accounted for a relative abundance of 45.9% and 21.3%, respectively, followed by Firmicutes, at 20.7%. When the diet of concentrate proportion increased from 55% to 90%, the abundance of Proteobacteria was increased significantly in the HCRU group than the NCRU group ($p < 0.05$) (Figure 4B). However, the proportion of Spirochaetes decreased dramatically in goats when the diet administered to goats was switched from 55% concentrate diet to 90% concentrate diet ($p < 0.01$) (Figure 4C). At the same time, the HC diet significantly dropped the abundance of Fibrobacteres ($p < 0.01$) (Figure 4D).

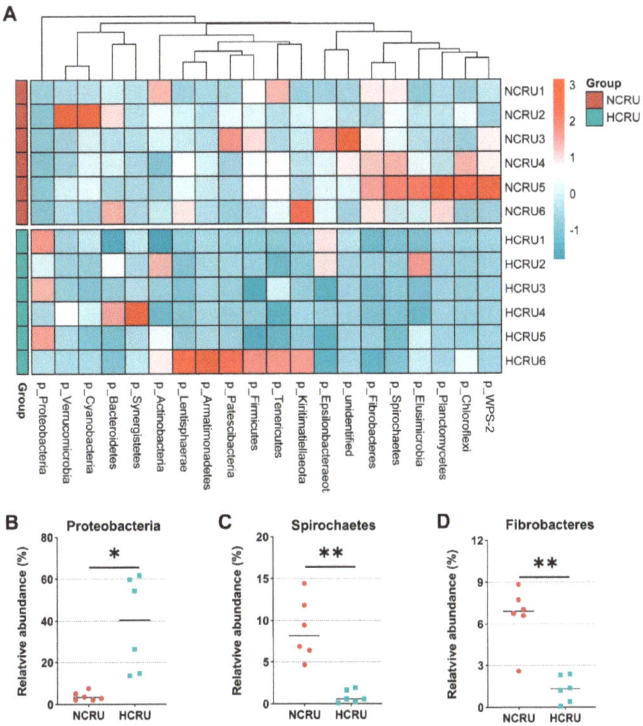

Figure 4. (**A**) Heatmap based on hierarchical clustering solution at phylum level (Bray-Curtis distance metric and complete clustering method) in rumen for goats fed a normal diet (NCRU) or a high-concentrate diet (HCRU), shown across columns. (**B–D**) Differential bacteria in the rumen at the phylum level. (*) $0.01 < p < 0.05$; (**) $p < 0.01$.

Within the bacterial population, the top 30 genera were identified across all samples (Figure 5A)-the genus Succinivibrionaceae_UCG-002 was the most abundant genera, accounting for 15.4%, followed by Prevotella_1, Rikenellaceae_RC9_gut_group, Treponema_2, and Fibrobacter as the predominant genera of rumen in the NCRU and HCRU groups. Among the genera, the percentage abundance of Succinivibrionaceae_UCG-002 was greater in the rumen of the HCRU group when compared to that in the NCRU group ($p < 0.05$) (Figure 5B), and Erysipelotrichaceae_UCG-004 tended to be lower by the HC diet ($p = 0.056$) (Figure 5G). In contrast, the relative abundance of Rikenellaceae_RC9_gut_group, Treponema_2 and Fibrobacter was markedly decreased in the rumen of the HCRU group than in the NCRU group ($p < 0.01$) (Figure 5C–E), and Ruminococcus_1 tended to decrease by the HC diet ($p = 0.058$) (Figure 5I). The Christensenellaceae_R-7_group and Ruminococcaceae_UCG-010 in rumen was notably decreased when a diet administered to steers was switched from 55% concentrate diet to 90% concentrate diet ($p < 0.05$) (Figure 5H,K).

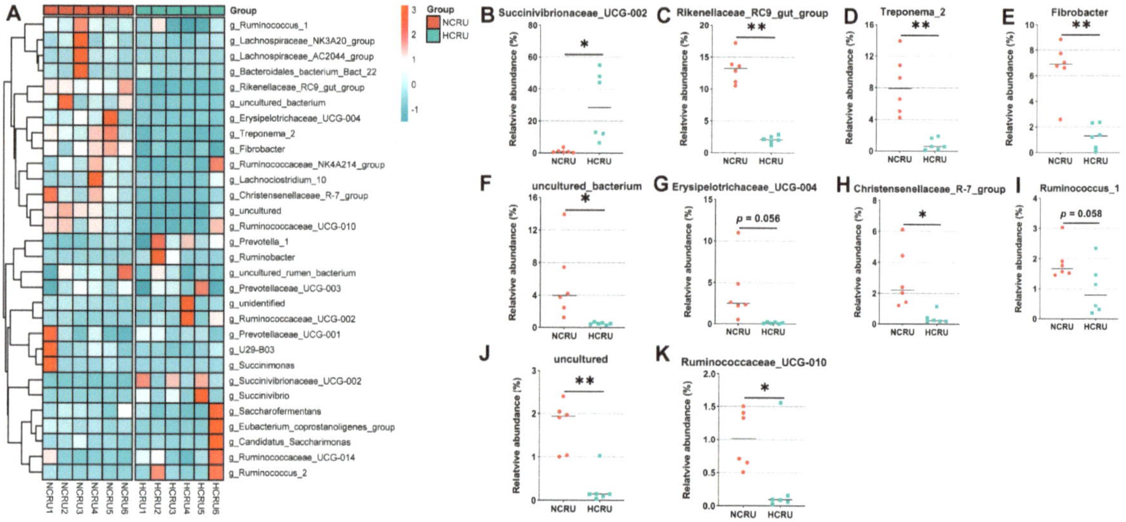

Figure 5. (**A**) Heatmap based on hierarchical clustering solution at genus level (Bray-Curtis distance metric and complete clustering method) in rumen for goats fed a normal diet (NCRU) or a high-concentrate diet (HCRU), shown across rows. (**B–K**) Differential bacteria in the rumen at the genus level. (*) $0.01 < p < 0.05$; (**) $p < 0.01$.

3.5. Function Prediction of Rumen Bacterial Community Using PICRUSt

Functional potentials of rumen microbiota were predicted using 16S rRNA marker gene sequences through PICRUSt against KEGG pathways. As showed in Figure 6, within the 20 most abundant level 2 KEGG pathways, ten, amino acid metabolism, carbohydrate metabolism, energy metabolism, the metabolism of cofactors and vitamins, nucleotide metabolism, glycan biosynthesis and metabolism, lipid metabolism, enzyme families, the metabolism of terpenoids and polyketides, and the metabolism of other amino acids, were associated with metabolism; and one, membrane transport, was associated with environmental information processing; four, including replication and repair, translation, folding, sorting and degradation, and transcription were associated with genetic information processing; one, cell motility was associated with cellular processes.

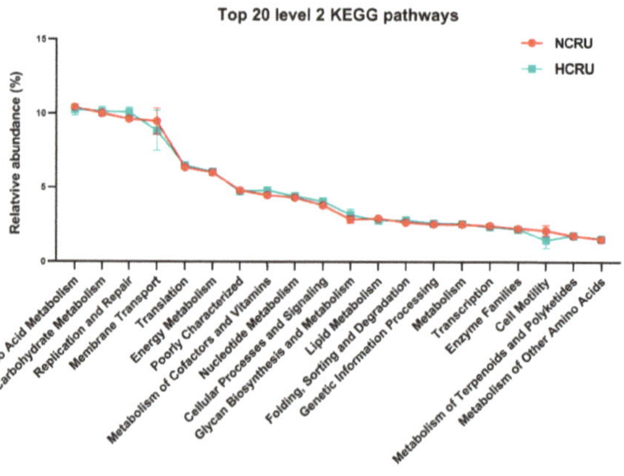

Figure 6. Top 20 predicted metagenomic functions at level 2 of the KEGG (Kyoto Encyclopedia of Genes and Genomes) pathways in rumen bacteria of goats fed the NC and HC diets.

In total, the related amino acid metabolism, carbohydrate metabolism and the metabolism of other amino acids individual pathways were predicted, and a comparison of different pathways (as shown in Table 2) revealed that the HC diet influenced the functional potentials of rumen microbiota. Specifically, compared to the NC group, the HC diet tended to decrease the abundance of histidine metabolism (ko00340) (p = 0.081), and the HC diet significantly decreased the abundance of lysine degradation (ko00310), tryptophan metabolism (ko00380), valine, leucine and isoleucine degradation (ko00280), and selenocompound metabolism (ko00450) ($p < 0.05$). In contrast, the HC diet significantly increased the abundance of the citrate cycle (TCA cycle) (ko00020) associated with carbohydrate metabolism (p = 0.028). Furthermore, the HC diet significantly increased the glutathione metabolism (ko00480) associated with the metabolism of other amino acids (p = 0.008).

Table 2. KEGG pathways that showed different abundances between ruminal digesta microbiota of NC- and HC-diet goats.

Level 2	Level 3	Pathway ID	NC [1]	HC [2]	p-Value
Amino acid metabolism	Histidine metabolism	ko00340	0.69 ± 0.03	0.65 ± 0.05	0.081
	Lysine degradation	ko00310	0.16 ± 0.02	0.13 ± 0.02	0.030
	Tryptophan metabolism	ko00380	0.19 ± 0.01	0.16 ± 0.02	0.003
	Valine, leucine and isoleucine degradation	ko00280	0.29 ± 0.02	0.25 ± 0.02	0.006
Carbohydrate metabolism	Citrate cycle (TCA cycle)	ko00020	0.73 ± 0.05	0.80 ± 0.04	0.028
Metabolism of other amino acids	Glutathione metabolism	ko00480	0.17 ± 0.004	0.24 ± 0.05	0.008
	Selenocompound metabolism	ko00450	0.38 ± 0.01	0.34 ± 0.02	0.002

[1] NC: normal-concentrate diet; [2] HC: high-concentrate diet.

4. Discussion

The diet's formulation and composition can potentially impact the growth performance and antioxidant capacity of animals, as stated in references [26–28]. Zhang et al. showed that SARA elevated systematic oxidative status and enhanced autophagy in the liver, and suppressed SIRT1 and FOXA2 may mediate enhanced oxidative damage and autophagy in the livers of dairy cows fed a HC diet [29]. Cows with SARA often develop complications or other diseases with physiologic immunosuppression and inflammation [30]. Our current study has verified that the HC diet could diminish the antioxidant capacity of goats, underscoring the potential adverse consequences of indiscriminately promoting weight gain in these animals. Dietary patterns can impact not only body weight but also bone density [31]. In addition to antioxidant capacity, anti-inflammatory and immunomodulatory properties are equally vital for the well-being of the host [32]. Cytokines, produced and released by immune cells that are activated, constitute a category of biologically active substances. Immune cells interact and mutually regulate each other, playing crucial parts in inflammation, immune responses, tissue healing, and hematopoietic functions. Promoting the proliferation of immune cells are several noteworthy immune mediators, namely IL-2, IL-4, IL-6, IL-10, and IL-12 [33–35]. Feeding goats with a high-concentrate (HC) diet resulted in lower nitric oxide (NO) production compared to the normal group. Typically, increased NO levels are associated with an enhanced pro-inflammatory response to lipopolysaccharides (LPS) [36], which correlates with elevated concentrations of various interleukins. Prior research has demonstrated that elevated IL-6 levels may enhance the permeability of the rumen epithelium and diminish its barrier function [37]. Furthermore, our research validated that the nitric oxide (NO) [38] mediated the control of IL-2 and IL-6 production in these goats through a negative feedback mechanism. These findings suggest that feeding goats with a high-concentrate (HC) diet during a state of subacute rumen acidosis (SARA) results in elevated plasma NO levels, which subsequently suppresses the production of cytokines, notably IL-6.

The activity of animal digestive enzymes [39] can be significantly influenced by the composition of the diet. The digestive enzymes produced by rumen microorganisms in ruminants help to break down and utilize the natural polymers found in their feed [40]. The high-carbohydrate diet resulted in an increase in AMY activity, in line with previous studies suggesting that higher intake of starch leads to elevated AMY activity in the contents of the pancreas and small intestine [41,42]. ALT and AST activity are reliable markers of liver function, released into the bloodstream when the liver is compromised. Typically, serum AST activity remains low under normal conditions but significantly elevates in the presence of cell damage. Increased AST and ALT values are indicative of varying degrees of liver impairment [43]. The HC group displayed elevated AST and ALT values in contrast to the NC group, implying that the HC diet may lead to liver damage in goats. ALP is an enzyme widely distributed in tissues and organs, including the liver, bones, and small intestine. It facilitates the hydrolysis of phosphate into phosphoric acid under alkaline conditions. Various forms of ALP, including small intestine ALP and bone-type ALP, contribute to the regulation of intestinal function, as well as calcium and phosphorus metabolism [44,45]. DiLorenzo and McCarthy, among others, have provided evidence suggesting that AMY may potentially accelerate the rate of ruminal carbohydrate degradation [46,47]. Dietary nutrients have the capacity to impact the morphology of small intestinal tissue and the digestive function of animals [48]. An imbalance in the interplay between gut microbiota and other factors can disrupt the homeostasis of the intestinal mucosa [49]. The intestinal epithelial barrier, acting as the first line of defense between the luminal environment and the host, can result in severe inflammation or other intestinal diseases if compromised [50]. This study additionally furnishes evidence that diets can impact the morphology of the rumen epithelium. Elevated enzyme levels (ALT, AST, and ALP) indicate potential damage or alterations in membrane permeability [51]. As a result, feeding goats an HC diet may have detrimental effects on their intestinal epithelium.

For this research, we employed 16S rRNA sequencing to investigate the bacteria present in rumen fluid. Our focus was on observing the changes in the composition and structure of rumen flora as goats progressed from a state of good health to one of acidosis. Our goal was to explore the reproduction patterns of rumen bacteria during disease and enhance the understanding of rumen microecology. Species accumulation curves are useful instruments for analyzing the species makeup of samples and forecasting species abundance. In biodiversity and community surveys, they are extensively used to assess the adequacy of sample size and estimate the diversity of species [52]. Thus, species accumulation curves not only assess sample size sufficiency but also predict species richness when the sample size is adequate. Utilizing the beta diversity distance matrix for hierarchical clustering analysis [53], the tree structure was constructed using the UPGMA algorithm, which helped with visual analysis. Our research validates the appropriateness of sequencing rumen samples from goats subjected to the NC and HC diets, revealing distinct clustering patterns for the two treatment groups. Rumen environmental imbalances often result in shifts in rumen flora. For example, Kim et al. induced subacute rumen acidosis (SARA) through a high-grain diet, which led to decreased rumen flora diversity and a reduction in the relative abundance of Prevotella [54]. Our study reveals that goats fed a HC diet display lower bacterial richness and diversity compared to those fed the NC diet, as evident from reductions in observed species, Shannon, Chao 1, and PD whole tree indices. This observation highlights the notably detrimental effect of the HC diet on the biodiversity of the goat rumen ecosystem.

Modifying dietary formulations can lead to shifts in the microbial composition of the animal gut and impact digestive function [55,56]. Zhao et al. indicated the importance of bacterial sphingolipids in maintaining hindgut symbiosis and homeostasis. Dietary supplementation with citrus flavonoid extract can decrease systemic inflammation by maintaining hindgut microbiota homeostasis and regulating sphingolipid metabolism in dairy cows fed a high-starch diet [57]. Environmental acid stress has a substantial impact on the growth and reproduction of microorganisms during microbial fermentation [58,59]. Previous studies have reported that Bacteroides is a major anaerobic genus in the rumen [60], and high-concentrate diets have been associated with a decrease in Bacteroides populations [61,62]. Prevotella, a member of the Bacteroides genus, is the predominant genus of rumen bacteria [63]. Prevotella is known for its high activity in protein degradation and its ability to utilize starch and pectin [60]. Surprisingly, our findings revealed no notable alteration in the prevalence of Prevotella in the rumen prior to and following SARA, which contradicts earlier research. Considering our experimental design, it is hypothesized that Prevotella may rapidly multiply when provided with ample starch substrates in a short timeframe, maintaining a stable relative abundance. The health and function of the rumen greatly depend on the diversity of microorganisms. During our investigation, we noticed a decrease in the ratios of the Rikenellaceae RC9 intestinal cluster, Treponema 2, Fibrobacter, and Christensenellaceae R-7 cluster when exposed to high-calorie feeding circumstances. It is worth mentioning that Treponema and Fibrobacter are acknowledged for their role in breaking down fibers [64,65]. Numerous studies on SARA have extensively documented the abundance of bacteria that break down starch and the decrease in bacteria that break down fiber [66,67]. Our current research aligns with these findings, confirming disruptions in rumen digestion and metabolic capabilities in the HC-fed goats. Succinivibrionaceae UCG-002 is a member of the Succinivibrionaceae family and has been associated with ruminal fatty acid metabolism [68]. As for the Rikenellaceae RC9 gut cluster, the Rikenellaceae family is a fairly recent taxonomic categorization, and there is a significant requirement for additional information regarding its metabolic role, despite the research conducted by Su et al.suggested a single isolate that produces acetate from the Rikenellaceae family [69]. Rikenellaceae RC9 was able to alter the digestibility of dietary NDF and ADF as well as the VFA concentration, as shown by Zhao et al. [70]. The Christensenellaceae R-7 cluster is most closely associated with Christensenella minuta [71], a species capable of generating acetate and butyrate through glucose [72]. In summary, these findings indicated that the HC goats may have impaired rumen digestion and metabolic capabilities.

Rumen bacteria play a crucial role in maintaining ruminant health [73–75]. In our study, we used PICRUSt to assess the potential functions of bacteria associated with digestion in the rumen. The clear disparities in anticipated functional pathways between the NC and HC groups suggest that alterations in bacterial populations can influence the functional potential of the rumen bacterial community. Within the HC group, we noticed a decline in pathways associated with the metabolism of amino acids, such as histidine, lysine, tryptophan, valine, leucine, and isoleucine breakdown. This indicates possible detrimental impacts on the metabolism of amino acids in the host resulting from prolonged consumption of an HC diet. Additionally, we noted a relative increase in pathways associated with the tricarboxylic acid (TCA) cycle and carbohydrate metabolism in the HC group, suggesting that the rumen microbiomes of HC-fed goats may have an enhanced capacity for energy extraction. Overall, these predicted findings suggest that long-term HC diet feeding may impact the bacterial function of the rumen microbiome in goats. However, it is important to note that while PICRUSt is a well-validated predictive tool, true metagenomic shotgun sequencing is necessary for a more comprehensive examination of the effects of HC diets on rumen-associated bacteria in goats.

5. Conclusions

In summary, SARA could alter the total antioxidant capacity and inflammatory factors levels of goats, and affect enzyme metabolism in their blood and rumen. Through our research, we have gained a deeper comprehension of the microorganisms present in the rumen, the presence of SARA reduced the number and variety of bacteria in the rumen and may affect the development and performance of bacteria in the rumen.

Author Contributions: Conceptualization, L.Y.; methodology, S.F.; formal analysis, S.F.; data curation, M.Z., A.R., H.M. and D.L.; writing—original draft preparation, S.F.; writing—review and editing, L.Y.; visualization, S.F.; supervision, L.Y.; project administration, L.Y. All authors have read and agreed to the published version of the manuscript.

Funding: This research project was funded by the Hunan Provincial Innovation Foundation for Postgraduate (Grant No. CX2017B368).

Institutional Review Board Statement: This study was conducted according to the guidelines of the Declaration of Helsinki, and approved by the Institutional Review Board of the Animal Care and Use Committee of Hunan Agricultural University.

Informed Consent Statement: Not applicable for studies not involving humans.

Data Availability Statement: The data presented in this study are available on request from the corresponding author. The availability of the data is restricted to investigators based in academic institutions.

Conflicts of Interest: The authors declare no conflict of interest.

References

1. Gozho, G.N.; Plaizier, J.C.; Krause, D.O.; Kennedy, A.D.; Wittenberg, K.M. Subacute ruminal acidosis induces ruminal lipopolysaccharide endotoxin release and triggers an inflammatory response. *J. Dairy Sci.* **2005**, *88*, 1399–1403. [CrossRef] [PubMed]
2. Gao, X.; Oba, M. Characteristics of dairy cows with a greater or lower risk of subacute ruminal acidosis: Volatile fatty acid absorption, rumen digestion, and expression of genes in rumen epithelial cells. *J. Dairy Sci.* **2016**, *99*, 8733–8745. [CrossRef] [PubMed]
3. Abdela, N. Sub-acute Ruminal Acidosis (SARA) and its Consequence in Dairy Cattle: A Review of Past and Recent Research at Global Prospective. *Achiev. Life Sci.* **2016**, *10*, 187–196. [CrossRef]
4. Kleen, J.L.; Hooijer, G.A.; Rehage, J.; Noordhuizen, J.P. Subacute ruminal acidosis (SARA): A review. *J. Vet. Med. A Physiol. Pathol. Clin. Med.* **2003**, *50*, 406–414. [CrossRef] [PubMed]
5. Bearson, S.; Bearson, B.; Foster, J.W. Acid stress responses in enterobacteria. *FEMS Microbiol. Lett.* **1997**, *147*, 173–180. [CrossRef]
6. Guan, N.; Liu, L. Microbial response to acid stress: Mechanisms and applications. *Appl. Microbiol. Biotechnol.* **2020**, *104*, 51–65. [CrossRef]
7. Monteiro, H.F.; Faciola, A.P. Ruminal acidosis, bacterial changes, and lipopolysaccharides. *J. Anim. Sci.* **2020**, *98*, skaa248. [CrossRef]
8. Guan, N.Z.; Li, J.H.; Shin, H.D.; Du, G.C.; Chen, J.; Liu, L. Microbial response to environmental stresses: From fundamental mechanisms to practical applications. *Appl. Microbiol. Biotechnol.* **2017**, *101*, 3991–4008. [CrossRef]
9. Mertens, D.R. Creating a system for meeting the fiber requirements of dairy cows. *J. Dairy Sci.* **1997**, *80*, 1463–1481. [CrossRef]
10. Li, C.; Beauchemin, K.A.; Yang, W. Feeding diets varying in forage proportion and particle length to lactating dairy cows: I. Effects on ruminal pH and fermentation, microbial protein synthesis, digestibility, and milk production. *J. Dairy Sci.* **2020**, *103*, 4340–4354. [CrossRef]
11. Yang, W.Z.; Beauchemin, K.A. Increasing physically effective fiber content of dairy cow diets through forage proportion versus forage chop length: Chewing and ruminal pH. *J. Dairy Sci.* **2009**, *92*, 1603–1615. [CrossRef] [PubMed]
12. He, B.; Zhu, R.; Yang, H.; Lu, Q.; Wang, W.; Song, L.; Sun, X.; Zhang, G.; Li, S.; Yang, J.; et al. Assessing the Impact of Data Preprocessing on Analyzing Next Generation Sequencing Data. *Front. Bioeng. Biotechnol.* **2020**, *8*, 817. [CrossRef]
13. Wang, K.; Ma, J.; Li, Y.; Han, Q.; Yin, Z.; Zhou, M.; Luo, M.; Chen, J.; Xia, S. Effects of essential oil extracted from Artemisia argyi leaf on lipid metabolism and gut microbiota in high-fat diet-fed mice. *Front. Nutr.* **2022**, *9*, 1024722. [CrossRef] [PubMed]
14. Liu, S.; Wang, K.J.; Lin, S.Z.; Zhang, Z.D.; Cheng, M.; Hu, S.S.; Hu, H.J.; Xiang, J.; Chen, F.M.; Li, G.H.; et al. Comparison of the Effects between Tannins Extracted from Different Natural Plants on Growth Performance, Antioxidant Capacity, Immunity, and Intestinal Flora of Broiler Chickens. *Antioxidants* **2023**, *12*, 441. [CrossRef] [PubMed]
15. Chen, F.M.; Wang, Y.S.; Wang, K.J.; Chen, J.Y.; Jin, K.; Peng, K.Q.; Chen, X.; Liu, Z.M.; Ouyang, J.; Wang, Y.; et al. Effects of Litsea cubeba essential oil on growth performance, blood antioxidation, immune function, apparent digestibility of nutrients, and fecal microflora of pigs. *Front. Pharmacol.* **2023**, *14*, 1166022. [CrossRef]
16. Zhang, R.Y.; Ye, H.M.; Liu, J.H.; Mao, S.Y. High-grain diets altered rumen fermentation and epithelial bacterial community and resulted in rumen epithelial injuries of goats. *Appl. Microbiol. Biotechnol.* **2017**, *101*, 6981–6992. [CrossRef]
17. Wang, K.; Ren, A.; Zheng, M.; Jiao, J.; Yan, Q.; Zhou, C.; Tan, Z. Diet with a High Proportion of Rice Alters Profiles and Potential Function of Digesta-Associated Microbiota in the Ileum of Goats. *Animals* **2020**, *10*, 1261. [CrossRef]
18. Wang, K.J.; Yan, Q.X.; Ren, A.; Zheng, M.L.; Zhang, P.H.; Tan, Z.L.; Zhou, C.S. Novel Linkages between Bacterial Composition of Hindgut and Host Metabolic Responses to SARA Induced by High-Paddy Diet in Young Goats. *Front. Vet. Sci.* **2022**, *8*, 791482. [CrossRef]
19. Wu, F.; Guo, X.; Zhang, J.; Zhang, M.; Ou, Z.; Peng, Y. Phascolarctobacterium faecium abundant colonization in human gastrointestinal tract. *Exp. Ther. Med.* **2017**, *14*, 3122–3126. [CrossRef]
20. Wang, H.; Hu, C.; Cheng, C.; Cui, J.; Ji, Y.; Hao, X.; Li, Q.; Ren, W.; Deng, B.; Yin, Y.; et al. Unraveling the association of fecal microbiota and oxidative stress with stillbirth rate of sows. *Theriogenology* **2019**, *136*, 131–137. [CrossRef]
21. Yin, L.; Li, J.; Wang, H.; Yi, Z.; Wang, L.; Zhang, S.; Li, X.; Wang, Q.; Li, J.; Yang, H.; et al. Effects of vitamin B6 on the growth performance, intestinal morphology, and gene expression in weaned piglets that are fed a low-protein diet. *J. Anim. Sci.* **2020**, *98*, skaa022. [CrossRef] [PubMed]

22. Wang, K.J.; Zheng, M.L.; Ren, A.; Zhou, C.S.; Yan, Q.X.; Tan, Z.L.; Zhang, P.; Yi, K. Effects of high rice diet on growth performance, nutrients apparent digestibility, nitrogen metabolism, blood parameters and rumen fermentation in growing goats. *Kafkas Univ. Vet. Fak. Derg.* **2019**, *25*, 749–755. [CrossRef]
23. Wang, K.J.; Peng, X.M.; Yang, A.Q.; Huang, Y.Q.; Tan, Y.X.; Qian, Y.J.; Lv, F.F.; Si, H.B. Effects of Diets with Different Protein Levels on Lipid Metabolism and Gut Microbes in the Host of Different Genders. *Front. Nutr.* **2022**, *9*, 940217. [CrossRef] [PubMed]
24. Wang, K.J.; Peng, X.M.; Lv, F.F.; Zheng, M.L.; Long, D.L.; Mao, H.X.; Si, H.B.; Zhang, P.H. Microbiome-Metabolites Analysis Reveals Unhealthy Alterations in the Gut Microbiota but Improved Meat Quality with a High-Rice Diet Challenge in a Small Ruminant Model. *Animals* **2021**, *11*, 2306. [CrossRef]
25. Langille, M.G.; Zaneveld, J.; Caporaso, J.G.; McDonald, D.; Knights, D.; Reyes, J.A.; Clemente, J.C.; Burkepile, D.E.; Vega Thurber, R.L.; Knight, R.; et al. Predictive functional profiling of microbial communities using 16S rRNA marker gene sequences. *Nat. Biotechnol.* **2013**, *31*, 814–821. [CrossRef] [PubMed]
26. Chen, F.M.; He, J.Y.; Wang, X.; Lv, T.; Liu, C.J.; Liao, L.P.; Li, Z.B.; Zhou, J.; He, B.S.; Qiu, H.J.; et al. Effect of Dietary Ramie Powder at Various Levels on the Growth Performance, Meat Quality, Serum Biochemical Indices and Antioxidative Capacity of Yanling White Geese. *Animals* **2022**, *12*, 2045. [CrossRef]
27. Ma, W.J.; Wei, S.S.; Peng, W.J.; Sun, T.L.; Huang, J.H.; Yu, R.; Zhang, B.K.; Li, W.Q. Antioxidant Effect of *Polygonatum sibiricum* Polysaccharides in D-Galactose-Induced Heart Aging Mice. *Biomed. Res. Int.* **2021**, *2021*, 6688855. [CrossRef]
28. Chen, J.Y.; Chen, F.M.; Peng, S.M.; Ou, Y.J.; He, B.S.; Li, Y.H.; Lin, Q. Effects of Artemisia argyi Powder on Egg Quality, Antioxidant Capacity, and Intestinal Development of Roman Laying Hens. *Front. Physiol.* **2022**, *13*, 902568. [CrossRef]
29. Zhang, H.; Xue, Y.; Xie, W.; Wang, Y.; Ma, N.; Chang, G.; Shen, X. Subacute ruminal acidosis downregulates FOXA2, changes oxidative status, and induces autophagy in the livers of dairy cows fed a high-concentrate diet. *J. Dairy Sci.* **2023**, *106*, 2007–2018. [CrossRef]
30. Sato, S. Subacute ruminal acidosis (SARA) challenge, ruminal condition and cellular immunity in cattle. *Jpn. J. Vet. Res.* **2015**, *63*, S25–S36.
31. Chen, Y.; Xiang, J.; Wang, Z.; Xiao, Y.; Zhang, D.; Chen, X.; Li, H.; Liu, M.; Zhang, Q. Associations of Bone Mineral Density with Lean Mass, Fat Mass, and Dietary Patterns in Postmenopausal Chinese Women: A 2-Year Prospective Study. *PLoS ONE* **2015**, *10*, e0137097. [CrossRef] [PubMed]
32. Li, J.; Yin, L.; Wang, L.; Li, J.; Huang, P.; Yang, H.; Yin, Y. Effects of vitamin B6 on growth, diarrhea rate, intestinal morphology, function, and inflammatory factors expression in a high-protein diet fed to weaned piglets. *J. Anim. Sci.* **2019**, *97*, 4865–4874. [CrossRef]
33. Zhu, J.F.; Paul, W.E. Peripheral CD4+ T-cell differentiation regulated by networks of cytokines and transcription factors. *Immunol. Rev.* **2010**, *238*, 247–262. [CrossRef] [PubMed]
34. Sun, X.; Gao, R.L.; Xiong, Y.K.; Huang, Q.C.; Xu, M. Antitumor and immunomodulatory effects of a water-soluble polysaccharide from Lilii Bulbus in mice. *Carbohydr. Polym.* **2014**, *102*, 543–549. [CrossRef] [PubMed]
35. Alexander, A.F.; Kelsey, I.; Forbes, H.; Miller-Jensen, K. Single-cell secretion analysis reveals a dual role for IL-10 in restraining and resolving the TLR4-induced inflammatory response. *Cell Rep.* **2021**, *36*, 109728. [CrossRef]
36. He, Z.; Tan, Z.; Sun, Z.; Beauchemin, K.A.; Tang, S.; Zhou, C.; Han, X.; Wang, M.; Wu, D. Unchanged interleukin 6 level of protein and energy restricted goats during late gestation: The role of elevated blood nitric oxide. *J. Endocrinol.* **2012**, *213*, 59–65. [CrossRef] [PubMed]
37. Trabi, E.B.; Seddik, H.E.; Xie, F.; Wang, X.F.; Liu, J.H.; Mao, S.Y. Effect of pelleted high-grain total mixed ration on rumen morphology, epithelium-associated microbiota and gene expression of proinflammatory cytokines and tight junction proteins in Hu sheep. *Anim. Feed Sci. Technol.* **2020**, *263*, 114453. [CrossRef]
38. Connelly, L.; Palacios-Callender, M.; Ameixa, C.; Moncada, S.; Hobbs, A.J. Biphasic regulation of NF-kappa B activity underlies the pro- and anti-inflammatory actions of nitric oxide. *J. Immunol.* **2001**, *166*, 3873–3881. [CrossRef]
39. Deng, Q.; Shao, Y.; Wang, Q.; Li, J.; Li, Y.; Ding, X.; Huang, P.; Yin, J.; Yang, H.; Yin, Y. Effects and interaction of dietary electrolyte balance and citric acid on growth performance, intestinal histomorphology, digestive enzyme activity and nutrient transporters expression of weaned piglets. *J. Anim. Physiol. Anim. Nutr.* **2021**, *105*, 272–285. [CrossRef]
40. Plaizier, J.C.; Khafipour, E.; Li, S.; Gozho, G.N.; Krause, D.O. Subacute ruminal acidosis (SARA), endotoxins and health consequences. *Anim. Feed Sci. Technol.* **2012**, *172*, 9–21. [CrossRef]
41. Wang, X.B.; Ogawa, T.; Suda, S.; Taniguchi, K.; Uike, H.; Kumagai, H.; Mitani, K. Effects of nutritional level on digestive enzyme activities in the pancreas and small intestine of calves slaughtered at same body weight. *Asian-Australas J. Anim. Sci.* **1998**, *11*, 375–380. [CrossRef]
42. Owsley, W.F.; Orr, D.E.; Tribble, L.F. Effects of age and diet on the development of the pancreas and the synthesis and secretion of pancreatic enzymes in the young pig. *J. Anim. Sci.* **1986**, *63*, 497–504. [CrossRef] [PubMed]
43. Guo, W.Z.; Ke, Y.T.; Shang, Y.Z.; Wang, J.H.; Wen, Y.Q.; Wei, J.T. Effects of different crude protein levels of formulated pellet diets on growth performance and blood biochemical parameters of Matou goats. *Chin. J. Anim. Sci.* **2019**, *55*, 108–110, 125.
44. Wang, Z.W.; Mou, X.L.; Yang, G.W.; Li, J.K.; Liu, D.S. Effects of dietary nutritional levels on calcium, phosphorus concentrations and ALK activity in serum and calcium, phosphorus contents in tibia for geese. *China Poult.* **2009**, *31*, 16–20.
45. Zhang, C.L.; Meng, H.J.; Zhu, S.X.; Wu, L.; Liu, Z. Effect of alkaline phosphatase on intestinal health of weaning piglets. *Swine Prod.* **2017**, *6*, 13–16.

46. DiLorenzo, N.; Smith, D.R.; Quinn, M.J.; May, M.L.; Ponce, C.H.; Steinberg, W.; Engstrom, M.A.; Galyean, M.L. Effects of grain processing and supplementation with exogenous amylase on nutrient digestibility in feedlot diets. *Livest. Sci.* **2011**, *137*, 178–184. [CrossRef]
47. McCarthy, M.M.; Engstrom, M.A.; Azem, E.; Gressley, T.F. The effect of an exogenous amylase on performance and total-tract digestibility in lactating dairy cows fed a high-byproduct diet. *J. Dairy Sci.* **2013**, *96*, 3075–3084. [CrossRef]
48. Chen, C.; Wang, Z.; Li, J.; Li, Y.; Huang, P.; Ding, X.; Yin, J.; He, S.; Yang, H.; Yin, Y. Dietary vitamin E affects small intestinal histomorphology, digestive enzyme activity, and the expression of nutrient transporters by inhibiting proliferation of intestinal epithelial cells within jejunum in weaned piglets. *J. Anim. Sci.* **2019**, *97*, 1212–1221. [CrossRef]
49. Luo, W.W.; Tian, L.; Tan, B.; Shen, Z.H.; Xiao, M.W.; Wu, S.; Meng, X.R.; Wu, X.; Wang, X.Y. Update: Innate Lymphoid Cells in Inflammatory Bowel Disease. *Dig. Dis. Sci.* **2022**, *67*, 56–66. [CrossRef]
50. Wu, J.; He, C.; Bu, J.; Luo, Y.; Yang, S.; Ye, C.; Yu, S.; He, B.; Yin, Y.; Yang, X. Betaine attenuates LPS-induced downregulation of Occludin and Claudin-1 and restores intestinal barrier function. *BMC Vet. Res.* **2020**, *16*, 75. [CrossRef]
51. Rashid, S.; Irshadullah, M. Evaluation of antioxidant and oxidant status of goats (*Capra aegagrus* hircus) naturally infected with Haemonchus contortus. *J. Helminthol.* **2020**, *94*, e36. [CrossRef] [PubMed]
52. Maughan, H.; Wang, P.W.; Diaz Caballero, J.; Fung, P.; Gong, Y.; Donaldson, S.L.; Yuan, L.; Keshavjee, S.; Zhang, Y.; Yau, Y.C.; et al. Analysis of the cystic fibrosis lung microbiota via serial Illumina sequencing of bacterial 16S rRNA hypervariable regions. *PLoS ONE* **2012**, *7*, e45791. [CrossRef] [PubMed]
53. Jiang, X.T.; Peng, X.; Deng, G.H.; Sheng, H.F.; Wang, Y.; Zhou, H.W.; Tam, N.F. Illumina sequencing of 16S rRNA tag revealed spatial variations of bacterial communities in a mangrove wetland. *Microb. Ecol.* **2013**, *66*, 96–104. [CrossRef] [PubMed]
54. Kim, Y.H.; Nagata, R.; Ohkubo, A.; Ohtani, N.; Kushibiki, S.; Ichijo, T.; Sato, S. Changes in ruminal and reticular pH and bacterial communities in Holstein cattle fed a high-grain diet. *BMC Vet. Res.* **2018**, *14*, 310. [CrossRef] [PubMed]
55. Ji, F.J.; Wang, L.X.; Yang, H.S.; Hu, A.; Yin, Y.L. Review: The roles and functions of glutamine on intestinal health and performance of weaning pigs. *Animal* **2019**, *13*, 2727–2735. [CrossRef]
56. Wang, K.; Zhou, M.; Gong, X.; Zhou, Y.; Chen, J.; Ma, J.; Zhang, P. Starch–protein interaction effects on lipid metabolism and gut microbes in host. *Front. Nutr.* **2022**, *9*, 1018026. [CrossRef]
57. Zhao, Y.; Yu, S.; Zhao, H.; Li, L.; Li, Y.; Liu, M.; Jiang, L. Integrated multi-omics analysis reveals the positive leverage of citrus flavonoids on hindgut microbiota and host homeostasis by modulating sphingolipid metabolism in mid-lactation dairy cows consuming a high-starch diet. *Microbiome* **2023**, *11*, 236. [CrossRef]
58. Akkermans, S.; Logist, F.; Van Impe, J.F. An interaction model for the combined effect of temperature, pH and water activity on the growth rate of E. coli K12. *Food Res. Int.* **2018**, *106*, 1123–1131. [CrossRef]
59. Ren, X.D.; Chen, X.S.; Tang, L.; Zeng, X.; Wang, L.; Mao, Z.G. Physiological mechanism of the overproduction of epsilon-poly-L-lysine by acidic pH shock in fed-batch fermentation. *Bioprocess Biosyst. Eng.* **2015**, *38*, 2085–2094. [CrossRef]
60. Li, Y.Q.; Xi, Y.M.; Zeng, H.F.; Zhang, L.; Chen, M.J.; Han, Z.Y. Diversity Changes of Bacterial Community in Rumen of Dairy Cows in Vitro Fermentation Analyzed by 16S rRNA Sequencing Technology. *Chin. J. Anim. Nutr.* **2018**, *30*, 4059–4070. [CrossRef]
61. Nagaraja, T.G.; Titgemeyer, E.C. Ruminal acidosis in beef cattle: The current microbiological and nutritional outlook. *J. Dairy Sci.* **2007**, *90* (Suppl. 1), E17–E38. [CrossRef]
62. Chiquette, J.; Allison, M.J.; Rasmussen, M.A. Prevotella bryantii 25A used as a probiotic in early-lactation dairy cows: Effect on ruminal fermentation characteristics, milk production, and milk composition. *J. Dairy Sci.* **2008**, *91*, 3536–3543. [CrossRef] [PubMed]
63. Bekele, A.Z.; Koike, S.; Kobayashi, Y. Genetic diversity and diet specificity of ruminal Prevotella revealed by 16S rRNA gene-based analysis. *FEMS Microbiol. Lett.* **2010**, *305*, 49–57. [CrossRef]
64. Bekele, A.Z.; Koike, S.; Kobayashi, Y. Phylogenetic diversity and dietary association of rumen Treponema revealed using group-specific 16S rRNA gene-based analysis. *FEMS Microbiol. Lett.* **2011**, *316*, 51–60. [CrossRef] [PubMed]
65. Ransom-Jones, E.; Jones, D.L.; McCarthy, A.J.; McDonald, J.E. The Fibrobacteres: An Important Phylum of Cellulose-Degrading Bacteria. *Microb. Ecol.* **2012**, *63*, 267–281. [CrossRef] [PubMed]
66. Plaizier, J.C.; Li, S.C.; Danscher, A.M.; Derakshani, H.; Andersen, P.H.; Khafipour, E. Changes in Microbiota in Rumen Digesta and Feces Due to a Grain-Based Subacute Ruminal Acidosis (SARA) Challenge. *Microb. Ecol.* **2017**, *74*, 485–495. [CrossRef] [PubMed]
67. Mickdam, E.; Khiaosa-ard, R.; Metzler-Zebeli, B.U.; Klevenhusen, F.; Chizzola, R.; Zebeli, Q. Rumen microbial abundance and fermentation profile during severe subacute ruminal acidosis and its modulation by plant derived alkaloids in vitro. *Anaerobe* **2016**, *39*, 4–13. [CrossRef]
68. Wang, X.; Martin, G.B.; Wen, Q.; Liu, S.; Li, Y.; Shi, B.; Guo, X.; Zhao, Y.; Guo, Y.; Yan, S. Palm oil protects alpha-linolenic acid from rumen biohydrogenation and muscle oxidation in cashmere goat kids. *J. Anim. Sci. Biotechnol.* **2020**, *11*, 100. [CrossRef]
69. Su, X.L.; Tian, Q.; Zhang, J.; Yuan, X.Z.; Shi, X.S.; Guo, R.B.; Qiu, Y.L. Acetobacteroides hydrogenigenes gen. nov., sp. nov., an anaerobic hydrogen-producing bacterium in the family Rikenellaceae isolated from a reed swamp. *Int. J. Syst. Evol. Microbiol.* **2014**, *64*, 2986–2991. [CrossRef]
70. Zhao, Y.; Xie, B.; Gao, J.; Zhao, G. Dietary supplementation with sodium sulfate improves rumen fermentation, fiber digestibility, and the plasma metabolome through modulation of rumen bacterial communities in steers. *Appl. Environ. Microbiol.* **2020**, *86*, e01412-20. [CrossRef]

71. Seshadri, R.; Leahy, S.C.; Attwood, G.T.; Teh, K.H.; Lambie, S.C.; Cookson, A.L.; Eloe-Fadrosh, E.A.; Pavlopoulos, G.A.; Hadjithomas, M.; Varghese, N.J.; et al. Cultivation and sequencing of rumen microbiome members from the Hungate1000 Collection. *Nat. Biotechnol.* **2018**, *36*, 359–367. [CrossRef] [PubMed]
72. Holman, D.B.; Gzyl, K.E. A meta-analysis of the bovine gastrointestinal tract microbiota. *FEMS Microbiol. Ecol.* **2019**, *95*, fiz072. [CrossRef] [PubMed]
73. Chen, X.; Su, X.; Li, J.; Yang, Y.; Wang, P.; Yan, F.; Yao, J.; Wu, S. Real-time monitoring of ruminal microbiota reveals their roles in dairy goats during subacute ruminal acidosis. *NPJ Biofilms Microbiomes* **2021**, *7*, 45. [CrossRef] [PubMed]
74. Fu, Y.; He, Y.; Xiang, K.; Zhao, C.; He, Z.; Qiu, M.; Hu, X.; Zhang, N. The role of rumen microbiota and its metabolites in subacute ruminal acidosis (SARA)-induced inflammatory diseases of ruminants. *Microorganisms* **2022**, *10*, 1495. [CrossRef]
75. Plaizier, J.C.; Mulligan, F.J.; Neville, E.W.; Guan, L.L.; Steele, M.A.; Penner, G.B. Invited review: Effect of subacute ruminal acidosis on gut health of dairy cows. *J. Dairy Sci.* **2022**, *105*, 7141–7160. [CrossRef]

Disclaimer/Publisher's Note: The statements, opinions and data contained in all publications are solely those of the individual author(s) and contributor(s) and not of MDPI and/or the editor(s). MDPI and/or the editor(s) disclaim responsibility for any injury to people or property resulting from any ideas, methods, instructions or products referred to in the content.

animals

MDPI

Article

The Effect of Replacing Whole-Plant Corn Silage with Daylily on the Growth Performance, Slaughtering Performance, Muscle Amino Acid Composition, and Blood Composition of Tan Sheep

Junli Zhang [1,†], Fen Li [2,†], Rina Na [2], Xue Bai [2], Yanfen Ma [2], Yuwei Yang [1], Yun Ma [2] and Xiuqin Wang [1,*]

1 Institute of Animal Science, Ningxia Academy of Agriculture and Forestry, Yinchuan 750002, China; zhangjl20000@163.com (J.Z.); yangyuwei0514@163.com (Y.Y.)
2 College of Animal Science and Technology, Ningxia University, Yinchuan 750002, China; lifen2843@126.com (F.L.); nrn_99@126.com (R.N.); baixue333work@163.com (X.B.); mayf@nxu.edu.cn (Y.M.); mayun@nxu.edu.cn (Y.M.)
* Correspondence: nxnkyw@163.com
† These authors contributed equally to this work.

Simple Summary: To date, few studies in China and abroad have investigated the utilization of daylily. Currently, there is no literature on the evaluation of the feed utilization value of daylily. By using biological fermentation technology, we developed a formula for replacing corn silage with daylily as a feed source for Tan sheep by analyzing coated silage production, diet formulation optimization, growth performance indicators, slaughter performance indicators, carcass quality, and blood biochemical indicators. Based on the feeding test and safety evaluation, the effect of daylily silage on the growth performance and slaughter performance of Tan sheep was evaluated, and the significance of daylily for feed utilization was determined, which provided a scientific basis for further scientific and rational utilization of daylily stems and leaves as feed resources.

Abstract: The shortage of high-quality coarse feed resources is the main factor that restricts the development of animal husbandry in many developing countries. The present study aimed to investigate the effects of replacing corn silage with daylily silage on the growth performance, slaughter performance, blood biochemical indicators, meat quality, and muscle amino acid composition of Tan sheep. A total of 72 healthy Tan sheep were randomly assigned to four groups. In each group, 0%, 20%, 40%, and 60% of corn silage were replaced with daylily silage (denoted as CON, HC20, HC40, and HC60, respectively). Tan sheep fed with daylily silage showed no significant adverse effects on their growth performance, meat quality, and muscle amino acid composition ($p > 0.05$). Some increase was observed in the carcass fat content value (GR-value, $p < 0.05$), thickness of backfat ($p < 0.05$), and the blood urea level ($p < 0.05$). These findings indicate that the utilization of daylily silage instead of whole-plant corn silage has no adverse effects on the growth performance and meat quality of Tan sheep, thus indicating that it can partially replace whole-plant corn feed as a feed resource for Tan sheep.

Keywords: daylily silage; tan sheep; growth performance; amino acids; blood biochemical indicators

Citation: Zhang, J.; Li, F.; Na, R.; Bai, X.; Ma, Y.; Yang, Y.; Ma, Y.; Wang, X. The Effect of Replacing Whole-Plant Corn Silage with Daylily on the Growth Performance, Slaughtering Performance, Muscle Amino Acid Composition, and Blood Composition of Tan Sheep. *Animals* **2023**, *13*, 3493. https://doi.org/10.3390/ani13223493

Academic Editors: Dingfu Xiao and Tiantian Meng

Received: 11 September 2023
Revised: 4 November 2023
Accepted: 6 November 2023
Published: 12 November 2023

Copyright: © 2023 by the authors. Licensee MDPI, Basel, Switzerland. This article is an open access article distributed under the terms and conditions of the Creative Commons Attribution (CC BY) license (https://creativecommons.org/licenses/by/4.0/).

1. Introduction

The shortage of high-quality roughage resources is the main factor that restricts the development of livestock industry in many developing countries [1]. Presently, several studies have reported that locally produced agricultural byproducts or herbaceous plants can replace whole-plant corn feed to some extent and promote the growth and health of livestock and poultry. Liu et al. [2] found that the addition of different proportions of wheat instead of corn in beef cattle diets can improve the apparent digestibility values of dry matter, organic matter, and crude protein and increase serum alanine aminotransferase levels but has no significant effect on rumen pH and fermentation. This finding indicates that an appropriate amount of wheat can be used to replace corn to meet the energy and fiber needs of beef cattle [2]. Silva observed that partial substitution of corn silage (CS) with whole soybean silage (SS) or black oat silage (OS) could reduce the intake and increase the rumination and chewing activities of cows without affecting their growth performance [3]. Zhang et al. reported that sweet straw and wheat straw can be used as cheaper substitutes for whole-plant corn silage (WPCS), without significantly reducing the economic benefits of fattening beef cattle [4]. Natalello et al. found that the addition of whole pomegranate byproducts instead of grain feed to sheep feed improved the antioxidant capacity of sheep meat through the presence of vitamin E in whole pomegranate byproducts [5]. Although many substitutes for WPCS in production have been reported, there is no relevant research on the use of daylily (*Hemerocallis citrina* Baroni) to replace WPCS.

Daylily, a perennial herb from the Liliaceae family, is suitable as both medicine and an edible product. It is named "zheng zhu cai" in China and is usually planted as a forage plant in southwest China. Young daylily stem contains abundant nutrients such as proteins, vitamin C, calcium, fats, carotene, amino acids, and other necessary nutrients beneficial for human health. Daylily is a healthy food with low fat and high protein content, and it is rich in minerals, flavonoids, and phenols, which are considered as active ingredients of daylily [6,7]. Daylily has several important medicinal properties, such as achieving hemostasis and anti-inflammatory effects, clearing heat, removing dampness, helping digestion, improving eyesight, and smoothing the nerves, and it can be used as a postpartum supplement [8]. Daylily can also be used as a high-quality forage food to feed livestock. Therefore, daylily is widely planted in northwest China (Yanchi County and the Hongsipu District of Wuzhong City in Ningxia) and has helped to form a local industry. Currently, there are no investigations on the use of daylily silage to replace WPCS for feeding Tan sheep. Hence, in the present study, we conducted a feeding experiment to determine how the addition of daylily silage to the diet of Tan sheep affects their growth performance, slaughter performance, meat quality, muscle amino acid composition, and blood biochemical indicators.

2. Materials and Methods

2.1. Crops and Silages

According to the local corn (milk-ripening period) and harvest time of daylily, the silage period is approximately 60 days from 20 September each year. Daylily silage was coated silage self-made by the study team. WPCS was self-made in the experimental field. Daylily and corn were cut into 2–3 cm pieces and placed in silage. A compactor was used to compact the plastic cloth on the back cover. Silage samples were collected from four different points in the silage cellar. Table 1 shows the nutritional components of daylily silage and corn silage feed.

The concentrated feed composition was as follows: soybean meal, rapeseed meal, cottonseed meal, DDGS (distiller's dried grains with solubles), stone powder, calcium hydrogen phosphate, soybean oil, sodium chloride, various trace elements, vitamins, and amino acids. The confirmed nutrient composition was as follows: crude protein $\geq$ 33.0%, crude fiber $\leq$ 18.0%, crude ash $\leq$ 20.0%, calcium 1.5–5.0%, total phosphorus $\geq$ 0.5%, sodium chloride 2.5–5.0%, lysine $\geq$ 1.1%, and moisture $\leq$ 13.0%.

2.2. Animals and Diets

The experiment was conducted at Ningxia Shuomu Yanchi Tan Sheep Breeding Co., Ltd. (Yinchuan, China). Based on random grouping, 72 healthy 3-month-old Tan sheep weighing approximately 24.4 $\pm$ 2.4191 kg were randomly assigned to 4 groups. Daylily silage was used to replace 20%, 40%, and 60% corn silage in the diet of Tan sheep as a total mixed diet (TMR). This experiment was conducted in groups, and the animals were fed twice a day (at 8 a.m. and 4 p.m.) after accurately weighing according to Table 2. The nutritional components of the feed are shown in Table 3.

Table 1. Nutrient composition of daylily silage and corn silage.

Items	Daylily Silage	Corn Silage
Crude protein (CP)	7.38%	6.04%,
Dry matter (DM)	31.7%	29.7%
Crude ash	13.1%	8.2%
Crude fat (EE)	2.34%	1.9%
Crude fiber (CF)	40.4%	24.8%
Neutral detergent fiber (NDF)	57.5%	51.0%
Acidity detergent fiber (ADF)	54.2%,	27.4%
Calcium	1.32%	2.80%
Phosphorus	0.12%	1.76%
pH	4.06	2.85
Ammoniacal nitrogen	0.21 g/kg	0.39 g/kg
Lactic acid	1.52 g/100 g	0.704 g/100 g
Acetic acid	2.94 g/kg	11.17 g/kg
Isobutyric acid	0.12 g/kg	0.12 g/kg
Butyrate	0.158 g/kg	0.025 g/kg

Table 2. Daylily feed ingredient addition for experimental sheep.

	Component (kg)	CON	HC20	HC40	HC60
Concentrate composition	Corn	8.6	8.6	8.6	8.6
	Concentrated material	2.5	2.5	2.5	2.5
Forage composition	Daylily silage	0	3.2	6.4	9.6
	Corn straw	0.8	0.8	0.8	0.8
	Corn silage	16	12.8	9.6	6.4
	Alfalfa straw	0.8	0.8	0.8	0.8
	Total (kg/d)	28.7	28.7	28.7	28.7

Notes: CON, HC20, HC40, and HC60 represent the replacement of 0, 20, 40, and 60% of corn silage with Daylily silage, respectively.

Table 3. Nutrient composition of the feed given to the four groups.

Raw Material	Abbreviation	CON	HC20	HC40	HC60
Dry matter	DM%	74.87	74.76	74.64	74.53
Crude protein	CP%	13.96	14.51	15.07	15.62
Digestible energy	DE (MJ/kg)	11.48	11.05	10.62	10.19
Metabolizable energy	ME (MJ/kg)	9.41	9.03	8.66	8.28
Neutral detergent fiber	NDF%	22.78	22.96	23.13	23.31
Acid detergent fiber	ADF%	13.03	14.14	15.25	16.36
Calcium	Ga%	0.24	0.32	0.34	0.35
Phosphorus	P%	0.23	0.23	0.24	0.25

Notes: DE = total energy (GE)—fecal energy (FE); ME = DE—urinary energy (UE)—gas energy (Eg). CON, HC20, HC40, and HC60 represent the replacement of 0, 20, 40, and 60% of corn silage with Daylily silage, respectively.

2.3. *Determination of Sheep Growth Performance and Slaughter Performance*

2.3.1. Determination of Sheep Growth Performance

During the experiment, we tracked and tested the growth performance of all 72 Tan sheep (18 sheep in each group). The amount of feed provided and rejected (10% provided) was recorded twice a day, and the animals were weighed to determine the feed intake for each animal. Average daily gain (ADG) and feed conversion rate (FCR) were calculated [9].

2.3.2. Determination of Sheep Slaughter Performance

At the end of the experiment, 3 sheep were randomly selected from each group for slaughter. According to the standard commercial procedure [10], Ministry of Agriculture, People's Republic of China), sheep were electrocuted and slaughtered humanely in the slaughterhouse. Before slaughtering, the animals were fasted for 24 h and stopped from drinking water for 2 h before weighing; the live weight of the animals was recorded before slaughtering.

Carcass weight: after slaughtering and releasing the blood, the test sheep were skinned, and the whole body (including kidney and surrounding fat) with head, hoof, and visceral tissue removed was weighed after standing for 30 min [11].

Net meat weight and bone weight: the weight of the meat after all bones were removed from the carcass was defined as net meat weight; the weight of the collected bones was defined as bone weight.

GR value: We measured the tissue thickness between the 12th and 13th ribs of a sheep carcass at a distance of 11 cm from the midline of the spine. We used this as an indicator to evaluate the fat content of the Tan sheep carcass. A vernier caliper (Booher-303215, Kunshan Kaipai Hardware, and Electrical Co., Ltd., Jiangsu, China.) was used to measure issue thickness.

Backfat thickness: A vernier caliper was used to measure the fat thickness just above the middle of the eye muscle between the 12th and 13th pairs of ribs.

Eye muscle area: The cross-sectional area of the longissimus dorsi muscle (LDM) at the interface of the 12th and 13th ribs was measured on a sulfur paper by using a planimeter (kp-21c, Koizumi Co., Ltd., Tsuchiura, Japan).

$$\text{Slaughter rate} = (\text{carcass weight}/\text{live weight before slaughter}) \times 100\%$$

$$\text{Net meat rate} = (\text{net meat mass}/\text{carcass mass}) \times 100\%$$

$$\text{Meat-to-bone ratio} = (\text{carcass net meat weight}/\text{carcass bone weight}) \times 100\%$$

2.3.3. Determination of Sheep Meat Quality

In the experiment, we slaughtered 3 Tan sheep in each group. We collected approximately 200 g of LDM samples from the left side of the 12th rib of the slaughtered Tan sheep. Then, we divided them into three sub samples to evaluate meat quality. The quality of meat was tested by measuring muscle water retention capacity (WRC), cooking loss, and shear force. Cooking loss and shear force were determined according to the method of Honikel (1998) [12].

The LDM sample was cooked in a temperature-controlled water bath in a plastic bag to an internal temperature of 75 °C, and the internal temperature of the sample was measured using a digital thermometer. The sample was weighed before and after cooking. Cooking loss was calculated based on the weight difference between raw and cooked samples and expressed as a percentage of the initial weight.

After removing the fat attached to the surface, the LDM was aged in a refrigerator at 4 °C for 24 h. Subsequently, the meat sample was kept at room temperature for 1 h, a thermometer was inserted into the center of the meat sample, and the sample was heated in an 80 °C constant temperature water bath until the temperature reached 70 °C. The meat sample was then removed, cooled in a refrigerator at 0~4 °C, and cut vertically in the direction of muscle fibers to obtain 50 cm-thick meat slices with a diameter of 1. A 27 cm circular sampler (SM-8007 type meat shear force measuring instrument, Dongguan Licheng Electronic Technology Co., Ltd., Dongguan, China) was used to sample and measure the shear force of the meat product along the direction of muscle fibers, and an average value was obtained through multiple measurements. The shear force was expressed in Newtons (N).

WRC was determined using the method described by Franco et al. [13]. Briefly, 10 g of meat samples was wrapped in double-layer gauze, and 18 layers of filter papers were placed on the top and bottom of the meat samples. The samples were then subjected to a pressure of 35 kg for 5 min and weighed immediately. WRC was calculated as follows:

$$\text{WRC} = ((\text{total muscle weight} - \text{muscle loss})/\text{total muscle moisture}) \times 100\%.$$

2.4. Determination of Amino Acid Composition

In the experiment, we collected approximately 100 g LDM samples from the left side of the 12th rib of the 3 slaughtered Tan sheep in each group. Then, we divided them into three sub samples to measure the amino acid composition of the meat. The content of various amino acids in the carcass was determined using an amino acid auto-analyzer. Before the experiment, excess oil was first removed, and the meat was then cut into small particles. The meat was then ground into a puree by using a food grinder, frozen, and stored. The amino acid composition was determined according to the method described in the national standard [14].

The amino acid composition of the LDM samples was determined using a high-performance liquid chromatography system (Waters 2695 HPLC system, Waters Technology, Taunton, MA, USA). Approximately 100 mg of minced muscle samples was hydrolyzed with 20 mL of HCl (6 mol/L) at 110 °C for 22 h in sealed evacuated tubes. The samples were then subjected to automatic precolumn derivatization. After derivatization, 10 μL of each sample was injected into the HPLC system to determine amino acid composition. Mobile phase A was 40 mM NaH_2PO_4 adjusted to pH 7.8 with NaOH, while mobile phase B was 45% acetonitrile, 45% methanol, and 10% deionized water. The chromatographic column temperature was set at 37 °C with a flow rate of 1 mL/min. The identity and quantity of the amino acids were assessed by comparison with the retention times and peak areas of standard amino acids.

2.5. Determination of Blood Biochemical Indicators

On the morning of the last day of the experiment, 10 sheep were randomly selected from each group, and 5 mL of fasting jugular vein blood was collected. The blood sample was left to stand for 30 min and then centrifuged at 2200× *g* rpm for 10 min to precipitate serum. The contents of total protein (TP), albumin (ALB), aspartate aminotransferase (AST), alanine transaminase (ALT), blood glucose (GLU), urea (UREA), creatinine (CREA), creatine kinase (CK), creatine kinase isoenzymes (CK-MB), lactate dehydrogenase (LDH), and cholinesterase (CHE) in serum were determined using Mindray BS-420 and Hwld DR-200BS automatic biochemical analyzer at Beijing Huaying Biotechnology Research Institute (Beijing, China).

2.6. Data Analysis

All data were processed using Microsoft Excel and analyzed using SAS version 9.3 (SAS Institute Inc., Cary, NC, USA). ANOVA and a graph-based test were used to analyze significant differences between the treatments. A *p*-value of <0.05 was considered statistically significant. Data are expressed as mean ± standard error.

3. Results

3.1. Effect of the Addition of Different Proportions of Daylily Silage on the Growth Performance of Tan Sheep

Table 4 shows the effect of the addition of different proportions of daylily silage instead of corn silage on the growth performance of Tan sheep. The HC20 group showed the highest final weight, and the weight was increased by 1.88%, 1.28%, and 4.10% as compared to those of the CON, HC40, and HC60 groups, respectively. The lowest FCR was noted for the HC20 group, and the FCR value for this group decreased by 11.02%, 12.27%, and 13.15% as compared to those for the CON, HC40, and HC60 groups, respectively. However, the initial weight, final weight, average daily gain, and feed-to-weight ratio were not significantly different between the groups ($p > 0.05$).

Table 4. Effect of daylily silage addition on the growth performance of Tan sheep.

Items	Treatments				SEM	*p*
	CON	HC20	HC40	HC60		
Initial weight (kg)	25.41 ± 1.71	25.18 ± 1.98	25.69 ± 1.95	24.87 ± 2.26	0.301	0.684
Final weight (kg)	35.60 ± 2.75	36.27 ± 2.54	35.81 ± 3.31	34.84 ± 3.44	0.516	0.604
Total weight gain (kg)	10.19 ± 1.99	11.09 ± 0.56	10.12 ± 1.36	9.97 ± 1.17	0.004	0.321
Average daily gain (kg)	0.17 ± 0.04	0.17 ± 0.03	0.16 ± 0.04	0.17 ± 0.03	0.265	0.338
Average daily feed intake (kg/day DM)	1.59 ± 0.01	1.60 ± 0.02	1.61 ± 0.02	1.62 ± 0.02	0.018	0.278
FCR (%)	4.90 ± 1.03	4.36 ± 0.40	4.97 ± 1.11	5.02 ± 1.16	0.439	0.211

Notes: CON, HC20, HC40, and HC60 represent the replacement of 0, 20, 40, and 60% of corn silage with Daylily silage, respectively. SEM represents standard error of mean, n = 18.

3.2. *Effect of the Addition of Different Proportions of Daylily Silage on the Slaughter Performance of Tan Sheep*

Table 5 shows the results of the slaughter performance analysis. No significant differences were noted in live weight, carcass weight, net meat weight, bone weight, fat thickness, and eye muscle area among the experimental groups. However, the GR values of the HC20 and HC40 groups were significantly higher than that of the CON group ($p < 0.05$). The backfat thickness of sheep fed with HC40 and HC60 daylily silage was also significantly lower than those fed with 100% WPCS ($p < 0.05$). Compared to the CON group, the net meat weight of the HC20, HC40, and HC60 groups was 6.57%, 4.85%, and 1.72% higher, respectively, and the net meat percentage of these groups was 8.26%, 6.45%, and 1.69% higher than that of the CON group, respectively.

Table 5. Effect of daylily silage addition on slaughter performance of Tan sheep.

Items	Treatments				SEM	*p*
	CON	HC20	HC40	HC60		
Pre-slaughter live weight (kg)	37.37 ± 0.15	37.07 ± 9.05	36.60 ± 0.44	37.83 ± 0.21	0.448	0.262
Carcass weight (kg)	21.00 ± 0.78	20.63 ± 7.58	20.87 ± 0.84	20.53 ± 0.75	0.187	0.859
Net meat weight (kg)	12.78 ± 0.43	13.62 ± 2.72	13.40 ± 0.38	13.00 ± 0.20	0.329	0.379
Bone weight (kg)	3.82 ± 0.34	3.72 ± 0.40	3.42 ± 0.21	4.18 ± 0.15	0.271	0.363
GR value (mm)	18.29 ± 2.15 ^{b}	22.86 ± 2.40 ^{a}	24.66 ± 3.44 ^{a}	18.73 ± 2.64 ^{ab}	2.706	<0.001
Thickness of backfat (mm)	7.35 ± 1.67 ^{a}	6.22 ± 0.88 ^{ab}	5.33 ± 1.08 ^{b}	4.78 ± 1.24 ^{b}	0.972	<0.001
Eye muscle area (cm^2)	12.32 ± 1.03	14.55 ± 1.42	15.73 ± 2.53	12.85 ± 0.89	1.357	0.088
Slaughter rate (%)	56.20 ± 1.93	55.75 ± 25.85	57.03 ± 2.67	54.27 ± 1.67	1.002	0.624
Net meat rate (%)	61.40 ± 3.39	66.47 ± 29.04	65.63 ± 1.99	63.79 ± 0.72	1.946	0.091
Meat to bone ratio (%)	3.43 ± 0.75	3.73 ± 1.18	4.00 ± 0.25	3.12 ± 0.31	0.329	0.325

Notes: CON, HC20, HC40, and HC60 represent the replacement of 0, 20, 40, and 60% of corn silage with Daylily silage, respectively. In the same row, mean values with different superscripts indicate significant differences ($p < 0.05$). SEM represents standard error of mean, n = 3.

3.3. *Effect of the Addition of Different Proportions of Daylily Silage on the Meat Quality of Tan Sheep*

Table 6 shows the results for the meat quality analysis. The HC40 group showed the highest water loss rate, which was 3.51%, 3.86%, and 9.07% higher than those of the CON, HC20, and HC60 groups, respectively. The HC40 group exhibited the highest cooked meat rate, which was 0.77%, 1.39%, and 0.92% higher than those of the CON, HC20, and HC60 groups, respectively. No significant difference in shear force values was observed among the groups.

Table 6. Effect of daylily silage addition on the carcass quality of Tan sheep.

Items	Treatments				SEM	*p*
	CON	HC20	HC40	HC60		
Water loss (%)	31.94 ± 1.94	31.83 ± 2.11	33.06 ± 2.59	30.31 ± 4.57	1.125	0.268
Cooked meat rate (%)	52.78 ± 1.29	52.42 ± 1.84	53.19 ± 0.30	52.70 ± 1.32	0.318	0.085
Shear force (N)	66.71 ± 2.34 ^{a}	52.42 ± 1.51 ^{b}	61.34 ± 2.18 ^{a}	52.69 ± 1.08 ^{b}	5.024	<0.001

Notes: CON, HC20, HC40, and HC60 represent the replacement of 0, 20, 40, and 60% of corn silage with Daylily silage, respectively. In the same row, mean values with different superscripts indicate significant differences ($p < 0.05$). SEM represents standard error of mean, n = 3.

3.4. Effect of the Addition of Different Proportions of Daylily Silage on Amino Acid Composition of Tan Sheep Meat

Seventeen types of amino acids were determined in the meat of each group. Table 7 shows the results of amino acid composition in the meat. There were 8 types of essential amino acids, and the total amino acid content reached 44.31%, 44.42%, 44.23%, and 44.11% for the CON, HC20, HC40, and HC60 groups, respectively. Semi-essential amino acids (conditionally essential amino acids) are those amino acids that can be synthesized in the body, but usually cannot meet normal needs and require additional supplements. Three types of semi-essential amino acids were detected in the meat, and their content reached 14.77%, 14.68%, 14.73%, and 14.78% in the CON, HC20, HC40, and HC60 groups, respectively; however, none of them reached a significant level. Six types of non-essential amino acids were detected in the meat, and their content was 40.92%, 40.90%, 41.03%, and 41.11% in the CON, HC20, HC40, and HC60 groups, respectively.

Table 7. Effect of daylily silage addition on the amino acid composition of Tan sheep.

Types of Amino Acids		Treatments				SEM	*p*
		CON	HC20	HC40	HC60		
Essential amino acids	Threonine	0.83 ± 0.02	0.79 ± 0.03	0.77 ± 0.05	0.77 ± 0.01	0.024	0.275
	Valine	0.93 ± 0.03	0.87 ± 0.03	0.86 ± 0.06	0.87 ± 0.00	0.028	0.209
	Isoleucine	0.85 ± 0.02	0.83 ± 0.03	0.78 ± 0.06	0.81 ± 0.00	0.026	0.239
	Leucine	1.46 ± 0.04	1.39 ± 0.04	1.36 ± 0.10	1.38 ± 0.00	0.038	0.309
	Phenylalanine	0.85 ± 0.03	0.81 ± 0.04	0.79 ± 0.06	0.78 ± 0.00	0.027	0.345
	Lysine	1.64 ± 0.05	1.56 ± 0.06	1.52 ± 0.11	1.53 ± 0.01	0.047	0.232
	Histidine	0.62 ± 0.04	0.58 ± 0.05	0.58 ± 0.05	0.55 ± 0.01	0.025	0.332
	Methionine	0.51 ± 0.01	0.49 ± 0.01	0.46 ± 0.03	0.48 ± 0.00	0.018	0.109
	Subtotal	7.72 ± 0.01	7.36 ± 0.02	7.16 ± 0.03	7.22 ± 0.00	0.217	0.271
Semi-essential amino acid	Arginine	1.14 ± 0.02	1.08 ± 0.03	1.06 ± 0.08	1.07 ± 0.01	0.031	0.238
	Glycine	0.77 ± 0.00	0.72 ± 0.02	0.72 ± 0.09	0.73 ± 0.02	0.021	0.186
	Tyrosine	0.64 ± 0.01	0.61 ± 0.02	0.59 ± 0.04	0.61 ± 0.01	0.018	0.289
	Subtotal	2.43 ± 0.05	2.43 ± 0.09	2.39 ± 0.23	2.42 ± 0.06	0.016	0.194
Non-essential amino acids	Aspartate	1.64 ± 0.05	1.56 ± 0.06	1.53 ± 0.11	1.53 ± 0.01	0.045	0.280
	Glutamate	2.92 ± 0.08	2.78 ± 0.10	2.70 ± 0.18	2.76 ± 0.02	0.081	0.552
	Serine	0.71 ± 0.02	0.68 ± 0.02	0.66 ± 0.05	0.66 ± 0.00	0.020	0.262
	Alanine	1.07 ± 0.02	1.01 ± 0.03	1.00 ± 0.07	1.01 ± 0.00	0.028	0.334
	Proline	0.69 ± 0.01	0.66 ± 0.02	0.66 ± 0.06	0.68 ± 0.01	0.013	0.743
	Cystine	0.06 ± 0.00	0.06 ± 0.00	0.06 ± 0.01	0.06 ± 0.01	0.000	0.723
	Subtotal	7.13 ± 0.03	6.77 ± 0.03	6.64 ± 0.60	6.73 ± 0.01	0.186	0.379

Notes: CON, HC20, HC40, and HC60 represent the replacement of 0, 20, 40, and 60% of corn silage with Daylily silage, respectively. SEM represents standard error of mean, n = 3.

The taste of amino acids is closely associated with the hydrophobicity of their side chain R groups. Amino acids with low hydrophobicity mainly have a sweet taste and are termed sweet amino acids, such as glycine, alanine, serine, lysine, and aspartate. Amino acids with high hydrophobicity mainly have a bitter taste and are termed bitter amino acids, such as leucine, isoleucine, phenylalanine, tyrosine, tryptophan, histidine, lysine, and arginine. For amino acids with an acidic side chain R group (such as COOH and SO_3H), the main flavor is acidity, such as aspartate and glutamate. These two amino acids are also important prerequisites for the formation of delicious substances and are also known as umami amino acids. The ANOVA analysis showed that their content and proportion were not significantly different among the different groups of umami amino acids, sweet amino acids, bitter amino acids, and total amino acids (Table 8).

Table 8. Effect of daylily silage addition on the functional amino acid content in Tan sheep muscle.

Groups	Umami Taste Amino Acid		Sweet Amino Acids		Bitter Amino Acid	
	Content (mg/g)	Percentage of Total Amino Acids (%)	Content (mg/g)	Percentage of Total Amino Acids (%)	Content (mg/g)	Percentage of Total Amino Acids (%)
CON	4.58 ± 0.14	26.26 ± 0.06	4.10 ± 0.08	23.52 ± 0.25	8.75 ± 0.30	50.22 ± 0.26
HC20	4.35 ± 0.16	26.25 ± 0.28	3.88 ± 0.14	23.44 ± 0.20	8.33 ± 0.33	50.31 ± 0.48
HC40	4.24 ± 0.30	26.21 ± 0.29	3.84 ± 0.35	23.68 ± 0.30	8.11 ± 0.63	50.11 ± 0.22
HC60	4.30 ± 0.02	26.27 ± 0.17	3.87 ± 0.04	26.68 ± 0.16	8.19 ± 0.02	50.05 ± 0.02
SEM	0.129	0.023	0.104	1.359	0.247	0.100
p	0.217		0.296		0.262	

Notes: CON, HC20, HC40, and HC60 represent the replacement of 0, 20, 40, and 60% of corn silage with Daylily silage, respectively. SEM represents standard error of mean, n = 3.

3.5. Effect of the Addition of Different Proportions of Daylily Silage on Blood Biochemical Indicators of Tan Sheep

Table 9 shows the effect of the addition of different proportions of daylily silage instead of corn silage on the blood biochemical indicators of Tan sheep. The results showed that the content of UREA in the HC60 group was significantly higher than that in the CON group ($p < 0.05$). The groups showed no significant differences in the levels of TP, ALB, AST, ALT, CREA, LDH, CHE, CK, CK-MB, and GLU.

Table 9. Effect of daylily silage addition on the blood biochemical indicators of Tan sheep.

Items	Treatments				SEM	*p*
	CON	HC20	HC40	HC60		
TP (g/L)	66.06 ± 3.19	65.09 ± 3.87	64.92 ± 2.92	65.14 ± 6.50	0.443	0.173
ALB (g/L)	24.44 ± 2.19	23.12 ± 2.60	22.53 ± 2.40	24.85 ± 2.35	0.945	0.121
AST (μ/L)	94.09 ± 7.79	83.64 ± 12.62	80.94 ± 17.21	93.18 ± 13.95	5.761	0.076
ALT (μ/L)	21.65 ± 3.92	19.50 ± 7.21	18.08 ± 6.03	23.16 ± 6.19	1.950	0.250
GLU (mmol/L)	1.10 ± 0.44	0.95 ± 0.34	1.01 ± 0.42	1.05 ± 0.35	0.055	0.850
UREA (mmol/L)	3.88 ± 1.18 ^{b}	4.39 ± 0.96 ^{ab}	4.44 ± 0.63 ^{ab}	5.77 ± 0.93 ^{a}	0.699	<0.001
CREA (μmol/L)	82.68 ± 8.69	80.84 ± 13.47	75.63 ± 8.36	77.01 ± 12.82	2.839	0.462
CK (μ/L)	262.98 ± 70.47	215.28 ± 51.92	262.39 ± 110.31	235.27 ± 53.83	19.997	0.432
CK-MB (μ/L)	119.56 ± 32.49	95.98 ± 12.14	119.77 ± 51.97	104.58 ± 24.03	10.158	0.350
LDH (μ/L)	539.23 ± 58.96	491.69 ± 91.19	495.55 ± 70.89	512.96 ± 42.22	18.756	0.400
CHE (μ/L)	2254.74 ± 262.15	2124.62 ± 448.11	2194.60 ± 309.11	2004.68 ± 348.87	93.016	0.427

Notes: TP, total protein; ALB, albumin; AST, aspartate aminotransferase; ALT, alanine transaminase; GLU, blood glucose; CREA, creatinine; CK, creatine kinase; CK-MB, creatine kinase isoenzymes; LDH, lactate dehydrogenase; CHE, cholinesterase. CON, HC20, HC40, and HC60 represent the replacement of 0, 20, 40, and 60% of corn silage with Daylily silage, respectively. In the same row, mean values with different superscripts indicate significant differences ($p < 0.05$). SEM represents standard error of mean, n = 10.

4. Discussion

Daylily silage and WPCS showed differences in their nutritional composition. Daylily silage showed higher CP, DM, EE, CF, NDF, and ADF content, but lower Ga and *p* content. Zhang et al. reported a positive correlation between the CP content in feed and body weight, ADG, and dry matter intake in ruminants [15]. Some studies have also shown that the protein content of feed can significantly affect the dietary intake, health, and growth performance of livestock [1,16–18]. The growth performance of sheep in the HC40 and

HC60 groups was not as high as that in the HC20 group, particularly for the HC60 group. This might be due to the poor quality of silage caused by excessive CP in HC40 and HC60 silage. Previous studies have shown that forage feed with a high protein content is not suitable for producing high-quality silage because of the high buffer capacity, pH value, and extensive proteolysis during ensiling [19,20].

Carcass trait is a key indicator that reflects the ideal livestock production in animal husbandry [3,21], which is closely associated with feed composition [22]. The experimental results showed that the use of daylily silage instead of 40% WPCS to feed sheep had higher GR values and lower backfat thickness ($p < 0.05$). The results showed that the body fat content of Tan sheep fed with Daylily silage increased, while the subcutaneous fat accumulation was less, with more edible parts. The slaughter performance of Tan sheep fed with daylily silage was better than that of Tan sheep fed with corn silage. This might be due to the reasonable balance of CP and ME concentrations in the diet. Wang et al. found that as the content of sweet sorghum silage feed increased, the CP content in the diet increased and the ME level decreased (which is consistent with the trend of dietary proportion in this experiment), resulting in less subcutaneous fat accumulation in Kurar sheep [1]. This finding was consistent with the results of the present study. The shear force of HC20, HC40, and HC60 lambs was smaller ($p < 0.05$), which may be attributed to the higher intramuscular fat deposition in the carcass. The increase in intramuscular fat will reduce the physical strength of muscle fibers; thus, making it easier for muscle bundles to separate. Consequently, muscles become softer and easier for consumers to chew. Intramuscular fat helps improve flavor, thereby increasing consumer acceptance of the product [23]. Previous studies have shown that lower shear forces can also be obtained through higher protein content [24] and appropriate energy levels [25] in the diet.

Amino acids are the most important nutrients in meat, and their content and composition are important indicators to evaluate the physiological value of food proteins [26]. The present study showed the presence of 17 types of total amino acids, and their content was balanced, with no significant difference. The ratio of essential amino acids to total amino acids in each group was above 44%. The contents of umami, sweet, and bitter amino acids in the meat were slightly different in each group; however, the differences among the groups were not significant. Ningxia meat is a delicious and nourishing food, and its glutamic acid and aspartic acid contents account for more than 25% of the total amino acid content [27]; this finding is consistent with the results of the present study. The content of various amino acids in the experimental sheep was generally relatively balanced, thus providing meat with rich and balanced amino acid content in each group.

Daylily contains various bioactive components, among which flavonoids and phenols are considered the active ingredients [6]. Flavonoids have anti-inflammatory, antiviral, and anticancer properties [28]. Phenolic compounds show potential antioxidant activity [29]. Although the present study did not explicitly explore the effects of daylily feed on the antioxidant and inflammatory properties in the meat of Tan sheep, the results confirmed that feeding daylily feed had no harmful effects on the health of Tan sheep. The normal levels of serum total protein, albumin, and globulin in sheep are 60–70 g/L, 24–30 g/L, and 35–57 g/L, respectively [30]. In the present study, all serum indices in the experimental groups were within the normal range. High serum total protein concentration is a manifestation of vigorous protein metabolism, which is conducive to promote animal growth and improve FCR. The serum UREA nitrogen content reflects the protein metabolism of the animal body and more accurately reflect the balance of animal protein metabolism and dietary amino acids [31,32]. The CP level of the experimental diet increased slightly following the addition of daylily silage, and the blood urine nitrogen content increased with the increase in daylily silage content in the diet. The serum UREA nitrogen content in the HC60 group was significantly higher than that in the CON, HC20, and HC40 groups. The substitution of WPCS with HC60 containing daylily silage significantly reduced the protein utilization rate of Tan sheep. These results indicated that the addition of 60% daylily silage to WPCS was not conducive to protein decomposition in Tan sheep. Enzyme activities in serum could be used as a determinant of various pathological diseases. Aminotransferase activities of the AST and ALT enzymes in sera are the key determinants for hepatocellular damage [33]. ALT and AST, the measures of liver function, are elevated in hepatocyte necrosis or alcoholism. In the present study, no significant change was observed in the AST level, and the AST content of the groups fed with daylily silage was reduced after WPCS was replaced. This finding indicated that feeding daylily silage had no adverse effect on the liver of Tan sheep.

5. Conclusions

The results of the present study suggest that replacing corn silage with daylily silage in the diet of Tan sheep has no negative effects on growth performance, slaughter performance, meat quality, amino acid composition, and blood biochemical indicators. Replacing a portion of corn silage with daylily silage can reduce the backfat thickness of Tan sheep, increase the GR value, and reduce shear force. Additionally, the serum urea content significantly increased. Thus, it is feasible to feed Tan sheep with dietary silage; however, our experiment only conducted preliminary exploration on small samples. In the future, it will be necessary to further explore the possibility of using Daylily silage feed to Tan sheep in large-scale experiments. The present study provides a theoretical basis for feeding fattening Tan sheep with daylily silage feed in the future.

Author Contributions: J.Z., Y.M. (Yanfen Ma) and X.W. designed the experiment. F.L., R.N. and X.B. collected samples and completed the manuscript. Y.M. (Yun Ma) and Y.Y. revised the manuscript. X.W. provided financial support. All authors have read and agreed to the published version of the manuscript.

Funding: Fund Project: the Ningxia Academy of Agricultural Sciences Science and Technology Innovation Project (project number: NYYZZ-G-19-08); the Ningxia Academy of Agricultural and Forestry Sciences "14th five-year" Innovation and Technology Project (project number: NGSB-2021-12-13); the National Modern Agriculture (Beef and Yak) Industrial Technology System Construction Special Project (project number: CARS-37).

Institutional Review Board Statement: All applicable international, national, and institutional guidelines for the care and use of animals were followed.

Informed Consent Statement: Not applicable.

Data Availability Statement: All the tables used to support the findings of the current study are included in the article.

Conflicts of Interest: The authors declare no conflict of interest.

References

1. Wang, J.; Yang, B.Y.; Zhang, S.J.; Amar, A.; Chaudhry, A.S.; Cheng, L.; Abbasi, I.H.R.; Al-Mamun, M.; Guo, X.F.; Shan, A.S. Using mixed silages of sweet sorghum and alfalfa in total mixed rations to improve growth performance, nutrient digestibility, carcass traits and meat quality of sheep. *Anim. Int. J. Anim. Biosci.* **2021**, *15*, 100246. [CrossRef] [PubMed]
2. Liu, Y.F.; Zhao, H.B.; Liu, X.M.; You, W.; Cheng, H.J.; Wan, F.C.; Liu, G.F.; Tan, X.W.; Song, E.L.; Zhang, X.L. Substitution of Wheat for Corn in Beef Cattle Diets: Digestibility, Digestive Enzyme Activities, Serum Metabolite Contents and Ruminal Fermentation. *Asian-Australas. J. Anim. Sci.* **2016**, *29*, 1424–1431. [CrossRef] [PubMed]
3. Silva, L.H.P.; Assis, D.E.F.; Estrada, M.M.; Assis, G.J.F.; Chizzotti, M.L.J.L. Carcass and meat quality traits of Nellore young bulls and steers throughout fattening. *Livest. Sci.* **2019**, *229*, 28–36. [CrossRef]
4. Zhang, H.; Zhang, L.; Xue, X.; Zhang, X.; Wang, H.; Gao, T.; Phillips, C. Effect of feeding a diet comprised of various corn silages inclusion with peanut vine or wheat straw on performance, digestion, serum parameters and meat nutrients in finishing beef cattle. *Anim. Biosci.* **2022**, *35*, 29–38. [CrossRef] [PubMed]
5. Natalello, A.; Priolo, A.; Valenti, B.; Codini, M.; Mattioli, S.; Pauselli, M.; Puccio, M.; Lanza, M.; Stergiadis, S.; Luciano, G. Dietary pomegranate by-product improves oxidative stability of lamb meat. *Meat Sci.* **2020**, *162*, 108037. [CrossRef]
6. Guo, S.; Qin, N.; Wang, X.; Zuo, Z.; Li, Q.; Wang, Y. Freeze-dried powder of daylily bud improves bromocriptine-induced lactation disorder in rats via JAK2/STAT5 pathway. *J. Ethnopharmacol.* **2023**, *313*, 116536. [CrossRef]
7. Hao, Z.; Liang, L.; Liu, H.; Yan, Y.; Zhang, Y. Exploring the Extraction Methods of Phenolic Compounds in Daylily (*Hemerocallis citrina* Baroni) and Its Antioxidant Activity. *Molecules* **2022**, *27*, 2964. [CrossRef]

8. Guo, A.; Li, S.; Yang, Y.; Hou, F.; Wu, J.; Gao, Y.; Xing, G.J.S.H. Lecithin extraction optimisation and synthesis in Hemerocallis citrina Baroni. *Sci. Hortic.* **2022**, *293*, 110682. [CrossRef]
9. do Sacramento Ribeiro, J.; Santos, L.L.; de Lima Júnior, D.M.; de Albuquerque Mariz, T.M.; Ladeira, M.M.; de Azevedo, P.S.; Lima, C.B.; Dos Santos Silva, M.J.M. Spineless cactus associated with Tifton hay or sugarcane bagasse may replace corn silage in sheep diets. *Trop. Anim. Health Prod.* **2017**, *49*, 995–1000. [CrossRef]
10. *NY 467-2001*; Health and Quarantine Regulations for Livestock and Poultry Slaughter Domestic—Industry Standards. Industry Standards—Agriculture CN NY: Beijing, China, 2001.
11. Wu, P.; Fu, X.; Wang, H.; Hou, M.; Shang, Z. Effect of Silage Diet (Sweet Sorghum vs. Whole-Crop Corn) and Breed on Growth Performance, Carcass Traits, and Meat Quality of Lambs. *Animals* **2021**, *11*, 3120. [CrossRef]
12. Honikel, K.O. Reference methods for the assessment of physical characteristics of meat. *Meat Sci.* **1998**, *49*, 447–457. [CrossRef] [PubMed]
13. Franco, D.; Bispo, E.; González, L.; Vázquez, J.A.; Moreno, T. Effect of finishing and ageing time on quality attributes of loin from the meat of Holstein-Fresian cull cows. *Meat Sci.* **2009**, *83*, 484–491. [CrossRef] [PubMed]
14. Li, Q.; Li, H.; Lu, G.; Zhou, H.; Xu, Z. Improvement of the National Standard Method for Determining 16 Amino Acids in Food. *Phys. Test. Chem. Anal. Part B Chem. Anal.* **2022**, *58*, 77–83.
15. Zhang, B.; Wang, C.; Liu, H.; Liu, J.; Liu, H. Effects of dietary protein level on growth performance and nitrogen excretion of dairy heifers. *Asian-Australas. J. Anim. Sci.* **2017**, *30*, 386–391. [CrossRef] [PubMed]
16. Zhang, H.B.; Wang, Z.S.; Peng, Q.H.; Tan, C.; Zou, H.W. Effects of different levels of protein supplementary diet on gene expressions related to intramuscular deposition in early-weaned yaks. *Anim. Sci. J. Nihon Chikusan Gakkaiho* **2014**, *85*, 411–419. [CrossRef] [PubMed]
17. Balasubramanian, B.; Park, J.H.; Shanmugam, S.; Kim, I.H. Influences of Enzyme Blend Supplementation on Growth Performance, Nutrient Digestibility, Fecal Microbiota and Meat-Quality in Grower-Finisher Pigs. *Animals* **2020**, *10*, 386. [CrossRef]
18. Wang, X.; Xu, T.; Zhang, X.; Geng, Y.; Kang, S.; Xu, S. Effects of Dietary Protein Levels on Growth Performance, Carcass Traits, Serum Metabolites, and Meat Composition of Tibetan Sheep during the Cold Season on the Qinghai-Tibetan Plateau. *Animals* **2020**, *10*, 801.
19. Fisher, D.S.; Burns, J.C.J.A.J. Quality Analysis of Summer-Annual Forages. II. Effects of Forage Carbohydrate Constituents on Silage Fermentation1. *Agron. J.* **1987**, *79*, 242–248. [CrossRef]
20. Mcdonald, P.; Henderson, A.R.; Heron, S.J.E.J.J.W. *The Biochemistry of Silage*; Chalcombe Publications: Bucks, UK, 1991.
21. Luo, J.; Zeng, D.; Cheng, L.; Mao, X.; Yu, J.; Yu, B.; Chen, D.J. Dietary β-glucan supplementation improves growth performance, carcass traits and meat quality of finishing pigs. *Anim. Nutr.* **2019**, *5*, 380–385. [CrossRef]
22. Francisco, A.E.; Janíček, M.; Dentinho, T.; Portugal, A.P.V.; Almeida, J.M.; Alves, S.P.; Fialho, L.; Jerónimo, E.; Bessa, R.J.B.; Santos-Silva, J. Effects of alfalfa particle size and starch content in diets on feeding behaviour, intake, rumen parameters, animal performance and meat quality of growing lambs—ScienceDirect. *Meat Sci.* **2020**, *161*, 107964. [CrossRef]
23. Kouba, M.; Bonneau, M. Compared development of intermuscular and subcutaneous fat in carcass and primal cuts of growing pigs from 30 to 140kg body weight. *Meat Sci.* **2009**, *81*, 270–274. [CrossRef] [PubMed]
24. Protes, V.M.; Costa, C.; Pariz, C.M.; Castilhos, A.M.; Meirelles, P.R.L.; Longhini, V.Z.; Roa, R.O.; Ricardo, H.A.; Melo, V.F.P. Effects of soybean silage on feeding behavior, performance, and meat quality of lambs. *Small Rumin. Res.* **2018**, *164*, 64–69. [CrossRef]
25. Abouelezz, K.F.M.; Wang, Y.; Wang, W.; Lin, X.; Mohammed, K.A.F. Impacts of Graded Levels of Metabolizable Energy on Growth Performance and Carcass Characteristics of Slow-Growing Yellow-Feathered Male Chickens. *Animals* **2019**, *9*, 461. [CrossRef] [PubMed]
26. Cai, Z.-W.; Zhao, X.-F.; Jiang, X.-L.; Yao, Y.-C.; Zhao, C.-J.; Wu, C.-X. Comparison of muscle amino acid and fatty acid composition of castrated and uncastrated male pigs at different slaughter ages. *Ital. J. Anim. Sci.* **2010**, *9*, 173–178.
27. Zhang, X.; Liu, C.; Kong, Y.; Li, F.; Yue, X. Effects of intramuscular fat on meat quality and its regulation mechanism in Tan sheep. *Front. Nutr.* **2022**, *9*, 908355. [CrossRef]
28. Serafini, M.; Peluso, I.; Raguzzini, A. Flavonoids as anti-inflammatory agents. *Proc. Nutr. Soc.* **2010**, *69*, 273–278. [CrossRef]
29. Pang, Y.; Ahmed, S.; Xu, Y.; Beta, T.; Zhu, Z.; Shao, Y.; Bao, J. Bound phenolic compounds and antioxidant properties of whole grain and bran of white, red and black rice. *Food Chem.* **2018**, *240*, 212–221. [CrossRef]
30. Duncan, J.R.; Prasse, K.W.J. *Veterinary Laboratory Medicine: Clinical Pathology*; John Wiley & Sons, Inc.: Canberra, Australian, 1977.
31. Lebzien, P.; Riemeier, A.; Flachowsky, G. Investigations on the effect of the ruminal N-balance on rumen metabolism, urea content in blood serum and milk as well as some liver parameters of lactating cows. *Arch. Anim. Nutr.* **2006**, *60*, 99–109. [CrossRef]
32. Giesecke, K.; Magnusson, I.; Ahlberg, M.; Hagenfeldt, L.; Wahren, J. Protein and amino acid metabolism during early starvation as reflected by excretion of urea and methylhistidines. *Metab. Clin. Exp.* **1989**, *38*, 1196–1200. [CrossRef]
33. Lv, X.; Chen, L.; Zhou, C.; Zhang, G.; Xie, J.; Kang, J.; Tan, Z.; Tang, S.; Kong, Z.; Liu, Z.; et al. Application of different proportions of sweet sorghum silage as a substitute for corn silage in dairy cows. *Food Sci. Nutr.* **2023**, *11*, 3575–3587. [CrossRef]

Disclaimer/Publisher's Note: The statements, opinions and data contained in all publications are solely those of the individual author(s) and contributor(s) and not of MDPI and/or the editor(s). MDPI and/or the editor(s) disclaim responsibility for any injury to people or property resulting from any ideas, methods, instructions or products referred to in the content.

animals

Article

Expansion Improved the Physical and Chemical Properties and In Vitro Rumen Digestibility of Buckwheat Straw

Xiaohui Cao [1], Sasa Zuo [1], Yanli Lin [2], Rui Cai [3], Fuyu Yang [2], Xuekai Wang [2,*] and Chuncheng Xu [1,*]

1 College of Engineering, China Agricultural University, Beijing 100083, China; s20193071205@cau.edu.cn (X.C.); zss@cau.edu.cn (S.Z.)
2 College of Grassland Science and Technology, China Agricultural University, Beijing 100093, China; lyl161@126.com (Y.L.); yfuyu@cau.edu.cn (F.Y.)
3 School of Environmental Science & Engineering, Huazhong University of Science and Technology, Wuhan 430074, China; chenmei@cau.edu.cn
* Correspondence: xkwang2016@163.com (X.W.); xucc@cau.edu.cn (C.X.)

Simple Summary: Straw has poor palatability, a high degree of lignification, and difficulty in compression, which limits its feed application. Expansion technology can make the hard straw soft and improve the physical and chemical properties of the straw. At present, there is no comprehensive characterization and evaluation of the physical and chemical properties and feed utilization value of expanded straw, and there is a lack of research on the texture properties of straw itself. The texture properties have important effects on the compressibility of straw during silage, the palatability during feeding, and even the energy conversion efficiency during digestion, which have important guiding significance for the further utilization of straw. We combined mechanical and chemical methods to investigate the hardness, elasticity, chewiness, compressive properties, and structural changes of expanded straw and verified the effect of the changes in mechanical properties and chemical structure of expanded straw on improving digestibility by in vitro digestion. This result is of great significance for the further effective application of expanded straw as a feed resource.

Abstract: The hard texture and poor palatability of straw are important factors that hinder its application in feed. Expansion is a technology that can improve the utilization of biomass, but few studies have comprehensively revealed how to change physicochemical characteristics to improve nutritional value. In this study, mechanical and chemical methods were combined to study the texture properties, rheological properties, and physicochemical structures of straw, and its utilization value was evaluated by in vitro rumen digestion. Expansion caused hemicellulose degradation, cellulose separation, and lignin redistribution, resulting in a decrease in crystallinity. The hardness and chewiness of expanded straw were reduced by 55% to 66%, significantly improving palatability. The compressive stress could be reduced by 54–73%, and the relaxation elasticity was reduced by 5% when expanded straw was compressed. The compression deformation of expanded straw was doubled compared to feedstock, and the compacting degree was improved. Expanded straw significantly improved digestibility and gas production efficiency, which was due to the pore structure increasing the attachment of rumen microorganisms; besides that, the reduction of the internal structural force of the straw reduced energy consumption during digestion. The lignin content decreased by 10%, the hardness decreased further in secondary expansion, but the digestibility did not improve significantly.

Keywords: buckwheat straw; expansion; structure; physio-chemical properties; in vitro rumen digestion

Citation: Cao, X.; Zuo, S.; Lin, Y.; Cai, R.; Yang, F.; Wang, X.; Xu, C. Expansion Improved the Physical and Chemical Properties and In Vitro Rumen Digestibility of Buckwheat Straw. *Animals* **2024**, *14*, 29. https://doi.org/10.3390/ani14010029

Academic Editors: Dingfu Xiao and Tiantian Meng

Received: 16 November 2023
Revised: 16 December 2023
Accepted: 18 December 2023
Published: 21 December 2023

Copyright: © 2023 by the authors. Licensee MDPI, Basel, Switzerland. This article is an open access article distributed under the terms and conditions of the Creative Commons Attribution (CC BY) license (https://creativecommons.org/licenses/by/4.0/).

1. Introduction

Buckwheat (*Fagopyrum esculentum*) belongs to the Polygonaceae family and is a nutritious coarse grain. Numerous studies have shown that buckwheat is rich in effective natural antioxidants and has gradually become a popular source of human food. However, buckwheat will produce a large amount of straw every year after a series of processing steps, and buckwheat straw has become a promising by-product that can no longer be ignored. The rational utilization of buckwheat straw for feed purposes could help reduce feed shortages in livestock production [1]. However, most cellulose is associated with lignin in the form of lignocellulose, restricting the enzymatic conversion due to the unique resistance structure, so removing lignin or increasing the accessibility of holocelluloses to microorganisms can enhance the nutritional value of straw [2,3]. Therefore, how to improve the physicochemical structure of buckwheat straw while retaining the nutrients is particularly important to improving its utilization value as a feed source.

Expansion is a kind of thermo-mechanical pretreatment that relies on the friction between the screw, straw, and inner wall of the machine to provide additional heat and energy, forming a high-temperature and high-pressure environment [4,5]. When the pressure is released instantaneously at the discharge port, the straw explodes, causing the depolymerization of macromolecules and enhancing their biodegradability [6]. The substrate simultaneously undergoes high shearing forces because of the protrusions of the screw during transport, and the twisting action of the screw is conducive to the full mixing of straw and water, thus leading to a better effect upon the explosion. This technique avoids the severe treatment conditions that produce a large number of by-products from steam explosions and liquid ammonia, as well as the limitations of grinding. Expansion is usually used for snack food, feed, and plastic products [7]. It could improve the expansion rate and porosity of oat bran, and the feasibility of oil extraction from oat bran was realized [8]. Regarding application in crop straw, some studies have shown that expansion pretreating corn stover or rice straw can significantly improve the sugar recovery rate and increase biogas production from grass [4,9]. The expansion process interferes with the biomass structure through the combination of thermal energy and mechanical energy, causing a series of physical and chemical changes in the material. However, few studies have comprehensively revealed how to change physicochemical characteristics in the assessment of its nutritional value as a feed source.

The change in mechanical properties of lignocellulosic biomass is closely related to its chemical composition and structure. The lignocellulosic biomass structure is composed of a highly complex matrix of cellulose, hemicellulose, and lignin. These components are interlinked by ether, ester, carbon-carbon, and hydrogen bonds [10]. Therefore, the relationship between stress and deformation of the samples can reflect the binding properties of the lignocellulosic biomass [11]. Generally, traditional chemical methods are usually used to evaluate the effect of pretreatment on the in vitro digestibility of straw feed, and there is a lack of evaluation of mechanical properties for biomass. However, Viamajala et al. [12] showed that the enzymatic hydrolysis system showed shear dilution characteristics through the relationship between rheological properties and initial solid concentration, pretreatment degree, and particle size in the enzymatic hydrolysis system of dilute acid pretreated corn straw. Liu and Chen [13] reported the heterogeneity and mechanical properties of corn stovers and their correlations with the efficiency of steam explosion and hydrolysis. Therefore, changes in mechanical properties may affect the conversion efficiency after the feed enters the rumen, which has important research significance for the improvement of the biological transformation process of the straw nutritional value. Besides, changes in mechanical properties can be used to improve the palatability and the degree of compaction during ensilage.

In this paper, the correlation between chemical composition and mechanical properties was investigated from the perspective of biomechanics, and the internal mechanism of the effect of expansion modification on physical and chemical properties was explored. Furthermore, the effects of cell wall structure and biomechanics on ruminal digestibility were explained, and a new idea was opened up for the evaluation of crop straw nutrition and utilization value.

2. Materials and Methods

2.1. Buckwheat Straw Preparation

The buckwheat straw was supplied by Fuxin City, Liaoning Province. The stubble height was about 20 cm when harvested and stored in a cool and dry place. Then, the buckwheat straw raw material was processed into a length of 5 cm with a cutter machine.

2.2. Different Degrees of Expansion Treatment

The cut buckwheat straw was expanded on the straw expander (Model 9P-150, Liaoning Xianghe agricultural animal husbandry Industry Co., Fuxin, China). The screw diameter of the expander is 150 mm, the spindle speed is 510 r/min, and the output is 1–1.5 t/h. The dried buckwheat straw was evenly put into the feeding port and conveyed by the screw auger to the expanding chamber. Through the extrusion, shearing, and friction of the screw and the screw sleeve, the pressure and temperature in the expanding chamber were continuously increased, and finally, the straw was instantly sprayed from the outlet (EO). The buckwheat straw of the EO was put into the expansion chamber again for secondary expansion (ES). Water was appropriately added to the expansion chamber through the nozzle during both processes to prevent carbonization. The expanding process lasted about 6 s, the temperature could reach above 120~150 °C, and the pressure was higher than 3.4 MPa. Retain part of the original buckwheat straw of three treatment groups in ziplock bags for mechanical analysis, and dry the rest at 65 °C for 48 h.

2.3. Chemical Composition

Dried samples were milled to pass through a 1-mm screen. The contents of neutral detergent fiber (NDF) and acid detergent fiber (ADF) were determined according to the improved Van Soet's methods [14,15], with the addition of alpha amylase but no sodium sulfite during the analysis of NDF. The content of lignin was determined by dissolving cellulose with sulphuric acid [16]. The hemicellulose (HCL) content was calculated by the difference between NDF and ADF, and the cellulose (CL) content was calculated as the difference between ADF and lignin, which were expressed on a dry matter (DM) basis. The crude protein (CP) content was analyzed by the Kjeldahl method [17]. The water-soluble carbohydrate (WSC) content was determined using the method of Owens et al. [18]. Crude ash content was determined by complete combustion of the samples to a constant weight at 550 °C in a muffle furnace until a constant weight was recorded [19].

2.4. Structure

2.4.1. Fourier Transform Infrared Spectroscopy

Infrared spectroscopy is widely studied in agricultural biomass materials. When the infrared light is irradiated on the sample, the molecules in the sample will absorb radiation of certain frequencies, causing vibration and rotation of the molecules, and the absorption spectrum is called the infrared spectrum. Each functional group has its own specific infrared spectral absorption region, so it is often used to characterize the functional groups of biomass.

The changes in functional groups of buckwheat straw before and after expansion treatment were conducted by an FTIR spectrometer (iS10, Nicolet Corporation, Madison, WI, USA). Buckwheat straw samples were dried at 105 °C for 4 h and stored in a desiccator before analysis. The sample spectra were obtained by scanning 32 times in the spectral range of 400–4000 cm^{-1}, with a spectral resolution of 4 cm^{-1}. The samples were analyzed by comparing the spectra for changes in chemical bonds.

2.4.2. X-ray Diffraction

X-rays are scattered when they are irradiated on crystalline cellulose, and the scattered X-ray is strengthened at some specific angles, thus showing the characteristic diffraction phenomenon of crystalline cellulose. The crystallinity of cellulose can be obtained by X-ray atlas, which refers to the proportion of crystalline cellulose. Crystallinity is an important index affecting the conversion rate of cellulose.

The cellulose crystallinity index (CrI) of Control, EO, and ES was obtained by an X-ray diffractometer (Purkinje, XD3, Beijing, China) using Cu/Kα irradiation at 36 kV and 20 mA. All buckwheat straw samples were scanned from 5° to 40° with a step size of 0.02°. The CrI was calculated using the method of Segal et al. [20].

2.4.3. Scanning Electron Microscopy

Changes in the surface morphology of buckwheat straw were visually examined using a scanning electron microscope (SEM). All dried samples were attached to the sample stage with conductive tape and then coated with a thin layer of gold with a vacuum coater. Finally, the samples were viewed using an SEM SU3500 (Hitachi, Tokyo, Japan) with an accelerating voltage of 15 kV.

2.5. Mechanical Properties

The mechanical properties directly affect the palatability, compression deformation, and even the flow properties of straw during digestion. The texture properties of straw can reveal changes in hardness, chewability, elasticity, and cohesiveness, and the rheological properties, including the stress relaxation and creep of straw, can reflect the resistance to internal structural deformation of straw under pressure and the deformation and flow properties under external forces.

2.5.1. Dynamic Compression Test on the Texture Properties

Dynamic mechanical experiments were selected to compare their mechanical properties using a Texture Analyzer (TAXT2i, Stable Micro Systems, Surrey, South East England, UK). Texture profile analysis (TPA) tests were conducted in two-cycle compression mode with a P/50 cylindrical probe at 1 mm/s speed to 75% of the original height of the samples. The probe stayed for 5 s after the first depression, then started an upward stroke at a speed of 1 mm/s, and the second compression-decompression cycle was performed later. To more accurately evaluate the effect of expansion on the texture characteristics of buckwheat straw, the remaining undried samples were used for the tests. The raw material was wetted to a moisture content of 38% before the compression test to ensure that the moisture content was at the same level and avoid being affected by the large difference in moisture content. Similarly, in order to eliminate the influence of compression density, 10 g samples were weighed and put into a 150 mL beaker (diameter about 50 mm, height 70 mm) before the experiments, and the height of each sample was guaranteed to be 50 mm. The load cell was calibrated with a 1 kg weight with a 1 g least count. Each compression test for each sample was repeated five times. Calculated TPA parameters related to hardness, springiness index, cohesiveness, and chewiness from the output force-time curves [21].

2.5.2. Static Compression Test on the Rheological Properties

Stress relaxation and creep tests were carried out with a P/50 probe at 1 mm/s test speed using a Texture Analyzer (TAXT2i, Stable Micro Systems, Surrey, South East England, UK), obtaining the curves of force-time and deformation-time through compression mode test, respectively [22]. The resulting force-time curves were developed for the calculation of compressive stress and relaxation elasticity, and deformation-time curves were recorded as the creep deformation.

2.6. In Vitro Rumen Digestibility

In vitro rumen digestibility was performed according to Tilley and Terry [23]. Fresh rumen fluid collected in the morning from three Angus bullocks fed with corn silage and a wheat straw-based diet was filtered through four layers of cheesecloth to remove large feed particles and mixed with the buffer solution in a 1:4 (v/v) ratio under continuous flushing with CO_2. The samples of different expanding degrees were weighed (approximately

500 mg of each sample) into filter bags, then placed in a Daisy II Incubator and incubated at 39 °C for 48 h. The filter bags were retrieved after digestion and rinsed with water until the water was clear. Then, the rinsed filter bag was gently squeezed and dried at 65 °C for 48 h. Digestibility was calculated based on changes in DM, NDF, ADF, CL, and HCL contents.

The air-dried samples were weighed (220 mg each) into a 100 mL glass syringe. Each syringe received 30 mL of the buffered rumen fluid and was incubated at 39 °C for 96 h to obtain gas production. It was recorded manually at 0, 2, 4, 6, 8, 10, 12, 16, 20, 24, 30, 36, 42, 48, 56, 64, 72, 84, and 96 h of incubation. The kinetics of gas production were fitted using the NLIN procedure of SAS version 9.4 (SAS Institute, Cary, NC, USA), according to France et al. [24].

$$\text{Model 1: } Y = B\,(1 - e - c\,(t - L)) \tag{1}$$

$$\text{Model 2: } Y = B_1\,(1 - e - c_1\,t_1) + B_2\,(1 - e - c_2\,t_2) \tag{2}$$

where Y (mL/g DM) is the volume of gas production at time t; B (mL/g DM) is the asymptotic gas production; B_1 and B_2 (mL/g DM) are the asymptotic gas production of the rapid or slow fermentation component; c (h^{-1}) is the rate of gas production; c_1 and c_2 (h^{-1}) are the rates of the rapid or slow fermentation component; and L (h) is the lag time prior to gas production.

2.7. Statistical Analysis

The statistical analysis was analyzed by one-way analysis of variance (ANOVA) followed by Tukey's multiple range tests and declared when $p < 0.05$ using IBM SPSS 25 for Windows (IBM SPSS Inc., Chicago, IL, USA). FTIR spectra were baseline corrected and normalized using OMNIC 32 (Nicolet, New York, NY, USA). Graph processing was performed using Origin 2023 for Windows (Origin Lab Corporation, Northampton, MA, USA). Pearson analysis was used for correlation analysis.

3. Results

3.1. Changes in Chemical Components

Expansion led to a downward trend in lignocellulose components, while WSC and CP showed an upward trend ($p < 0.05$, Table 1). There was a significant difference in NDF content between EO and ES ($p < 0.05$), which was mainly caused by the decrease in lignin and hemicellulose content. The lignin content decreased in EO but was not significant, while it decreased significantly in ES ($p < 0.05$). Hemicellulose also had an obvious loss in both expansion processes ($p < 0.05$). The crude ash content did not change significantly.

Table 1. Chemical compositions of buckwheat straw in different treatments.

Items (g/kg DM)	Control	EO	ES	SEM [1]	*p*-Value [2]
Neutral detergent fiber (NDF)	735.8 a	705.8 b	688.6 c	7.01	<0.001
Acidity detergent fiber (ADF)	584.7 a	554.9 b	542.9 b	6.93	0.004
Lignin	84.9 a	81.9 a	76.5 b	1.39	0.007
Cellulose (CL)	498.0	472.3	463.7	6.87	0.086
Hemicellulose (HCL)	154.1 a	151.5 ab	145.7 b	1.36	0.017
Water-soluble carbohydrate (WSC)	16.7 b	21.5 a	21.6 a	0.94	0.017
Crude protein (CP)	32.1 c	39.5 b	43.7 a	1.71	<0.001
Crude ash	92.9	96.8	95.5	1.31	0.541

Control, untreated; EO, once expansion; ES, secondary expansion. [1] SEM represents the standard error of the mean, n = 3. [2] Significant differences in each row are indicated by different letters ($p < 0.05$).

3.2. Analysis in Structure

3.2.1. FTIR Spectra

Figure 1a indicates the variation of the absorption intensity of chemical bonds with different pretreatments, and Table S1 reveals the changes in the absorption peak position of chemical bonds. The absorption peaks near 1737 cm^{-1} and 898 cm^{-1} weakened by expansion. Compared to Control, the absorption peak at 1629 cm^{-1} and 3418 cm^{-1} was significantly blue-shifted. The absorption peaks at 1506 cm^{-1} and 1425 cm^{-1} related to the vibration of the benzene ring showed a decrease in EO and ES. The positions of 1319 cm^{-1} and 1245 cm^{-1}, which represent the structural units of syringyl and guaiacyl, respectively, were shifted to different degrees. The vibration frequency decreased after pretreatment to 2920 cm^{-1}.

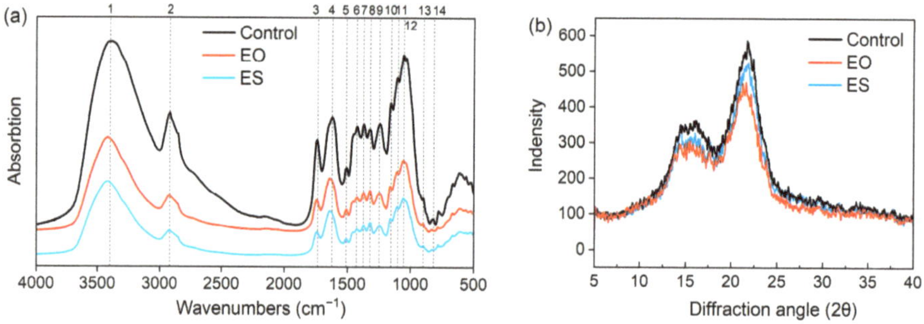

Figure 1. Structural changes in chemical bonding with FTIR (**a**) and crystallinity with XRD (**b**). Control, untreated; EO, once expansion; ES, secondary expansion.

3.2.2. Diffraction Pattern

The XRD curves in Figure 1b displayed that the expanded samples had crystalline regions and amorphous regions of cellulose around 2θ = 22° and 2θ = 18°, respectively. It could be seen that the diffraction intensity of the Control, ES, and EO decreased successively. Using the Segal formula to calculate the crystallinity index of different treatments from the corresponding XRD patterns and the CrI of the Control, the EO, and the ES were 56.58%, 51.21%, and 52.79%, respectively.

3.3. *Analysis of Mechanical Properties*

3.3.1. Texture Properties Based on Texture Profile Analysis

Texture profile analysis (TPA) includes two compression cycles as shown in Figure 2a, and the hardness, chewiness, springiness index, and cohesiveness of materials were calculated according to the curves (Figure 2b,c). The results showed that hardness was reduced by about 50% in EO compared with the Control and decreased by about 66% in ES. Chewiness showed a similar trend to hardness. The highest springiness index and cohesiveness were obtained in Control, while they decreased by 25% and 18%, respectively, after expansion and had no significant difference in EO and ES.

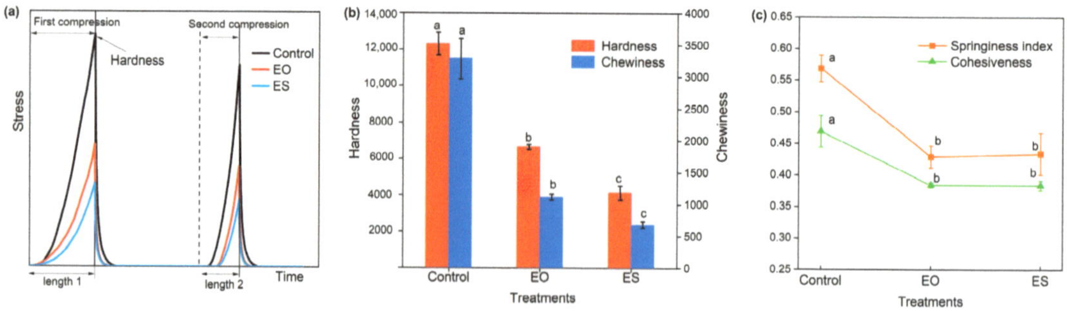

Figure 2. Compressive curves of TPA and calculation results of the texture properties of different treatments. (**a**) Two cyclic compression curves in texture analysis; (**b**) Hardness and chewiness calculated

from compression curves; (**c**) Springiness index and cohesiveness calculated from compression curves. Control, untreated; EO, once expansion; ES, secondary expansion. Significant differences between treatments are indicated by different letters ($p < 0.05$).

3.3.2. Rheological Properties Based on Stress Relaxation and Creep Tests

Figure 3 displayed the stress relaxation and creep curves, which showed the force and deformation of the sample over time, respectively. The compressive stress decreased by about 54% in EO and 73% in ES ($p < 0.001$, Table S2). The relaxation elasticity of the expanded samples decreased by about 5%, and there was no significant difference between the EO and ES. Under the same load condition, the creep deformation increased from about 6 mm in Control to 10 mm in EO and 13 mm in ES ($p < 0.001$).

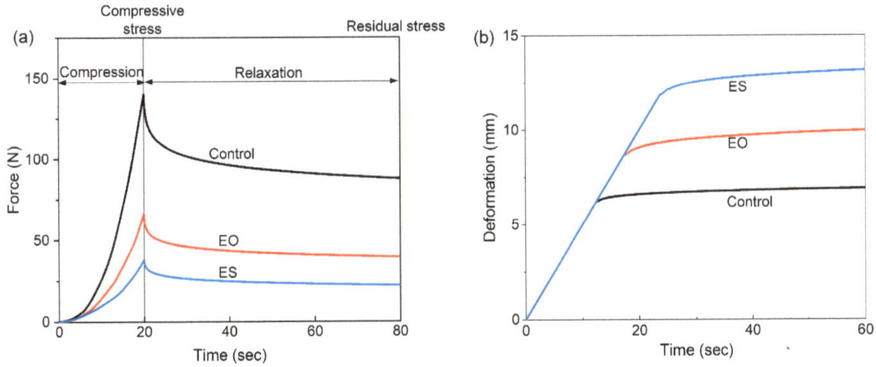

Figure 3. Compressive curves of stress relaxation and creep deformation curves of different treatments. (**a**) Representation of compressive stress and relaxed elasticity; (**b**) Representation of creep deformation. Control, untreated; EO, once expansion; ES, secondary expansion.

3.4. *In Vitro Rumen Digestibility*

After 48 h incubation, expanded samples showed significantly higher digestibility of DM, NDF, ADF, CL, and HCL compared to the Control (Figure 4a, $p < 0.05$). The EO and ES increased the digestibility of DM, NDF, ADF, CL, and HCL by 3.8–5.3%, 4–5.1%, 3.7–5%, 4.6–6.0%, and 5.3%, respectively. Figure 4b showed that the Control had the lowest gas production, and there was a gas production delay period. There was no significant difference in gas production between EO and ES after 96 h of incubation. The results of gas production model fitting showed that model 1 was the best fit for Control, while model 2 was the best fit for EO and ES. The maximum theoretical gas production of EO and ES was composed of fast fermentation components and slow fermentation components, and the EO was slightly higher than the ES, whereas the gas production rates in ES were higher than those in EO (Table 2).

Table 2. Kinetic parameters of in vitro gas production models.

Treatment [1]	Y [2] (mL/g DM)	B [3] (mL/g DM)		c [4] (h^{-1})		L [5] (h)
		B_1	B_2	c_1	c_2	
Control	170.1	172.64	172.64	0.0398	0.0398	0.2038
EO	180.1	153.65	78.96	0.0523	0.0044	/
ES	178.2	123.54	75.67	0.0697	0.0139	/

[1] Control, untreated; EO, once expansion; ES, secondary expansion. [2] Y, gas production at time t. [3] B, theoretical gas production; B_1 and B_2 represent theoretical maximum gas production of the rapid and slow fermentation component. [4] c, gas production rate; c_1 and c_2 represent gas production rates of the rapid and slow fermentation component. [5] L, lag phase.

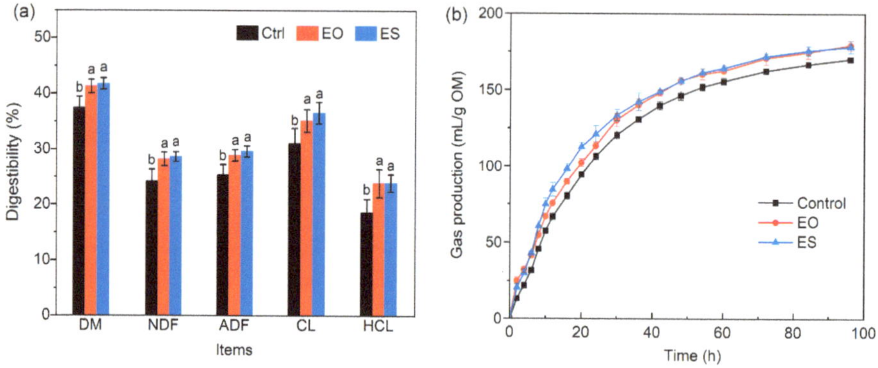

Figure 4. In vitro digestibility of the raw and treated buckwheat straw. (**a**) Degradation of each component after incubation for 48 h; (**b**) Gas production during incubation for 96 h. SEM represents standard error of the mean, n = 6. Significant differences between treatments are indicated by different letters ($p < 0.05$).

3.5. Changes in Morphological Composition

Figure 5a illustrates the smooth surface and ordered fibers of the Control. However, the exposed fibers of the EO were rough with deposits (Figure 5b), and the fiber structure of the ES was further destroyed and disordered, even displaying a honeycomb shape (Figure 5c). After in vitro rumen fermentation, cracks and holes could also be observed on the surface of the Control, but there was still an obvious lignified structure (Figure 5d). However, the fiber in the EO curled and the pore structure greatly increased (Figure 5e), and the samples of the ES had more deposits on the surface and had no visible complete fiber structure (Figure 5f).

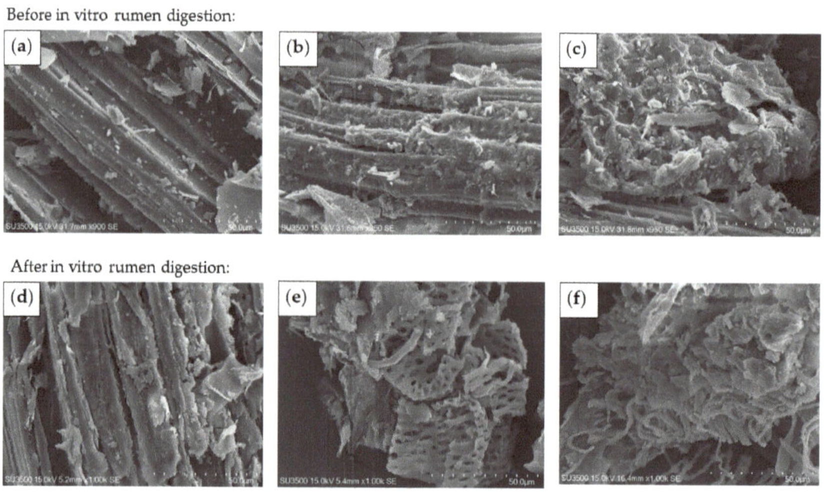

Figure 5. Scanning electron microscopy (SEM) of buckwheat straw before and after in vitro rumen digestion. (**a–c**) Untreated, once-expanded, and secondary-expanded buckwheat straw before digestion; (**d–f**) Untreated, once-expanded, and secondary-expanded buckwheat straw after digestion.

3.6. Correlation of Mechanical Properties, Chemical Characteristics, and Gas Production Rate

The internal relationship between the change law of mechanical properties and chemical properties, as well as the efficiency of the gas production process during the expansion, is shown in Figure 6. Hardness, chewiness, springiness index, cohesiveness, compression stress, and relaxation elasticity all had a strong positive correlation with NDF content. Hardness and compression resistance were strongly correlated with lignin ($p < 0.01$), chewiness had a strong correlation with cellulose ($p < 0.01$), and relaxation elasticity was correlated with hemicellulose ($p < 0.01$). Hardness, chewiness, compression stress, and relaxation elasticity were significantly positively correlated with crystallinity ($p < 0.001$), while negatively correlated with gas production rate ($p < 0.001$). The deformation was negatively correlated with fiber composition and crystallinity while positively correlated with gas production rate ($p < 0.001$). There was a significant negative correlation between crystallinity and gas production rate ($p < 0.001$). The content of CP and WSC was negatively correlated with the mechanical properties of straw except for creep deformation, while positively correlated with the gas production rate.

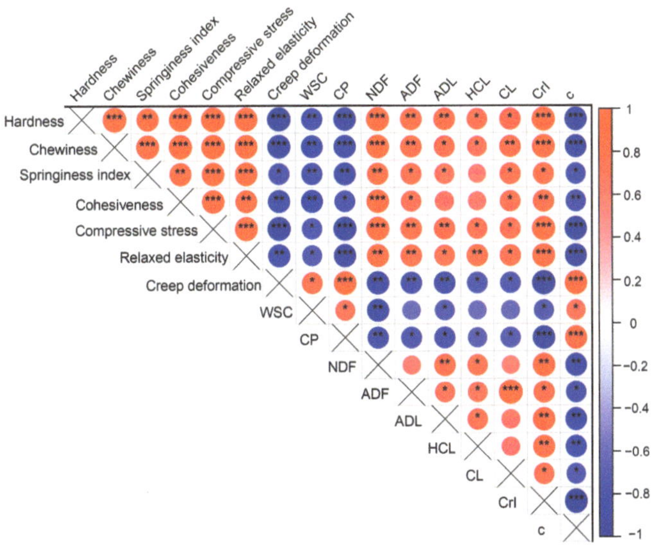

Figure 6. Correlation of mechanical properties, chemical characteristics, and gas production rate. Positive correlations were shown in red, and negative correlations were shown in blue. The dot color and size were proportional to the correlation coefficient, and Pearson's correlation coefficient and the corresponding color were shown on the ruler on the right side. × means that correlations between the same items are ignored. * $p \leq 0.05$, ** $p \leq 0.01$, *** $p \leq 0.001$.

4. Discussion

4.1. Changes in Chemical Components

The EO did not significantly contribute to reducing lignin content, but the ES significantly reduced the lignin content of buckwheat straw. Lignin has been reported to have a lower melting point of approximately 140 °C, and it becomes soft and sometimes melts when biomass is heated [25]. Since the extrusion process is shorter in time and the lignin structure is hard, the fast rotating speed of the screw and the rise in temperature during the EO might lead to the softening of the lignin and the destruction of the structure to a greater extent. As the structure of the straw had been preliminarily destroyed in the EO, more lignin was removed in the ES when the softened straw was placed in the high-temperature and high-pressure environment again. Hemicellulose content was degraded less in EO, which might be attributed to the shorter expanding time of the raw material, and it was discharged from the discharge port before being degraded in large quantities. However, hemicellulose was largely degraded in ES due to its amorphous state and low polymerization degree under high temperature and high pressure. Some studies have also shown that the thermally unstable acetyl division of hemicellulose can release acetic acid, which can catalyze the hydrolysis of lignocellulosic biomass [26]. The plant cell wall mainly consists of the intercellular layer, primary wall, and secondary wall. The main chemical components of the intercellular layer are lignin and pectin. The primary wall contains a lot of hemicellulose and lignin, and the secondary wall mainly consists of cellulose and hemicellulose. Through the analysis of the changes in lignocellulose components, it can be inferred that expansion mainly destroys the intercellular layer and primary wall structure. The increase in WSC content after expansion was possibly due to the fact that expansion broke the seal of the lignin, and partial hemicellulose or cellulose was hydrolyzed under acidic conditions and elevated temperatures to oligo- or monosaccharides [27]. However, WSC content did not increase significantly in ES compared to EO, probably because the acidic environment could result in the formation of some inhibitors. The principle of the Kjeldahl nitrogen determination method for detecting crude protein content is to measure the element content of nitrogen in straw and multiply it by 6.25 to obtain the crude protein content. Nitrogen is impossible to change during the expansion process, so the absolute content of crude protein is also impossible to change. However, due to the action of high temperature, thermal mechanical energy, and water steam, some soluble substances such as starch, fat, and sugar would be lost in the expansion process, resulting in an increase in the proportion of nitrogen elements.

4.2. Analysis in the Chemical Structure

FTIR spectra reflected modifications of the chemical bond structure of lignocellulose. The signal at 1629 cm^{-1} represents the stretching vibration of C=O in the conjugated carbonyl group. The peak position was blue-shifted in EO and ES, suggesting that expansion could destroy the structure of the conjugated carbonyl group and that lignin is easier to oxidize. The expanded samples showed a decrease in the aromatic skeleton vibrational intensity at 1507/1506 cm^{-1} and in the aromatic vibration intensity at 1425/1424 cm^{-1}, indicating that aliphatic and aromatic groups in lignin were degraded. Moreover, the shift of characteristic peaks and changes in absorption strength of lignin structural units, including guaiacyl (1245/1246 cm^{-1}), syringyl (1323~1319 cm^{-1}), and *p*-hydroxyphenyl (831~828 cm^{-1}), demonstrated the modification of lignin structure by expansion. It has been reported that the lignin extraction rate of expanded rice straw is high, and expansion is considered to be an effective method to improve the contribution of lignin separation [28]. In the present study, changes in chemical bond structure also demonstrated that although the EO had no significant effect on the degradation of lignin, it destroyed the tight connection between lignocellulose. The reduction of the absorption peak at 1737 cm^{-1} indicated that the expansion effectively broke the ester bond of lignin and hemicellulose, and the red-shifted of peak position in ES indicated that the damage degree was more serious, which was consistent with the decrease of hemicellulose. Declines of peaks intensities at 1374/1373 cm^{-1} represented C-H deformation of crystalline cellulose and hemicellulose, and 898 cm^{-1} represented C-O-C stretching of amorphous cellulose [29] suggested that both cellulose and hemicellulose structure were destroyed during expansion. The changes at 3418 cm^{-1} indicated that the degree of O-H association was reduced [30] and the hydrogen bond was broken during expansion. Chemical bonds determine the reaction rate of components since the different types and quantities of chemical bonds connected between molecular polymers determine the energy required [13]. Therefore, the destruction of chemical bonds and the reduction of the number of chemical bonds would promote the reaction rate of the digestive system.

The crystallinity index (CrI) value reflected the sample's crystallinity and sensitivity to other components, including amorphous cellulose, hemicellulose, and lignin, which together contribute to the amorphous portion [31]. Buckwheat straw was extruded, sheared, and rubbed during expansion, then exploded instantly at the discharge port, resulting in a size reduction of lignocellulose as well as depolymerization [32,33]. With the secondary expansion, high-pressure steam further produced mechanical separation and tearing of fibers [34], destroying the polarity of cellulose molecules and the interaction between molecular chains. However, the ES resulted in a decrease in amorphous structures such as lignin and hemicellulose, which increased the proportion of crystalline regions, leading to a slightly higher crystallinity compared to the EO. The partial expansion of amorphous regions might improve the nutritional availability of pretreated feeds since this loosely agglomerate structure enables water and enzymes to access the hydrolytic reaction [35].

4.3. Analysis of Mechanical Properties

The mechanical properties of lignocellulosic materials are important factors affecting the conversion efficiency, which affects energy consumption and material transfer efficiency of the digestive system. TPA mainly reflects the hardness, elasticity, chewability, and cohesiveness of the straw itself, and stress relaxation and creep deformation can reflect the rheological characteristics of the system after straw compression. Hardness is an indicator used to measure the soft and hard texture of samples, and chewiness is a description of solid samples used to simulate the energy expended when the sample is chewed to a state that can be swallowed. During the expansion process, the straw was blasted into 1–2 cm fragments under the dual action of high-temperature effect and mechanical effect, and the cell wall disintegrated [36], resulting in a decrease in the hardness of the straw. Compressive stress refers to the force required to resist the deformation of the internal structure of the biomass material, and the expanded samples required less force to be compressed. Creep deformation represents the deformation of the samples under the same load condition, reflecting the degree of difficulty of the sample being compressed. The expansion broke the ordered structure of cell wall, resulting in larger deformation under the same load. Springiness index indicates the recovery ability of deformation. Expanded straw had the lower springiness index, indicating that expanded biomass was not easy to rebound after compression, which is conductive to the silage compaction of biomass. Cohesiveness refers to the ratio of compression work between the second cycle and the first cycle and reflects the inner structure force of the sample. The cohesiveness of expanded buckwheat straw decreased. The in-depth study of physical properties provides a theoretical basis for the improvement of the palatability and compression performance of silage.

4.4. Analysis of In Vitro Rumen Digestibility

It was found that the expansion significantly improved the utilization rate of nutrients through the analysis of in vitro digestibility and gas production. The findings were also supported by chemical and physical property modifications. The reduction of the mechanical properties of the expanded straw increased the gas production rate. The disruption of the lignocellulose structure and separation of fibers increased the contact area of enzymes. However, the further decrease in hardness and compressive stress in ES did not lead to further improvement in digestibility and gas production. It was speculated that the high deposition density of lignin droplets covered the fiber surface and blocked part of the pore structure [37], resulting in limited space for the attachment of microorganisms. SEM images seemed to demonstrate the opinion, which was due to the gradual separation of tightly bound cellulose microfibers and the interaction of thermal expansion and water, eventually causing lignin to be melted, solubilized, and recondensed on the surface [38]. Interestingly, the gas production rate was further increased along with the decrease in hardness. It was inferred that the weakening of the mechanical properties of the expanded straw reduced the energy consumption required to overcome the internal structural force during digestion in rumen, thus promoting the improvement of conversion efficiency. Although some similar pretreatment methods could achieve higher digestion rates, such as steam explosion, which increased the degradation rates of NDF, ADF, CL, and HC by 10.53%, 7.84%, 5.79%, and 13.04%, respectively, severe treatment conditions sacrificed more than 50% of hemicellulose and other nutrients and even produced by-product inhibitors such as furfural and hydroxymethylfurfural [39]. There is also a study indicating that the loss of a large amount of fiber components during steam explosion treatment of wheat, rice, and corn straw results in no significant change in the in vitro gas production after 96 h compared to the raw materials, only accelerating the gas production rate [40]. Therefore, expansion can improve the utilization value of buckwheat straw without causing a lot of nutrient loss.

4.5. Correlation Analysis of Mechanical and Chemical Characteristics and Gas Production Efficiency

Different chemical components cause different mechanical properties owing to different contents and unique chemical bonds. It proved that hardness, chewiness, compressive stress, and deformation were strongly correlated with lignin and cellulose, indicating lignin and cellulose play the main role in hardness. Some studies have shown that reductions in the lignin content caused the stem mechanical strength to decrease, suggesting that lignin is closely tied to cell wall mechanical strength [41,42]. Cellulose, a glucose homopolymer composed of β-(1,4)-glucan chains, is the main component of the cell wall and exhibits strong rigidity due to intermolecular and intramolecular hydrogen bonding [43]. The degradation of lignin and cellulose caused by expansion led to a reduction in mechanical strength. Relaxation elasticity reflects the energy required by the sample to resist relaxation deformation when being compressed, which is positively correlated with hemicellulose. This was not only related to the hydrophilicity of hemicellulose but also to its role as a coupling agent between matrix and hard cellulose fibrils due to the spatial organization and structural diversity of hemicellulose [44]. The above analysis suggested that structural carbohydrates play important roles in the mechanical strength of straw. The decline in the hardness and relaxation elasticity of expanded straw improves the palatability and chewiness. Besides, these findings are beneficial to compression during silage, and the resilience of the feed is weakened, which reduces the infiltration of air and is beneficial to improving the aerobic stability of the feed.

Notably, crystallinity is a key factor affecting the digestibility and resistance of biomass. The crystallinity decreased, accompanied by a decrease in the amount of crystalline cellulose and degree of densification, thereby reducing the work to overcome the internal structural force of straw. Besides, the moisture content of the straw increased from 5% to about 38% after expansion. Some studies have shown that moisture can enter the cell wall and interact with the macromolecules by hydrogen bond and van der Waals force [30,45], changing the internal structure force of plant biomass. The swelling effect of cellulose can alter the mechanical properties of plant biomass [13].

Changes in physicochemical properties would lead to different chemical potential energy and reaction activation energy in the reaction process, resulting in different reaction rates. Mechanical properties and CrI have a stronger correlation with gas production rate. In vitro digestion process mainly relies on enzymes secreted by rumen microorganisms for enzymatic hydrolysis. Liu and Chen [13] proposed that the high cohesiveness of the enzymatic hydrolysis system affected the efficiency of enzymatic hydrolysis. This was due to the incompatibility between liquid and solid, which needs to overcome the resistance and surface tension between interfaces. The highly cohesive system would increase energy consumption during digestion. This study proved that the gas production rate increased with the decrease in cohesiveness. The decrease in cell wall stiffness caused by expansion supported by physical properties led to a decrease in internal structural forces. It could reduce the energy consumption required to overcome the internal structural forces during digestion, thus promoting gas production efficiency. The decrease in cellulose crystallinity was also conducive to in vitro rumen fermentation. The decrease in lignin content and crystallinity is accompanied by an increase in grain dislocation, which promotes the accessibility of catalysts to reach reactive sites in cellulose and thus improves digestibility [46,47]. In addition, the denatured protein was more easily hydrolyzed and better absorbed, which was consistent with the very significant positive correlation between crude protein and gas production rate. The results of this study showed that the modification of the physiochemical properties of straw by expansion had a favorable effect on digestibility. Given expanded straw utilization value enhancement, in the future we will further analyze expansion pretreatment in feed for the application of microbial metabolism and function [48].

5. Conclusions

Expansion effectively improved the physicochemical properties of buckwheat straw to enhance its utilization as a feed source. Expansion resulted in a 4–7% decrease in NDF content, partial hydrolysis of hemicellulose, and redistribution of lignin. Compared with chemical composition, the improvement of mechanical properties more reflected the bioavailability of straw as fermented feed. The significantly reduced hardness and elasticity of the expanded straw improved the chewiness and palatability, which were related to the lower CrI and chemical bond energy. The weakening of mechanical properties reduced the energy required to overcome internal structural forces during digestion, thereby increasing in vitro gas production efficiency. The secondary expansion might have only a small component modification effect on the damaged lignocellulose structure and could not significantly increase digestibility, but it could cause a significant decrease in hardness. Expansion pretreatment significantly improved the utilization value of straw and solved the problems of high lignification degree, poor palatability, and difficult compression of the dry straw. This study lays a theoretical foundation for research on expansion applied in silage fermentation or compression transportation so as to further improve the nutritional value of straw as ruminant feed.

Supplementary Materials: The following supporting information can be downloaded at: https://www.mdpi.com/article/10.3390/ani14010029/s1, Table S1: Signal assignments and relative intensities in FTIR spectra; Table S2: Changes in rheological properties.

Author Contributions: Conceptualization, X.C.; methodology, X.C. and X.W.; software, S.Z. and R.C.; writing—original draft, X.C.; data curation, X.W.; validation, F.Y. and Y.L.; writing—review and editing, C.X.; supervision, C.X. and F.Y. All authors have read and agreed to the published version of the manuscript.

Funding: This research was supported by the Modern Agro-industry Technology Research System of China (No. CARS-07-E-3), Major Science and Technology Projects of Inner Mongolia Autonomous Region (2021SZD0017-4), and the China Postdoctoral Science Foundation (No. 2022M723420).

Institutional Review Board Statement: Digestion test was approved by the Institutional Animal Care and Use Committee of China Agricultural University (DK 996 No. 20144-2), Beijing, China.

Data Availability Statement: All the tables and figures used to support the findings of the current study are included in the article.

Acknowledgments: The authors wish to thank Dongze Niu for his advice in improving the manuscript.

Conflicts of Interest: The authors declare no conflict of interest.

References

1. Nie, D.C.; Yao, L.Y.; Xu, X.K.; Zhang, Z.; Li, Y.L. Promoting corn stover degradation via sequential processing of steam explosion and cellulase/lactic acid bacteria-assisted ensilage. *Bioresour. Technol.* **2021**, *337*, 125392. [CrossRef]
2. Wu, D.; Wei, Z.M.; Mohamed, T.A.; Zheng, G.R.; Qu, F.T.; Wang, F.; Zhao, Y.; Song, C.H. Lignocellulose biomass bioconversion during composting: Mechanism of action of lignocellulase, pretreatment methods and future perspectives. *Chemosphere* **2022**, *286*, 131635. [CrossRef] [PubMed]
3. Zhang, Y.L.; Chen, X.H.; Gu, Y.; Zhou, X.F. A physicochemical method for increasing methane production from rice straw: Extrusion combined with alkali pretreatment. *Appl. Energy* **2015**, *160*, 39–48. [CrossRef]
4. Souza, M.F.; Devriendt, N.; Willems, B.; Guisson, R.; Biswas, J.K.; Meers, E. Techno-economic Feasibility of Extrusion as a Pretreatment Step for Biogas Production from Grass. *Bioenerg. Res.* **2022**, *15*, 1232–1239. [CrossRef]
5. Guiao, K.S.; Gupta, A.; Tzoganakis, C.; Mekonnen, T.H. Reactive extrusion as a sustainable alternative for the processing and valorization of biomass components. *J. Clean. Prod.* **2022**, *355*, 131840. [CrossRef]
6. Duque, A.; Manzanares, P.; Ballesteros, M. Extrusion as a pretreatment for lignocellulosic biomass: Fundamentals and applications. *Renew. Energy* **2017**, *114*, 1427–1441. [CrossRef]
7. Karunanithy, C.; Muthukumarappan, K. Influence of Extruder Temperature and Screw Speed on Pretreatment of Corn Stover while Varying Enzymes and Their Ratios. *Appl. Biochem. Biotechnol.* **2010**, *162*, 264–279. [CrossRef] [PubMed]
8. Liu, J.; Jin, S.Y.; Song, H.D.; Huang, K.; Li, S.; Guan, X.; Wang, Y. Effect of extrusion pretreatment on extraction, quality and antioxidant capacity of oat (*Avena sativa* L.) bran oil. *J. Cereal Sci.* **2020**, *95*, 102972. [CrossRef]

9. Wang, Z.C.; He, X.J.; Yan, L.M.; Wang, J.P.; Hu, X.L.; Sun, Q.; Zhang, H.R. Enhancing enzymatic hydrolysis of corn stover by twin-screw extrusion pretreatment. *Ind. Crops Prod.* **2020**, *143*, 111960. [CrossRef]
10. Sundaram, V.; Muthukumarappan, K.; Kamireddy, S.R. Effect of ammonia fiber expansion (AFEX (TM)) pretreatment on compression behavior of corn stover, prairie cord grass and switchgrass. *Ind. Crops Prod.* **2015**, *74*, 45–54. [CrossRef]
11. Sokhansanj, S.; Mani, S.; Bi, X.; Zaini, P.; Tabil, L. Binderless pelletization of biomass. In Proceedings of the 2005 ASAE Annual Meeting, Tampa, FL, USA, 17–20 July 2005; p. 1.
12. Viamajala, S.; McMillan, J.D.; Schell, D.J.; Elander, R.T. Rheology of corn stover slurries at high solids concentrations—Effects of saccharification and particle size. *Bioresour. Technol.* **2009**, *100*, 925–934. [CrossRef] [PubMed]
13. Liu, Z.H.; Chen, H.Z. Mechanical property of different corn stover morphological fractions and its correlations with high solids enzymatic hydrolysis by periodic peristalsis. *Bioresour. Technol.* **2016**, *214*, 292–302. [CrossRef] [PubMed]
14. Van Soest, P.J.; Robertson, J.B.; Lewis, B.A. Methods for dietary fiber, neutral detergent fiber, and nonstarch polysaccharides in relation to animal nutrition. *J. Dairy Sci.* **1991**, *74*, 3583–3597. [CrossRef] [PubMed]
15. Cai, R.; Cui, X.Y.; Cao, X.H.; Jiang, X.; Wang, R.Z.; Xu, C.C. Crushing combined with high-frequency turning can promote material degradation of sheep manure compost on the Qinghai-Tibet Plateau by improving the microbial metabolic function. *J. Environ. Chem. Eng.* **2023**, *11*, 109535. [CrossRef]
16. Van Soest, P.J. Collaborative Study of Acid-Detergent Fiber and Lignin. *J. Assoc. Off. Anal. Chem.* **1973**, *56*, 781–784. [CrossRef]
17. AOAC. *Official Methods of Analysis*, 15th ed.; Association of Official Analytical Chemists: Arlington, VA, USA, 1990.
18. Owens, V.N.; Albrecht, K.A.; Muck, R.E.; Duke, S.H. Protein degradation and fermentation characteristics of red clover and alfalfa silage harvested with varying levels of total nonstructural carbohydrates. *Crop Sci.* **1999**, *39*, 1873–1880. [CrossRef]
19. Cai, R.; Zuo, S.S.; Cao, X.H.; Xu, C.C. Carbon and nitrogen transformation mechanism and functional microorganisms of sheep manure composting on the Qinghai-Tibet Plateau under different moisture content. *J. Environ. Chem. Eng.* **2023**, *11*, 111341. [CrossRef]
20. Segal, L.; Creely, J.J.; Martin, A.E.; Conrad, C.M. An Empirical Method for Estimating the Degree of Crystallinity of Native Cellulose Using the X-Ray Diffractometer. *Text. Res. J.* **1959**, *29*, 786–794. [CrossRef]
21. Bourne, M.C.; Kenny, J.F.; Barnard, J. Computer-Assisted Readout of Data from Texture Profile Analysis Curves. *J. Texture Stud.* **1978**, *9*, 481–494. [CrossRef]
22. Sozer, N.; Dalgic, A.C. Modelling of rheological characteristics of various spaghetti types. *Eur. Food Res. Technol.* **2007**, *225*, 183–190. [CrossRef]
23. Tilley, J.; Terry, D.R. A two-stage technique for the in vitro digestion of forage crops. *Grass Forage Sci.* **1963**, *18*, 104–111. [CrossRef]
24. France, J.; Dijkstra, J.; Dhanoa, M.S.; Lopez, S.; Bannink, A. Estimating the extent of degradation of ruminant feeds from a description of their gas production profiles observed in vitro: Derivation of models and other mathematical considerations. *Br. J. Nutr.* **2000**, *83*, 143–150. [CrossRef] [PubMed]
25. van Dam, J.E.G.; van den Oever, M.J.A.; Teunissen, W.; Keijsers, E.R.P.; Peralta, A.G. Process for production of high density/high performance binderless boards from whole coconut husk—Part 1: Lignin as intrinsic thermosetting binder resin. *Ind. Crops Prod.* **2004**, *19*, 207–216. [CrossRef]
26. Zhao, X.; Wang, L.; Lu, X.; Zhang, S. Pretreatment of corn stover with diluted acetic acid for enhancement of acidogenic fermentation. *Bioresour. Technol.* **2014**, *158*, 12–18. [CrossRef] [PubMed]
27. Singh, J.; Suhag, M.; Dhaka, A. Augmented digestion of lignocellulose by steam explosion, acid and alkaline pretreatment methods: A review. *Carbohydr. Polym.* **2015**, *117*, 624–631. [CrossRef] [PubMed]
28. Wu, M.; Pang, J.H.; Lu, F.C.; Zhang, X.M.; Che, L.; Xu, F.; Sun, R.C. Application of new expansion pretreatment method on agricultural waste. Part I: Influence of pretreatment on the properties of lignin. *Ind. Crops Prod.* **2013**, *50*, 887–895. [CrossRef]
29. Ma, J.; Yue, H.; Li, H.; Zhang, J.; Liu, G.Q. Selective delignification of poplar wood with a newly isolated white-rot basidiomycete Peniophora incarnata T-7 by submerged fermentation to enhance saccharification. *Biotechnol. Biofuels* **2021**, *14*, 1–15. [CrossRef]
30. Luo, R.H.; Liao, Q.; Xia, A.; Deng, Z.C.; Huang, Y.; Zhu, X.Q.; Zhu, X. Synergistic Treatment of Alkali Lignin via Fungal Coculture for Biofuel Production: Comparison of Physicochemical Properties and Adsorption of Enzymes Used As Catalysts. *Front. Energy Res.* **2020**, *8*, 575371. [CrossRef]
31. Chen, L.; Li, J.; Lu, M.; Guo, X.; Zhang, H.; Han, L. Integrated chemical and multi-scale structural analyses for the processes of acid pretreatment and enzymatic hydrolysis of corn stover. *Carbohydr. Polym.* **2016**, *141*, 1–9. [CrossRef]
32. Zhai, Q.L.; Li, F.L.; Wang, F.; Feng, J.F.; Jiang, J.C.; Xu, J.M. Ultrafine grinding of poplar biomass: Effect of particle morphology on the liquefaction of biomass for methyl glycosides and phenolics. *Cellulose* **2019**, *26*, 3685–3701. [CrossRef]
33. Chen, X.H.; Zhang, Y.L.; Gu, Y.; Liu, Z.G.; Shen, Z.; Chu, H.Q.; Zhou, X.F. Enhancing methane production from rice straw by extrusion pretreatment. *Appl. Energy* **2014**, *122*, 34–41. [CrossRef]
34. Jacquet, N.; Maniet, G.; Vanderghem, C.; Delvigne, F.; Richel, A. Application of Steam Explosion as Pretreatment on Lignocellulosic Material: A Review. *Ind. Eng. Chem. Res.* **2015**, *54*, 2593–2598. [CrossRef]
35. Hahor, W.; Thongprajukaew, K.; Nuntapong, N.; Saekhow, S.; Rungruangsak-Torrissen, K.; Dumrongrittamatt, T.; Phonchai, A. Partial pretreatment of ingredient mixture effectively improved feed chemical composition, physicochemical properties and in vitro digestibility. *Anim. Feed Sci. Technol.* **2022**, *285*, 115216. [CrossRef]
36. Brummell, D.A.; Dal Cin, V.; Crisosto, C.H.; Labavitch, J.M. Cell wall metabolism during maturation, ripening and senescence of peach fruit. *J. Exp. Bot.* **2004**, *55*, 2029–2039. [CrossRef] [PubMed]

37. Selig, M.J.; Viamajala, S.; Decker, S.R.; Tucker, M.P.; Himmel, M.E.; Vinzant, T.B. Deposition of lignin droplets produced during dilute acid pretreatment of maize stems retards enzymatic hydrolysis of cellulose. *Biotechnol. Prog.* **2007**, *23*, 1333–1339. [CrossRef] [PubMed]
38. Chen, H.Z.; Liu, Z.H. Steam explosion and its combinatorial pretreatment refining technology of plant biomass to bio-based products. *Biotechnol. J.* **2015**, *10*, 866–885. [CrossRef] [PubMed]
39. Zhao, S.; Li, G.; Zheng, N.; Wang, J.; Yu, Z. Steam explosion enhances digestibility and fermentation of corn stover by facilitating ruminal microbial colonization. *Bioresour. Technol.* **2018**, *253*, 244–251. [CrossRef]
40. He, L.W.; Meng, Q.X.; Li, D.Y.; Wang, F.; Ren, L.P. Effect of steam explosion on gas production kinetics and rumen fermentation profiles of three common straws. *Ital. J. Anim. Sci.* **2015**, *14*, 4076. [CrossRef]
41. Hussain, S.; Liu, T.; Iqbal, N.; Brestic, M.; Pang, T.; Mumtaz, M.; Shafiq, I.; Li, S.X.; Wang, L.; Gao, Y.; et al. Effects of lignin, cellulose, hemicellulose, sucrose and monosaccharide carbohydrates on soybean physical stem strength and yield in intercropping. *Photochem. Photobiol. Sci.* **2020**, *19*, 462–472. [CrossRef]
42. Zhao, D.Q.; Luan, Y.T.; Xia, X.; Shi, W.B.; Tang, Y.H.; Tao, J. Lignin provides mechanical support to herbaceous peony (Pall.) stems. *Hortic. Res.* **2020**, *7*, 213. [CrossRef]
43. Li, Q.; Fu, C.F.; Liang, C.L.; Ni, X.J.; Zhao, X.H.; Chen, M.; Ou, L.J. Crop Lodging and The Roles of Lignin, Cellulose, and Hemicellulose in Lodging Resistance. *Agronomy* **2022**, *12*, 1795. [CrossRef]
44. Burgert, I. Exploring the micromechanical design of plant cell walls. *Am. J. Bot.* **2006**, *93*, 1391–1401. [CrossRef] [PubMed]
45. Roberts, K.M.; Lavenson, D.M.; Tozzi, E.J.; McCarthy, M.J.; Jeoh, T. The effects of water interactions in cellulose suspensions on mass transfer and saccharification efficiency at high solids loadings. *Cellulose* **2011**, *18*, 759–773. [CrossRef]
46. Pihlajaniemi, V.; Sipponen, M.H.; Liimatainen, H.; Sirvio, J.A.; Nyyssola, A.; Laakso, S. Weighing the factors behind enzymatic hydrolyzability of pretreated lignocellulose. *Green Chem.* **2016**, *18*, 1295–1305. [CrossRef]
47. Shen, F.; Sun, S.; Zhang, X.; Yang, J.R.; Qiu, M.; Qi, X.H. Mechanochemical-assisted production of 5-hydroxymethylfurfural from high concentration of cellulose. *Cellulose* **2020**, *27*, 3013–3023. [CrossRef]
48. Cai, R.; Zuo, S.S.; Cao, X.H.; Jiang, X.; Xu, C.C. Effects of turning frequency on fermentation efficiency and microbial community metabolic function of sheep manure composting on the Qinghai-Tibet Plateau. *Bioresour. Bioprocess.* **2023**, *10*, 53. [CrossRef]

Disclaimer/Publisher's Note: The statements, opinions and data contained in all publications are solely those of the individual author(s) and contributor(s) and not of MDPI and/or the editor(s). MDPI and/or the editor(s) disclaim responsibility for any injury to people or property resulting from any ideas, methods, instructions or products referred to in the content.

animals

MDPI

Article

Enhancing Rumen Fermentation and Bacteria Community in Sika Deer (*Cervus nippon*) through Varying Levels of Dragon Fruit Peel Polyphenolic Extract: An In Vitro Study

Chao Ban [1], Xingzhou Tian [2,3], Qi Lu [2,3], Pipat Lounglawan [1,*] and Guilan Wen [2,*]

1 School of Animal Technology and Innovation, Institute of Agricultural Technology, Suranaree University of Technology, Nakhon Ratchasima 30000, Thailand; ban_chao@outlook.com
2 College of Animal Science, Guizhou University, Guiyang 550025, China; qlu@gzu.edu.cn (Q.L.)
3 Key Laboratory of Animal Genetics, Breeding and Reproduction in the Plateau Mountainous Region, Ministry of Education, College of Animal Science, Guizhou University, Guiyang 550025, China
* Correspondence: pipat@sut.ac.th (P.L.); glwen@gzu.edu.cn (G.W.)

Simple Summary: Dragon fruit peel polyphenolic extract is a potential ruminal modulator endorsed by consumers for its inherent safety and acceptability. This in vitro study demonstrated that dragon fruit peel polyphenolic extract increased the concentration of volatile fatty acids in the rumen, reduced gas production, and enhanced the rumen bacterial abundances. These findings suggest that dragon fruit peel polyphenol extract has the potential to improve ruminal fermentation of sika deer and provides a theoretical basis for dragon fruit peel polyphenol extract to be used as a functional feed additive for Sika deer.

Abstract: The aim of this study is to investigate the effect of dragon fruit peel polyphenolic extract (DFPE) on gas production, rumen fermentation, and bacterial communities in sika deer using an in vitro technique. Three treatments with different DFPE levels (DFPE0, base diet; DFPE5, base diet + 5 mg/g DFPE; DFPE10, base diet + 10 mg/g DFPE, respectively; n = 6) were implemented. The phenolic composition of DFPE, gas production (GP), ammonia nitrogen (NH_3-N), volatile fatty acid (VFA), and bacteria communities was evaluated after 24 h of incubation. The results showed that GP and NH_3-N were reduced by DFPE supplementation. Total VFA, isovaleric acid, and valeric acid were increased ($p < 0.05$) by the addition of DFPE. No changes ($p > 0.05$) were observed in pH, acetic acid, propionic acid, isobutyric acid, butyric acid, and the ratio of acetic acid to propionic acid. Additionally, the alpha indexes, including Sobs, Shannon, and Ace, were increased by DFPE supplementation. Moreover, at the phylum level, DFPE supplementation increased ($p = 0.01$) *Bacteroidota* but reduced ($p < 0.01$) *Firmicutes*. At the genus level, compared to DFPE0, the DFPE10 had increased relative abundances of *Rikenellaceae_RC9_gut_group* ($p < 0.01$), *norank_f_Muribaculaceae* ($p = 0.01$), *Lachnospiraceae_NK3A20_group* ($p < 0.01$), *Christensenellaceae_R-7_group* ($p < 0.01$), and *NK4A214_group* ($p < 0.01$), decreased relative abundances of *Streptococcus* ($p < 0.01$), *Oribacterium* ($p = 0.01$), and *Enterococcus* ($p < 0.01$). Compared to DFPE0, DFPE5 had no change ($p > 0.05$) in all bacteria at the genus level except for decreased relative abundance of *Enterococcus* ($p < 0.01$). These results indicated that DFPE may be able to be used as a feed additive to enhance fermentation parameters and improve ruminal bacteria communities in Sika deer.

Keywords: dragon fruit peel; in vitro gas production; polyphenolic extract; rumen fermentation; sika deer; 16S rRNA sequencing

Citation: Ban, C.; Tian, X.; Lu, Q.; Lounglawan, P.; Wen, G. Enhancing Rumen Fermentation and Bacteria Community in Sika Deer (*Cervus nippon*) through Varying Levels of Dragon Fruit Peel Polyphenolic Extract: An In Vitro Study. *Animals* **2024**, *14*, 1139. https://doi.org/10.3390/ani14081139

Academic Editors: Dingfu Xiao and Tiantian Meng

Received: 5 March 2024
Revised: 6 April 2024
Accepted: 7 April 2024
Published: 9 April 2024

Copyright: © 2024 by the authors. Licensee MDPI, Basel, Switzerland. This article is an open access article distributed under the terms and conditions of the Creative Commons Attribution (CC BY) license (https://creativecommons.org/licenses/by/4.0/).

1. Introduction

The structure of a ruminal microbial community is influenced by various factors, including health status, physiological status (e.g., pregnancy, lactation) [1], individual variation, and diet composition. Of these, diet composition exerts a dominant influence over the composition of ruminal microbiota [2,3]. In addition to diet, plant secondary metabolites, especially phenolic compounds, play a crucial role in the formation of ruminal microbial ecosystems [4]. For instance, naringin, which is a flavonoid (a subclass of polyphenols) compound commonly found in fruit peel, has been shown to reduce methane production to improve feed efficiency and increase the ruminal butyrate ratio while altering the composition of ruminal bacteria [5]. Tian et al. [6] found that polyphenols-rich sorghum stalks, which contain various flavonoids, including naringin, could reduce ruminal gas production and increase propionic acid proportions. Thus, enhancing rumen fermentation by utilizing polyphenols represents an effective approach to safeguarding the health of ruminant animals, increasing animal product yield and improving product quality.

Dragon fruit (*Hylocereus polyrhizus*) peel polyphenolic extract can be considered an antioxidant source that is non-toxic and biologically safe, serving as the functional feed additive that has the potential to improve rumen fermentation. Previous research has shown that dietary dragon fruit peel could increase microbial protein supply in Holstein crossbred bulls [7]. Furthermore, dragon fruit peel could increase the content of volatile fatty acids and reduce methane production [8].

Sika deer (*Cervus nippon*) is a special livestock that is widely distributed in Asian countries, including China, Vietnam, Korea, and Japan [9]. The velvet antler of Sika deer serves as the primary product, which is an exceedingly coveted natural resource with substantial market potential [10]. Understanding the bacterial ecology within the rumen of Sika deer is crucial for optimizing nutrient digestion efficiency and boosting velvet antler productivity [11]. However, there is a scarcity of research on the utilization of phenolic substances in Sika deer, particularly in relation to their impact on rumen microorganisms, which remains largely unexplored. Therefore, we hypothesized that supplementing dragon fruit peel polyphenolic extract could enhance rumen fermentation and the microbial population. The current study aimed to investigate dragon fruit peel polyphenolic extract on gas production, ruminal fermentation characteristics, and bacteria communities in Sika deer using an in vitro technique.

2. Materials and Methods

2.1. Animal Care

The animal study underwent a thorough review, ensuring that all procedures related to the care and treatment of experimental animals were carefully examined and assessed. Reviewing department: Animal Welfare and Experimental Animal Ethics of Guizhou University (EAE-GZU-2024-E001), Guizhou, China.

2.2. Experimental Design

Three treatments were administered involving the supplementation of polyphenols from dragon fruit peel extract (DFPE was commercial extract, it was purchased from Ningshan Guosheng Biotechnology Co., Ltd., Ankang, China). These treatments were distinguished by varying levels of dragon fruit peel extract: 0, 5, and 10 mg/g of fermentation substrate based on dry matter (DM), respectively. Each treatment was 6 repetitions. The in vitro fermentation substrate composition is shown in Table 1.

Table 1. The composition and nutrient level of the basal substrate (g/kg of DM) for in vitro fermentation.

Ingredients	Content
Napier grass	456.54
Peanut vine	43.46
Peanut shell	159.43
Garlic peel	106.34
Soybean waste	106.34
Soybean meal	37.06
Corn	72.73

Table 1. *Cont.*

Ingredients	Content
Wheat bran	7.41
Calcium bicarbonate	2.59
Salt	5.34
Premix [1]	2.76
Total	1000
Nutrient level	
Gross energy, MJ/kg of DM	13.06
Dry matter, g/kg	956.97
Ash	98.95
Crude protein	143.68
Ether extract	42.87
Neutral detergent fiber	363.08
Acid detergent fiber	281.47
Calcium	14.62
Phosphorus	3.26

[1] 1 kg premix consists of vitamin premix 700,000 IU, Fe 10 g, Zn 3 g, Cu 0.75 g, Mn 3 g, Se 0.01 g, Co 0.02 g.

2.3. Simulating Rumen Fermentation Processes In Vitro

Two male Sika deer (BW = 78.41 ± 6.82 kg, mean ± SD) at 2 years old were selected from Guizhou Shengchangyao Ecological Agriculture Comprehensive Development Co., Ltd. (Bijie, China) as the donors of rumen fluid.

All Sika deer were housed in pens with free access to fresh drinking water and were fed twice a day at 8:00 am and 4:00 pm. The feed had a concentrate-to-roughage ratio of 30:70, and the dietary nutrient requirements were sourced from the National Research Council [12]. The roughage was corn stalk, and the composition of concentrate was (based on DM) 21.6% corn, 16.6% rice bran, 38.8% soybean meal, wheat bran 20.5%, 1% salt, 1% calcium bicarbonate, and 0.5% premix (the premix consists of vitamin premix 700 IU/g, Fe 10 mg/g, Zn 3 mg/g, Cu 0.75 mg/g, Mn 3 mg/g, Se 0.01 mg/g, Co 0.02 mg/g).

The rumen fluid was collected after 12 h of fasting, and the deer were anesthetized with 2 mL xylazine hydrochloride injection (Jilin Huamu Animal Health Products Co., Ltd., Changchun, China) using an anesthetic gun. Rumen fluids were obtained via mouth by a stomach tube connected to a vacuum pump. The rumen fluids were immediately placed into a prefilled CO_2 thermos in equal proportions and transferred to the laboratory. Subsequently, rumen fluids were filtered through four layers of gauze while being continuously flushed with CO_2. According to the method of Menke and Steingass [13], the artificial saliva was configured as follows: the 1 L of artificial saliva consisted of 520.2 mL of distilled water, 0.1 mL of micromineral solution (1 L of micromineral solution containing 132 g of $CaCl_2{\cdot}2H_2O$, 100 g of $MnCl_2{\cdot}4H_2O$, 10 g of $CoCl_2{\cdot}6H_2O$ and 80 g of $FeCl_3{\cdot}6H_2O$), 208.1 mL of buffer solution (1 L of buffer solution containing 35 g of $NaHCO_3$ and 4 g of NH_4HCO_3), 208.1 mL of macromineral solution (1 L of micromineral solution containing 5.7 g of Na_2HPO_4, 6.2g of KH_2PO_4, and 0.6 g of $MgSO_4$), 1 mL of 0.1% resazurin solution, and 62.4 mL of reduction solution (added 570 mg of $Na_2S{\cdot}7H_2O$ to 100 mL of 0.04 N of NaOH). The artificial rumen fluid was prepared by mixing rumen fluid with artificial saliva at a ratio of 1:2.

According to the method of Menke and Steingass [13], 200 mg of substrate was accurately weighed and placed at the bottom of a glass gas-tight syringe (120 mL, Changzhou yangming glass production Co., Ltd., Changzhou, China), and 30 mL of artificial rumen fluid was injected, sealed with a deoxidizer, then placed into incubator shaker (IS-RDH1, Suzhou JieMEI Electronics Co., Ltd., Suzhou, China) and incubated at 39 °C. There was a total of 21 glass gas-tight syringes, including 6 repetitions per treatment and 3 repetitions in blank. The gas production was measured by directly observing the scale on the glass gas-tight syringe at 3, 5, 7, 10, 20, 22, and 24 h after incubation, and the net gas production value was calculated by subtracting the average of gas generated in the blank.

At 24 h of incubation time, fermentation was halted by placing the glass gas-tight syringe into an ice bath. Subsequently, the pH value was promptly measured using a pH meter (Sartorius AR company, Gottingen, Germany). The rumen digesta was stored at −80 °C for future DNA extraction. The rumen fluid was then filtered through 4 layers of gauze, and 1 mL of rumen fluid was mixed with 200 μL of 25% metaphosphoric acid. Next, through vortexing and shaking, the mixture underwent centrifugation at 10,000× *g* at 4 °C for 10 min. The supernatant was meticulously filtered using a 13 mm nylon syringe filter with a pore size of 0.45 μm for VFA analysis. The VFA concentrations, including acetic acid (AA), propionic acid (PA), isobutyric acid (isoBA), butyric acid (BA), isovaleric acid (isoVA), and valeric acid (VA), were determined by gas chromatography (CP-3800, Varian Medical Systems Company, Palo Alto, CA, USA) with a column size of 30 m × 0.32 mm × 0.15 μm (DB-FFAP, Onlysci, China), in triplicate. The GC condition was based on the method referred to by Yang et al. [14]; in brief, the column was initially heated to a temperature of 80 °C for a duration of 2 min. Subsequently, the temperature was raised to 150 °C at a controlled rate of 10 °C/min for an additional 2 min. Following this, the temperature was further increased to 180 °C at a rate of 15 °C/min, maintaining this temperature for 5 min. The entire process lasted for a total of 18 min. The carrier gas used was helium, maintained at a consistent flow rate of 3 mL/min.

2.4. Chemical Analysis

Approximately 1 g of DFPE was extracted using 10 mL of extraction reagent (1% HCl-Methanol) by vortexing for 1 min, in triplicate. The mixture was then placed in an ultrasonic ice-water bath for 30 min. Subsequently, the extracted solution underwent filtration using a 0.22 μm organic phase filter membrane and was subsequently preserved at −20 °C for future analysis. A total of 15 kinds of phenolics were determined. The standard phenolics, including gallic acid, chlorogenic acid, catechin, epicatechin, protocatechuic acid, caffeic acid, syringate, rutin, coumaric acid, naringin, vanillic acid, myricetin, luteolin, quercetin, and kaempferol, were obtained from Merck (Merck Co., Ltd., Darmstadt, Germany). The phenolic compounds of all extract solutions were determined using an Agilent 1260 HPLC (Agilent Technologies, Santa Clara, CA, USA) coupled to an Agilent 6420 triple quadrupole mass spectrometer with a column of ZORBAX Eclipse Plus C_{18} (3.5 μm, 2.1 × 150 mm) operated at 35 °C, in triplicates.

All the feed (substrate) samples were passed through a 1 mm screen for analysis of chemical composition, in triplicate. The dry matter (DM), ash, crude protein (CP), ether extract (EE), calcium (Ca), and phosphorus (P) were analyzed according to the method of AOAC [15], and the gross energy (GE) was determined by using a calorimeter (Parr 6200, Moline, IL, USA). The rumen fluid amino nitrogen (NH_3-N) was analyzed according to the method of Tian et al. [16]. The acid detergent fiber (ADF) and neutral detergent fiber (NDF) were determined according to the method of Van Soect et al. [17].

2.5. DNA Extract and Detection

The DNeasy® PowerSoil® Pro Kit (QIAGEN, Hilden, Germany) was used to extract the total microbial DNA from 18 rumen fluid samples according to the kit instructions. The concentration and purity of extracted DNA were determined by a NanoDrop 2000 spectrophotometer (Thermo Fisher Scientific, Waltham, MA, USA).

Bacterial 16S rRNA gene fragments (V3–V4) were amplified from the extracted DNA using primers Ba9F (5′-GAGTTTGATCMTGGCTCAG-3′) and Ba515Rmod1R (5′-CCGCGGCKGCTGGCAC-3′). The PCR condition was based on an earlier method reported by Zhang et al. [18] with minor modifications. Briefly, the process begins with an initial denaturation at 95 °C for 3 min, followed by 27 cycles of denaturation at 95 °C for 30 s, annealing at 55 °C for 30 s, and elongation at 72 °C for 45 s, concluding with a final extension at 72 °C for 10 min.

The PCR reactions were conducted using the following components: 4 μL of 5× TransStart FastPfu buffer, 2 μL of 2.5 mM deoxynucleoside triphosphates (dNTPs), 0.8 μL of each primer (5 μM), 0.4 μL of TransStart FastPfu DNA Polymerase, and 10 ng of the extracted DNA. Subsequently, distilled deionized water (ddH_2O) was added to bring the total volume to 20 μL. Next, agarose gel electrophoresis (2%) was carried out to confirm the amplicon's size. Finally, the amplicons were subjected to paired-end sequencing on the Illumina MiSeq sequencing platform (using PE300) through Majorbio Bio-Pharm Technology Co., Ltd. in Shanghai, China.

The raw sequences underwent quality control using FASTP software (https://github.com/OpenGene/fastp, version 0.19.6, accessed on 18 January 2024) and were merged using FLASH software (https://ccb.jhu.edu/software/FLASH/index.shtml, version 1.2.11, accessed on 18 January 2024). Specifically, the reads are filtered based on quality scores, removing bases with quality scores < 20 at the tail end. A 50 bp window was applied, and if the average quality value within the window was <20, the back-end bases were truncated. Reads < 50 bp after quality control (QC) were excluded, and reads containing N bases were eliminated. Pairs of reads were merged into a single sequence based on the overlap relationship between PE reads, requiring a minimum overlap length of 10 bp. The maximum allowable mismatch ratio in the overlap region of the merged sequence was 0.2, and non-conforming sequences were discarded. Samples were distinguished by barcode and primers at the sequence's start and end. The sequence orientation was adjusted accordingly.

The operational taxonomic unit (OTU) clustering was performed at 97% sequence identity using the UPARSE software (http://drive5.com/uparse/, version 11, accessed on 18 January 2024) and compared with the Silva 16S rRNA database (https://www.arb-silva.de/, version 138, accessed on 18 January 2024), with a comparison threshold of 70%.

2.6. Statistical Analysis

In this study, the effect of DFPE on gas production and rumen fermentation parameters was performed by SPSS 27 (Chicago, IL, USA) using one-way analysis of variance (ANOVA), and Tukey's test was used to compare the differences among the 3 treatment groups. The effects of DFPE on ruminal bacteria parameters were investigated using nonparametric tests (Kruskal–Wallis test). The significance level was $p < 0.05$. The correlation between rumen fermentation parameters and bacterial communities was analyzed using the Spearman test.

3. Results

3.1. The Phenolic Composition of Dragon Fruit Peel Extract

The top five phenolics with the highest content in dragon fruit peel extract were naringin, caffeic acid, vanillic acid, quercetin, and protocatechuic acid, respectively. The bottom five phenolics with the lowest content were luteolin, catechin, gallic acid, epicatechin, and chlorogenic acid, respectively (Table 2).

Table 2. Composition of phenolic compounds in DFPE (ng/g of DM).

Item	Mean	SD
Gallic acid	30.88	1.02
Chlorogenic acid	47.42	3.62
Catechin	27.88	2.99
Epicatechin	35.57	1.50
Protocatechuic acid	1363.07	109.56
Caffeic acid	2682.22	250.81
Syringate	199.99	51.93
Rutin	174.90	24.15
Coumaric acid	731.49	45.76
Naringin	9977.35	317.84
Vanillic acid	2649.25	206.65
Myricetin	314.84	19.93
Luteolin	10.94	1.62
Quercetin	2187.91	133.37
Kaempferol	49.68	2.74

DFPE, dragon fruit peel polyphenolic extract; SD, standard deviation; values represent the mean of three replicates.

3.2. DFPE on Gas Production and Rumen Fermentation

The gas production curve tended to be flat after 24 h of incubation in all treatments, and the supplementation of DFPE significantly reduced the cumulative gas production at 24 h (Figure 1). No significant differences ($p > 0.05$) were observed in pH, propionic acid, isobutyric acid, butyric acid, and the ratio of acetic acid to propionic acid among all treatments. DFPE0 had a significantly lower ($p < 0.05$) ammonia nitrogen, total volatile fatty acid, isovaleric acid, and valeric acid, whereas DFPE0 showed a significantly higher ($p < 0.01$) acetic acid than the DFPE5 and DFPE10, see Table 3 below.

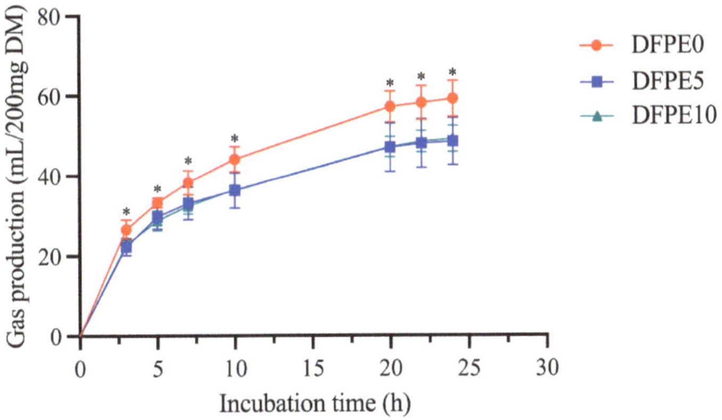

Figure 1. Effect of DFPE on gas production in vitro. DFPE0, base diet; DFPE5, base diet + 5 g/kg DFPE; DFPE10, base diet + 10 g/kg DFPE; DM, dry matter, "*" indicated significant difference.

Table 3. Effect of DFPE on gas production and rumen fermentation.

Item	DFPE0	DFPE5	DFPE10	SEM	p-Value
GP_{24h}, mL/g of DM	295.83 a	242.50 b	245.83 b	7.85	<0.01
pH	6.17	6.12	6.15	0.02	0.53
NH_3-N, mg/dL	34.80 a	30.90 b	31.57 b	0.44	<0.01
TVFA, mmol/L	10.78 b	11.83 a	11.95 a	0.20	0.02
Molar proportion (mmol/100 mmoL)					
AA	58.88 a	57.69 b	57.48 b	0.22	<0.01
PA	25.22	24.85	24.04	0.26	0.17
isoBA	0.61	0.71	0.65	0.03	0.49
BA	11.82	12.04	12.56	0.15	0.09
isoVA	0.73 b	0.94 a	1.02 a	0.04	<0.01
VA	2.75 b	3.77 a	4.25 a	0.19	<0.01
AA/PA	2.34	2.33	2.39	0.03	0.58

SEM, standard error of the mean; GP, gas production, g/mL of DM; NH_3-N, ammonia nitrogen; TVFA, total volatile fatty acid; AA, acetic acid; PA, propionic acid; isoBA, isobutyric acid; BA, butyric acid; isoVA, isovaleric acid; VA, valeric acid; AA/PA, the ratio of acetic acid to propionic acid. Different letters indicated significant difference.

3.3. Rumen Bacterial Richness, Diversity, and Composition

After removing low-quality or chimeric sequences, there was a total of 1,081,480 high-quality sequences, and the number of total effective bases was 523,605,765 obtained from the 18 samples. A total of 2921 OTUs were identified and categorized into 15 phyla, 23 classes, 61 orders, 100 families, 230 genera, and 485 species. The rarefaction curve based on Sobs and Shannon indexes revealed that (Figure 2A,B) as the number of sequences increased, the curves gradually leveled off, signifying saturated coverage and sufficient sequencing depth in the present experiment.

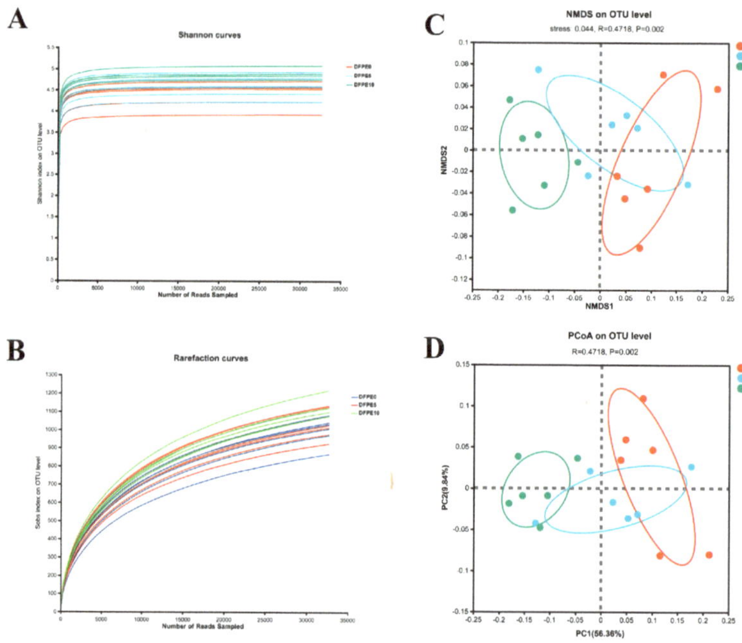

Figure 2. Effect of DFPE on ruminal bacterial diversity. (**A**) The Shannon curves. (**B**) The rarefaction curves. (**C**) NMDS analysis. (**D**) PCoA based on the Bray–Curtis distance.

As shown in Table 4, the coverage index for all samples was greater than 99%. The indexes of Sobs, Shannon, and Ace were higher ($p < 0.05$) in the DFPE10 than DFPE5 and DFPE0 groups. No difference ($p > 0.05$) was observed in the Chao index.

Table 4. Analysis of alpha diversity index in the present study.

Item	DFPE0	DFPE5	DFPE10	SEM	*p*-Value
Sobs	999.67 b	1033.17 b	1118.50 a	19.95	0.03
Shannon	4.43 b	4.59 b	4.89 a	0.07	0.01
Ace	1218.43 b	1228.20 b	1340.36 a	0.01	0.04
Chao	1183.92	1189.80	1287.96	22.59	0.07

Different letters indicated significant difference.

The non-metric multidimensional scaling (NMDS) revealed a microbial community and distinct structure. Based on the distance calculation of Bray–Curtis, the stress of NMDS at the OTU level was 0.044 (Figure 2C), indicating the criteria for obtaining a dependable representation of the alterations in the bacterial community structure. The PCoA noted based on Bray–Curtis, with the contribution values of principal component (PC) 1 and PC2 for sample differences, were 56.36% and 9.84%, respectively (Figure 2D ANOSIM; $p = 0.002$).

In the Venn diagram shown (Figure 3), each ellipse symbolizes a group, and the intersecting region among the ellipses signifies the common OTU shared between the groups, with the numerical value indicating the quantity of OTUs. The DFPE0, DFPE5, and DFPE10 groups had 1863, 1955, and 2116 OTUs, respectively. A total of 1308 OTUs were observed across all groups, and specifically, 335, 379, and 502 OTUs were detected for the DFPE0, DFPE5, and DFPE10 groups, respectively.

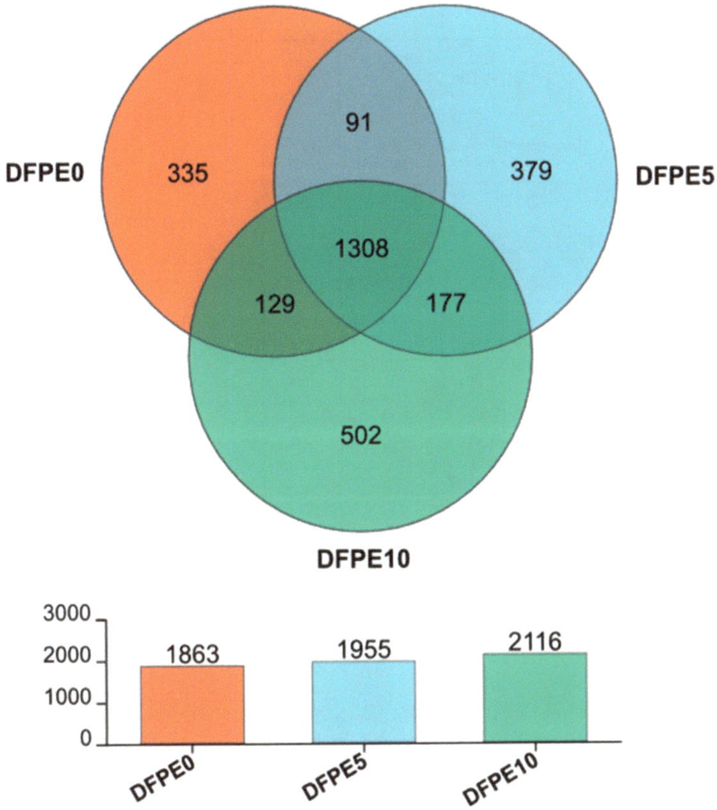

Figure 3. The Venn diagram.

3.4. Rumen Bacteria Composition at Phylum Level

At the phylum level (Figure 4A and Table 5), we found that *Firmicutes* (DFPE0, DFPE5, and DFPE10 were 70.21%, 66.30%, and 62.13%, respectively) was the predominant bacteria phylum, followed by *Bacteroidota* (DFPE0, DFPE5, and DFPE10 were 27.21%, 30.84%, and 35.06%, respectively), and *Firmicutes* and *Bacteroidota* together accounted for more than 97% of the total bacterial abundance at the phylum level. Furthermore, followed by *unclassified_k_norank_bacteria* (DFPE0, DFPE5, and DFPE10 were 0.68%, 0.86%, and 1.01%, respectively), the ratio of *Firmicutes* to *Bacteroidetes* in DFPE0, DFPE5, and DFPE10 groups were 2.65, 2.21, and 1.79, respectively.

We employed the nonparametric Kruskal–Wallis test to assess the relative richness at the phylum level (at least one treatment was relative abundance > 0.5%) among the three groups. There were no differences ($p > 0.05$) between the DFPE0 and DFPE5 groups in the relative abundances of each bacterium at the phylum level. In comparison to the DFPE0 group, the DFPE10 group exhibited a significant rise in the relative abundances of the *Bacteroidota* and *unclassified_k_norank_bacteria* phyla, accompanied by a notable decrease ($p < 0.05$) in the relative abundance of the *Firmicutes* phylum, as well as a reduced ratio of *Firmicutes* to *Bacteroidota*. Furthermore, in contrast to the DFPE10 group, the DFPE5 group demonstrated an increase ($p < 0.05$) in the ratio of *Firmicutes* to *Bacteroidota*.

Table 5. Effect of DFPE on ruminal microbiota at the phylum level (relative abundance > 0.5).

Item	DFPE0	DFPE5	DFPE10	SEM	*p*-Value
Firmicutes	70.21 b	66.3 ab	62.13 a	1.195	<0.01
Bacteroidota	27.21 b	30.84 ab	35.06 a	1.176	0.01
unclassified_k_norank_d_Bacteria	0.68 b	0.86 ab	1.01 a	0.043	<0.01
Firmicutes: Bacteroidota	2.65 b	2.21 ab	1.79 a	0.130	0.01

Different letters indicated significant difference.

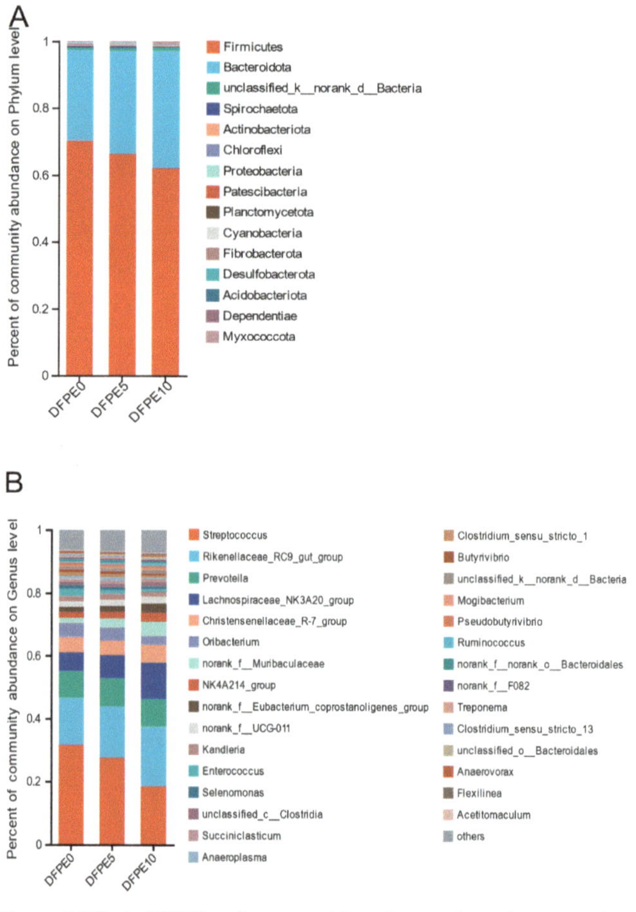

Figure 4. Effect of DFPE on the composition of rumen bacteria community. (**A**) Top 15 ruminal bacterial communities at the phylum level. (**B**) Top 30 ruminal bacterial communities at the genus level.

3.5. Rumen Bacteria Composition at Genus Level

A total of 181, 193, and 197 bacterial genera were observed in the DFPE0, DFPE5, and DFPE10 groups, respectively (Figure 4B). The top 30 relative abundances of bacteria at the genus level with a total of 17 genera were found with an abundance greater than 1% (Table 6). In the DFPE0 group, *Streptococcus* (31.71%) and *Rikenellaceae_RC9_gut_group* (14.93%) were the dominant genera, followed by *Prevotella* (8.40%), *Lachnospiraceae_NK3A20_group* (5.95%), *Christensenellaceae_R-7_group* (4.88%), and *Oribacterium* (4.45%). Combined, these taxa made up more than 70% of the overall bacterial composition. In the DFPE5, *Streptococcus* (27.5%) and *Rikenellaceae_RC9_gut_group* (16.52%) were predominant genera, followed by *Prevotella* (8.70%), *Lachnospiraceae_NK3A20_group* (7.35%), *Christensenellaceae_R-7_group* (4.69%), and *Oribacterium* (4.14%); collectively, these taxa comprised roughly 65% of the overall bacterial composition. In DFPE10, *Rikenellaceae_RC9_gut_group* (18.89%), *Streptococcus* (18.52%), and *Lachnospiraceae_NK3A20_group* (11.55%) were the predominant genera, followed by *Prevotella* (8.75%), *Christensenellaceae_R-7_group* (5.74%), and *norank_f_Muribaculaceae* (4.60%); collectively, these taxa comprised roughly 65% of the overall bacterial composition.

Table 6. Effect of DFPE on ruminal microbiota at the genus level (relative abundance in at least one treatment > 1%).

Genus	DFPE0	DFPE5	DFPE10	SEM	*p*-Value
Rikenellaceae_RC9_gut_group	14.93 b	16.52 ab	18.89 a	0.57	<0.01
Prevotella	8.40	8.70	8.75	0.21	0.58
norank_f_Muribaculaceae	1.60 b	2.90 ab	4.60 a	0.44	0.01
Streptococcus	31.71 b	27.50 b	18.52 a	1.71	<0.01
Lachnospiraceae_NK3A20_group	5.95 b	7.35 b	11.55 a	0.73	<0.01
Christensenellaceae_R-7_group	4.88 b	4.69 b	5.74 a	0.15	<0.01
norank_f_Eubacterium_coprostanoligenes_group	1.73 b	1.96 b	3.02 a	0.16	<0.01
NK4A214_group	1.88 b	2.09 ab	2.94 a	0.14	<0.01
Oribacterium	4.45 b	4.14 b	2.72 a	0.25	0.01
norank_f_UCG-011	1.83	2.00	2.16	0.08	0.29
Kandleria	1.73	1.89	1.53	0.11	0.44
unclassified_c_Clostridia	0.95 b	1.03 b	1.22 a	0.04	0.01
Succiniclasticum	1.02	0.90	1.02	0.03	0.05
Selenomonas	1.29	0.99	0.98	0.10	0.68
Enterococcus	2.17 b	1.18 a	0.77 a	0.18	<0.01
Anaeroplasma	0.98	1.24	0.57	0.12	0.05
unclassified_k_norank_d_Bacteria	0.68 b	0.86 b	1.01 a	0.04	<0.01

Different letters indicated significant difference.

We employed the nonparametric Kruskal–Wallis test to assess the relative richness at the genus level (at least one treatment was relative abundance > 1%) among the three groups. The relative abundances of genera *Kandleria, Prevotella, Succiniclasticum, norank_f_UCG-011, Selenomonas,* and *Anaeroplasma* showed no significant differences ($p > 0.05$) among all the treatments. The relative abundances of *Lachnospiraceae_NK3A20_group* ($p < 0.01$), *Christensenellaceae_R-7_group* ($p < 0.01$), *norank_f_Eubacterium_coprostanoligenes_group* ($p < 0.01$), *unclassified_c_Clostridia*, and *unclassified_k_norank_d_Bacteria* ($p = 0.01$) were elevated in the DFPE10 group compared to the DFPE0 and DFPE5 groups, with no discernible differences ($p > 0.05$) between the DFPE0 and DFPE5 groups. Conversely, the relative abundances of *Streptococcus* ($p < 0.01$) and *Oribacterium* ($p = 0.01$) were lower in the DFPE10 group than in the DFPE0 and DFPE5 groups, with no significant distinctions observed between the DFPE0 and DFPE5 groups. The relative abundance of *Enterococcus* ($p < 0.01$) was decreased in the DFPE5 and DFPE10 groups. The relative abundances of *Rikenellaceae_RC9_gut_group* ($p < 0.01$), *norank_f_Muribaculaceae* ($p = 0.01$), *NK4A214_group* ($p < 0.01$) were lower in the DFPE10 group than in the DFPE0 group, and there was no significant difference in the DFPE5 group when compared to either the DFPE0 or DFPE10 groups.

3.6. Spearman Correlations between Bacterial Communities and Rumen Fermentation Parameters

The correlation between rumen fermentation parameters and bacterial communities is presented in Figure 5. The relative abundance of *Streptococcus* was negatively ($p < 0.001$, $r < -0.70$) correlated with TVFA, BA, isoVA, and VA; however, it showed a positive ($p < 0.01$, $r = 0.66$) correlation with PA. The relative abundances of *g__Rikenellaceae_RC9_gut_group*, *g__Lachnospiraceae_NK3A20_group*, *g__norank_f__Muribaculaceae*, and *g__NK4A214_group* were positively ($p < 0.05$, $r > 0.66$) correlated with BA, isoVA, and VA. However, it showed a negative ($p < 0.05$, $r < 0.62$) correlation with PA. The relative abundance of *g__norank_f__Eubacterium_coprostanoligenes_group* showed a positive ($p < 0.01$, $r > 0.62$) correlated with TVFA, isoVA, and VA. Additionally, the relative abundance of *g__Oribacterium* was negatively ($p < 0.001$, $r < -0.71$) correlated with VA. Moreover, TVFA had a positive ($p < 0.001$, $r > 0.75$) correlation with the relative abundances of *g__Lachnospiraceae_NK3A20_group*, and *g__norank_f__Eubacterium_coprostanoligenes_group*, whereas it showed a negative ($p = 0.001$, $r < 0.70$) correlation with the relative abundance of *g__Streptococcus*.

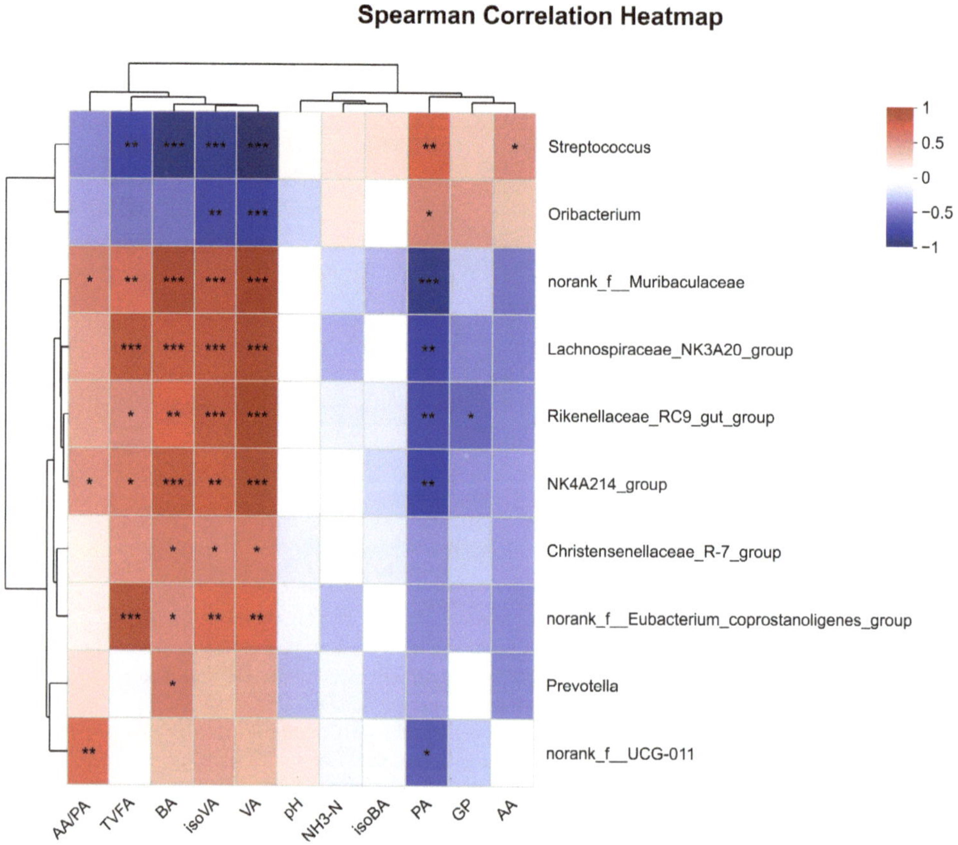

Figure 5. Spearman correlation analysis of the top 10 bacteria genera with gas production and ruminal fermentation parameters. (* $p < 0.05$, ** $p < 0.01$, *** $p < 0.001$).

4. Discussion

4.1. Effect of DFPE on Gas Production and Rumen Fermentation

GP serves as a crucial indicator of carbohydrate digestion, exhibiting a negative correlation with NDF content and a positive correlation with starch levels [19]. An excess of GP can contribute to impaired ruminal function, particularly under highly concentrated feeding conditions, leading to the likelihood of cattle and sheep experiencing a subacute rumen bloat state [20]. Polyphenols demonstrate potential in preventing rumen bloat [21]. As anticipated, our finding suggests that DFPE has the ability to reduce GP in vitro. A previous study [22] has reviewed that these compounds possess the capability to reduce GP and CH_4. Yu et al. [23] reported that naringin could reduce CH_4 production in in vitro rumen fermentation. This reduction in GP may be attributed to the presence of phenolic compounds in DFPE, including naringin, caffeic acid, vanillic acid, and quercetin (Table 2), which may contribute to this reduction in GP. Furthermore, we observed that supplement levels did not influence GP. One possible reason may be that some gas-producing bacteria (e.g., methanogens) have varying degrees of sensitivity to DFPE, which necessitates further investigation in the future.

On the other hand, phenolic compounds can induce the precipitation of soluble proteins by forming hydrogen bonds [24]. The correlation of this phenomenon with alleviating tissue swelling has been substantiated [25]. A previous study demonstrated that 10 g/kg of polyphenolic extract from *Eucommia ulmoides* leaf reduced ruminal NH_3-N levels from 175.6 to 129.6 mg/L [26]. Yu et al. [23] reported that naringin (20 mg/kg of diet) had decreased NH_3-N from 13.05 mg/dL to 11.41 mg/dL in dairy cows. Conversely, the protein-binding property of phenolic compounds can mitigate ruminal fermentation, thereby enhancing nitrogen use efficiency [27]. In the present study, our observations indicate that DFPE reduced NH_3-N levels, implying that DFPE has the ability to mitigate the degradation of dietary protein in the rumen and has the potential to enhance protein utilization.

Ruminal pH serves as a valuable indicator for assessing both ruminal health and optimal function. Consistent with the findings reported by Matra et al. [7], no significant differences in pH were observed in this study, suggesting that DFPE has the capability to maintain an appropriate acid–base status without inducing rumen acidosis. In contrast, the pH observed in this experiment ranges from 6.12 to 6.17, which differs from the optimal pH range (6.2–7.2) for microbial activity. As stated by Jackson et al. [28], this deviation may be attributed to constraints inherent in the in vitro fermentation system used in this experiment, wherein fermentation by-products such as VFA cannot be absorbed by the artificial rumen, leading to their accumulation and subsequent pH reduction. Nevertheless, the pH values remained within the normal range [29].

VFA are end-products of carbohydrates and provide the main source of energy for ruminant metabolism. Polyphenols have the potential to increase VFA by modifying the rumen bacteria population and increasing bacterial richness and diversity [30,31]. Gao et al. [32] reported that feeding beef bulls with polyphenol-rich red cabbage extract (120 g/d/head) resulted in a significant increase in total VFA from 83.06 mmol/L to 89.19 mmol/L. In the present study, we found that DFPE significantly increased total VFA. Similarly, previous studies [33,34] have reported that quercetin can increase ruminal VFA and decrease GP in vitro. In contrast, the present study revealed a comparatively lower total VFA content (10.78 to 11.95 mmol/L), while Sun et al. reported rumen total VFA ranges from 13.17 to 23.59 mmol/L (in vivo) [35]. This may be related to the lower microbial concentration in vitro than in vivo [36].

Butyric acid accounts for about 70% of the daily energy metabolism of ruminants [37]. Higher butyrate concentrations also enhance ruminal structure and function, enhance nutrient absorption, promote gut health, and improve health benefits for ruminants [38]. Remling et al. [39] noted that butyric acid also stimulated the growth of the ruminal papilla, thereby increasing the absorptive surface area of ruminants. In the present study, we found that DFPE had no effect on acetic acid, propionic acid, or butyric acid. A similar result was found by Silva et al. [40]. On the other hand, branched-chain VFA (BCVFA), such as isobutyric acid, isovaleric acid, and valeric acid, is utilized by ruminal microbes as a source of carbon skeleton for synthesizing branched-chain amino acids [41]. In the present study, we found higher concentrations of isovaleric acid and valeric acid from DFPE treatments. This indicates that DFPE has the potential to augment the population of cellulolytic bacteria and improve fiber digestibility.

4.2. Effect of DFPE on Bacteria Communities In Vitro

The well-being of the rumen is, in part, contingent upon the appropriate richness, diversity, and stability exhibited by the ruminal microbiome. The alpha diversity indexes of Sobs, Shannon, and Ace were increased by DFPE supplementation. In the present study, we examined the rumen microbiota using 16S rRNA sequencing analysis. As previously elucidated and documented in Sika deer [42], we observed that the predominant bacteria belonged to the phyla *Firmicutes* and *Bacteroidota*, with a high relative ratio of *Firmicutes* to *Bacteroidetes* in this study. This occurrence may be attributed to the fact that Sika deer typically inhabit cold and high-altitude regions [43], where a high ratio of *Firmicutes* to *Bacteroidetes* is advantageous for fat deposition, helping animals adapt to severe cold climates [44], thereby enhancing their resilience to these challenging conditions. Polyphenols are regarded as candidate compounds for prebiotics due to their interactions with gut microbiota [45]. The observations from the present study revealed that the relative abundance of *Bacteroidetes* increased with the inclusion of DFPE supplementation, while *Firmicutes* exhibited a significant decrease. In agreement with our study, Sutoyo et al. [46] reported that dietary intake of phenolic compounds is associated with a reduced proportion of *Firmicutes* and an increased proportion of *Bacteroidetes*.

Bacteroidetes are primarily involved with the degradation of carbohydrates, fats, and proteins. At the *Bacteroidetes* phylum, *Rikenellaceae_RC9_gut_group*, *Prevotella*, and *norank_f_Muribaculaceae* emerged as the predominant bacterial genera in our study, consistent with findings from a previous study by Si et al. [47]. The family *Rikenellaceae* participates in the degradation of structural carbohydrates, produces succinate and propionate as fermentation end products, and promotes lipid metabolism [48], while the family *Muribaculaceae* has been demonstrated to exhibit probiotic effects and is associated with the innate immune system and the absorption and utilization of fat [49]. A previous report indicated that the relative abundance of *Muribaculaceae* is negatively correlated with fat deposition and is affected by the digestion of carbohydrates [50]. In the present study, we observed an increase in the relative abundance of *Rikenellaceae_RC9_gut_group* and *norank_f_Muribaculaceae* with DFPE supplementation. Meanwhile, these bacterium abundances were positively correlated with butyric acid, isovaleric acid, and valeric acid. Similarly, a previous study showed that feeding Hu sheep with the polyphenol-rich fruit *Nitraria tangutorum* could increase the abundance of *Rikenellaceae_RC9_gut_group* and *Christensenellaceae_R-7_group* [51]. In this study, a positive correlation was identified between *Rikenellaceae_RC9_gut_group* and butyric acid content. Loubet Filho et al. [52] observed that cyanidin-3-glucoside led to an augmentation in the abundance of *Muribaculaceae*, suggesting that the bacteria of the *Muribaculaceae* family may be the target of cyanidin-3-glucoside, potentially contributing to the positive effects of cyanidin-3-glucoside on metabolism and inflammation.

Firmicutes exhibit the capability to break down fibers into short-chain fatty acids. Within the Firmicutes phylum, *Christensenellaceae_R-7_group* is commonly found in the intestinal tract and mucosa of both humans and animals, and it plays a crucial role in preserving the health of the host [53]. Previous studies indicated that *Christensenellaceae_R-7_group* has the potential to stimulate the development of the rumen, improving the absorption and digestion of nutrients. They have also exhibited a positive correlation with the breakdown of dietary protein, a negative correlation with diseases, such as inflammation [54–56], and a positive correlation with nipple width, epithelial thickness, and stratum corneum thickness [57]. Certainly, it is worth noting that microorganisms with different taxonomic features may serve the same function. Conversely, microorganisms with the same taxonomic characteristics may fulfill distinct functions. For example, *NK4A214_group* (*Oscillospiraceae* family) and *Christensenellaceae_R-7_group* (*Christensenellaceae* family), despite not belonging to the same family, may contribute individually or cooperatively to the promotion of ruminal biohydrogenation [58]. *Lachnospiraceae_NK3A20_group* (*Lachnospiraceae* family) plays an important role in rumen development and the promotion of rumen fermentation by increasing butyrate production [57]. *Oribacterium* belongs to the *Lachnospiraceae* family and has been recognized as one of the predominant bacteria in the rumen of cows that are fed forage-based diets [59], but its role in the rumen has not been reported yet. In the present study, we found that polyphenols from DFPE increased the abundance of *Lachnospiraceae_NK3A20_group*, *Christensenellaceae_R-7_group*, and *NK4A214_group*, indicating that polyphenols from DFPE can regulate rumen fermentation and potentially promote ruminal epithelial development, which may explain the rise in butyrate. A similar result was found in a report by Xu et al. [60]. Polyphenols could potentially play a crucial role in safeguarding microorganisms from oxidative stress [16]. This seems to explain the increased abundance of multiple bacteria caused by DFPE.

Streptococcus serves as the primary lactic acid fermenter in the rumen, converting starch and lactic acid into acetic acid and propionic acid. Since lactic acid is a stronger acid than

VFA, the accumulation of lactate frequently causes a decrease in ruminal pH (5.0 or less) [61]. *Streptococcus* displays a notable tolerance to acidity, enabling them to thrive and multiply at low pH levels [62]. This characteristic may hinder the growth of other bacteria during such conditions [62]. On the other hand, the *Streptococcus* genus is recognized as one of the primary pathogens responsible for inducing mastitis [63]. In the present study, we reveal that DFPE had the ability to reduce the relative abundance of *Streptococcus*. The reason may be that polyphenols have the capability to engage with bacterial proteins, resulting in the inhibition of bacterial nucleic acid synthesis, the modulation of cell membrane function and fluidity, adjustments to cell wall integrity and synthesis, an impact on cell metabolism, and the prevention of biofilm formation [64]; in addition, polyphenols have the ability to increase the production of short-chain fatty acids and inhibit the growth of pathogenic bacteria [65]. Similarly, we also found *Streptococcus* had a negative correlation with the butyric acid content. *Enterococcus* is considered a probiotic, which contributes to the maintenance of lactic acid-utilizing bacteria activity, stimulates the growth of rumen microorganisms, and enhances the supply of gluco-propionic acid-producing energy for the host ruminants [66]. In the present study, we found that *Enterococcus* was decreased by DFPE. The mechanism for this is still unclear and requires future studies. A potential reason might be the increased relative abundance of certain microbes indirectly influencing the growth of other microbes [16].

5. Conclusions

The current study indicates that DFPE enhanced in vitro rumen fermentation serving as a potential ruminal modulator due to its ability to maintain ruminal pH, reduce gas production, decrease the concentration of NH_3-N and acetic acid, and increase the concentrations of total VFA, isovaleric acid, and valeric acid. It also indicates that DFPE enhanced in regulation of rumen bacteria in Sika deer. However, further studies are needed to elucidate how DFPE enhances tumen fermentation in in vivo feeding trials.

Author Contributions: Conceptualization, P.L.; data curation, C.B. and Q.L.; investigation, X.T. and G.W.; methodology, P.L. and G.W.; resources, C.B.; software, Q.L.; supervision, P.L.; writing—original draft, C.B.; writing—review and editing, X.T. and P.L. All authors have read and agreed to the published version of the manuscript.

Funding: This research received no external funding.

Institutional Review Board Statement: The study was conducted in accordance with the Declaration of Helsinki, and approved by the Rules of Animal Welfare and Experimental Animal Ethics of Guizhou University (EAE-GZU-2024-E001; 4 January 2024).

Informed Consent Statement: Not applicable.

Data Availability Statement: Raw data associated with each sample have been submitted to the National Center for Biotechnology Information under study accession number PRJNA1083787.

Conflicts of Interest: The authors declare no conflicts of interest.

References

1. Shults, M.A. *Maternal Reproductive Microbiome Influences on Pregnancy Status, Stage of Gestation and the Developing Rumen Microbiome in Cattle*; University of Wyoming: Laramie, WY, USA, 2022.
2. Henderson, G.; Cox, F.; Ganesh, S.; Jonker, A.; Young, W.; Global Rumen Census Collaborators; Janssen, P.H. Rumen microbial community composition varies with diet and host, but a core microbiome is found across a wide geographical range. *Sci. Rep.* **2015**, *5*, 14567. [CrossRef]
3. Vasta, V.; Daghio, M.; Cappucci, A.; Buccioni, A.; Serra, A.; Viti, C.; Mele, M. Invited review: Plant polyphenols and rumen microbiota responsible for fatty acid biohydrogenation, fiber digestion, and methane emission: Experimental evidence and methodological approaches. *J. Dairy Sci.* **2019**, *102*, 3781–3804. [CrossRef] [PubMed]
4. Vasta, V.; Luciano, G. The effects of dietary consumption of plants secondary compounds on small ruminants' products quality. *Small Rumin. Res.* **2011**, *101*, 150–159. [CrossRef]

5. Sun, H.; Zhao, F.; Hou, F.; Jin, Y.; Zhang, X.; Ma, Y.; Zhang, Y.; Fan, Y.; Yang, Z.; Wang, H. Influences of naringin supplementation on ruminal fermentation, inflammatory response, antioxidant capacity and bacterial community in high-concentrate diet of fattening goats. *Ital. J. Anim. Sci.* **2022**, *21*, 1498–1507. [CrossRef]
6. Tian, X.; Qin, J.; Luo, Q.; Xu, Y.; Xie, S.; Chen, R.; Wang, X.; Lu, Q. Differences in Chemical Composition, Polyphenol Compounds, Antioxidant Activity, and In Vitro Rumen Fermentation among Sorghum Stalks. *Animals* **2024**, *14*, 415. [CrossRef] [PubMed]
7. Matra, M.; Totakul, P.; Viennasay, B.; Phesatcha, B.; Wanapat, M. Dragon fruit (*Hylocereus undatus*) peel pellet as a rumen enhancer in Holstein crossbred bulls. *Anim. Biosci.* **2021**, *34*, 594. [CrossRef] [PubMed]
8. Matra, M.; Wanapat, M.; Cherdthong, A.; Foiklang, S.; Mapato, C. Dietary dragon fruit (*Hylocereus undatus*) peel powder improved in vitro rumen fermentation and gas production kinetics. *Trop. Anim. Health Prod.* **2019**, *51*, 1531–1538. [CrossRef] [PubMed]
9. Goodman, S.J.; Tamate, H.B.; Wilson, R.; Nagata, J.; Tatsuzawa, S.; Swanson, G.M.; Pemberton, J.M.; McCullough, D.R. Bottlenecks, drift and differentiation: The population structure and demographic history of sika deer (*Cervus nippon*) in the Japanese archipelago. *Mol. Ecol.* **2001**, *10*, 1357–1370. [CrossRef] [PubMed]
10. Hung, Y.-K.; Ho, S.-T.; Kuo, C.-Y.; Chen, M.-J. In vitro effects of velvet antler water extracts from Formosan Sambar deer and red deer on barrier integrity in Caco-2 cell. *Int. J. Med. Sci.* **2021**, *18*, 1778. [CrossRef]
11. Li, Z.P.; Liu, H.L.; Li, G.Y.; Bao, K.; Wang, K.Y.; Xu, C.; Yang, Y.F.; Yang, F.H.; Wright, A.-D.G. Molecular diversity of rumen bacterial communities from tannin-rich and fiber-rich forage fed domestic Sika deer (*Cervus nippon*) in China. *BMC Microbiol.* **2013**, *13*, 151. [CrossRef]
12. Committee on Nutrient Requirements of Small Ruminants. *Nutrient Requirements of Small Ruminants: Sheep, Goats, Cervids, and New World Camelids*; China Legal Publishing House: Beijing, China, 2007.
13. Kh, M. Estimation of the energetic feed value obtained from chemical analysis and in vitro gas production using rumen fluid. *Anim. Res. Dev.* **1988**, *28*, 7–55.
14. Yang, K.; Deng, X.; Jian, S.; Zhang, M.; Wen, C.; Xin, Z.; Zhang, L.; Tong, A.; Ye, S.; Liao, P.; et al. Gallic acid alleviates gut dysfunction and boosts immune and antioxidant activities in puppies under environmental stress based on microbiome–metabolomics analysis. *Front. Immunol.* **2022**, *12*, 813890. [CrossRef] [PubMed]
15. Horwitz, W. *Official Methods of Analysis of AOAC International. Volume I, Agricultural Chemicals, Contaminants, Drugs*; Horwitz, W., Ed.; AOAC International: Gaithersburg, MD, USA, 2010.
16. Tian, X.-Z.; Li, J.-X.; Luo, Q.-Y.; Zhou, D.; Long, Q.-M.; Wang, X.; Lu, Q.; Wen, G.-L. Effects of purple corn anthocyanin on blood biochemical indexes, ruminal fluid fermentation, and rumen microbiota in goats. *Front. Vet. Sci.* **2021**, *8*, 715710. [CrossRef] [PubMed]
17. Van Soest, P.; Robertson, J.B.; Lewis, B.A. Methods for dietary fiber, neutral detergent fiber, and nonstarch polysaccharides in relation to animal nutrition. *J. Dairy Sci.* **1991**, *74*, 3583–3597. [CrossRef] [PubMed]
18. Zhang, J.; Yang, Y.; Lei, X.; Wang, Y.; Li, Y.; Yang, Z.; Yao, J. Active dry yeast supplementation benefits ruminal fermentation, bacterial community, blood immunoglobulins, and growth performance in young dairy goats, but not for intermittent supplementation. *Anim. Nutr.* **2023**, *13*, 289–301. [CrossRef]
19. De Boever, J.; Aerts, J.; Vanacker, J.; De Brabander, D. Evaluation of the nutritive value of maize silages using a gas production technique. *Anim. Feed Sci. Technol.* **2005**, *123*, 255–265. [CrossRef]
20. Clarke, R.; Reid, C. Foamy bloat of cattle. A review. *J. Dairy Sci.* **1974**, *57*, 753–785. [CrossRef] [PubMed]
21. Tian, X.-Z.; Paengkoum, P.; Paengkoum, S.; Thongpea, S.; Ban, C. Comparison of forage yield, silage fermentative quality, anthocyanin stability, antioxidant activity, and in vitro rumen fermentation of anthocyanin-rich purple corn (*Zea mays* L.) stover and sticky corn stover. *J. Integr. Agric.* **2018**, *17*, 2082–2095. [CrossRef]
22. Jafari, S.; Ebrahimi, M.; Goh, Y.M.; Rajion, M.A.; Jahromi, M.F.; Al-Jumaili, W.S. Manipulation of rumen fermentation and methane gas production by plant secondary metabolites (saponin, tannin and essential oil)—A review of ten-year studies. *Ann. Anim. Sci.* **2019**, *19*, 3–29. [CrossRef]
23. Yu, S.; Zhao, Y.; Li, L.; Zhao, H.; Liu, M.; Jiang, L. Flavonoids from citrus peel display potential synergistic effects on inhibiting rumen methanogenesis and ammoniagenesis: A microbiome perspective. *Environ. Sci. Pollut. Res.* **2024**, *31*, 21208–21223. [CrossRef]
24. Dai, S.; Lian, Z.; Qi, W.; Chen, Y.; Tong, X.; Tian, T.; Lyu, B.; Wang, M.; Wang, H.; Jiang, L. Non-covalent interaction of soy protein isolate and catechin: Mechanism and effects on protein conformation. *Food Chem.* **2022**, *384*, 132507. [CrossRef] [PubMed]
25. McMahon, L.R.; McAllister, T.A.; Berg, B.P.; Majak, W.; Acharya, S.N.; Popp, J.D.; Coulman, B.E.; Wang, Y.; Cheng, K.-J. A review of the effects of forage condensed tannins on ruminal fermentation and bloat in grazing cattle. *Can. J. Plant Sci.* **2000**, *80*, 469–485. [CrossRef]
26. Liu, H.; Li, K.; Zhao, J.; Deng, W. Effects of polyphenolic extract from Eucommia ulmoides Oliver leaf on growth performance, digestibility, rumen fermentation and antioxidant status of fattening lambs. *Anim. Sci. J.* **2018**, *89*, 888–894. [CrossRef] [PubMed]
27. Correddu, F.; Lunesu, M.F.; Buffa, G.; Atzori, A.S.; Nudda, A.; Battacone, G.; Pulina, G. Can agro-industrial by-products rich in polyphenols be advantageously used in the feeding and nutrition of dairy small ruminants? *Animals* **2020**, *10*, 131. [CrossRef] [PubMed]
28. Jackson, P.G.; Cockcroft, P.D.; Elmhurst, S. *Clinical Examination of Farm Animals*; Wiley Online Library: Hoboken, NJ, USA, 2002; Volume 331.

29. Banakar, P.; Shashank, C.; Lakhani, N. Physically effective fibre in ruminant nutrition: A review. *J. Pharmacogn. Phytochem.* **2018**, *7*, 303–308.
30. De Nardi, R.; Marchesini, G.; Li, S.; Khafipour, E.; Plaizier, K.J.; Gianesella, M.; Segato, S. Metagenomic analysis of rumen microbial population in dairy heifers fed a high grain diet supplemented with dicarboxylic acids or polyphenols. *BMC Vet. Res.* **2016**, *12*, 29. [CrossRef] [PubMed]
31. Lee, S.J.; Kim, H.S.; Eom, J.S.; Choi, Y.Y.; Jo, S.U.; Chu, G.M.; Lee, S.S. Effects of Olive (*Olea europaea* L.) leaves with antioxidant and antimicrobial activities on in Vitro ruminal fermentation and methane emission. *Animals* **2021**, *11*, 2008. [CrossRef]
32. Gao, J.; Cheng, B.; Liu, Y.; Li, M.; Zhao, G. Effects of red cabbage extract rich in anthocyanins on rumen fermentation, rumen bacterial community, nutrient digestion, and plasma indices in beef bulls. *Animal* **2022**, *16*, 100510. [CrossRef]
33. Nørskov, N.P.; Battelli, M.; Curtasu, M.V.; Olijhoek, D.W.; Chassé, É.; Nielsen, M.O. Methane reduction by quercetin, tannic and salicylic acids: Influence of molecular structures on methane formation and fermentation in vitro. *Sci. Rep.* **2023**, *13*, 16023. [CrossRef]
34. Lourenço, M.; Cardozo, P.W.; Calsamiglia, S.; Fievez, V. Effects of saponins, quercetin, eugenol, and cinnamaldehyde on fatty acid biohydrogenation of forage polyunsaturated fatty acids in dual-flow continuous culture fermenters. *J. Anim. Sci.* **2008**, *86*, 3045–3053. [CrossRef]
35. Sun, W.; Shi, H.; Gong, C.; Liu, K.; Li, G. Effects of Different Yeast Selenium Levels on Rumen Fermentation Parameters, Digestive Enzyme Activity and Gastrointestinal Microflora of Sika Deer during Antler Growth. *Microorganisms* **2023**, *11*, 1444. [CrossRef] [PubMed]
36. Martínez-Fernández, G.; Abecia, L.; Martín-García, A.I.; Ramos-Morales, E.; Hervás, G.; Molina-Alcaide, E.; Yáñez-Ruiz, D.R. In vitro–in vivo study on the effects of plant compounds on rumen fermentation, microbial abundances and methane emissions in goats. *Animal* **2013**, *7*, 1925–1934. [CrossRef] [PubMed]
37. Bergman, E. Energy contributions of volatile fatty acids from the gastrointestinal tract in various species. *Physiol. Rev.* **1990**, *70*, 567–590. [CrossRef] [PubMed]
38. Alabi, J.O.; Okedoyin, D.O.; Anotaenwere, C.C.; Wuaku, M.; Gray, D.; Adelusi, O.O.; Ike, K.A.; Olagunju, L.K.; Dele, P.A.; Anele, U.Y. Essential Oil Blends with or without Fumaric Acid Influenced In Vitro Rumen Fermentation, Greenhouse Gas Emission, and Volatile Fatty Acids Production of a Total Mixed Ration. *Ruminants* **2023**, *3*, 373–384. [CrossRef]
39. Remling, N.; Riede, S.; Meyer, U.; Beineke, A.; Breves, G.; Flachowsky, G.; Dänicke, S. Influence of fumaric acid on ruminal parameters and organ weights of growing bulls fed with grass or maize silage. *Animal* **2017**, *11*, 1754–1761. [CrossRef] [PubMed]
40. e Silva, S.; Chabrillat, T.; Kerros, S.; Guillaume, S.; Gandra, J.; de Carvalho, G.; da Silva, F.; Mesquita, L.; Gordiano, L.; Camargo, G.; et al. Effects of plant extract supplementations or monensin on nutrient intake, digestibility, ruminal fermentation and metabolism in dairy cows. *Anim. Feed Sci. Technol.* **2021**, *275*, 114886. [CrossRef]
41. Yu, S.; Li, L.; Zhao, H.; Tu, Y.; Liu, M.; Jiang, L.; Zhao, Y. Characterization of the Dynamic Changes of Ruminal Microbiota Colonizing Citrus Pomace Waste during Rumen Incubation for Volatile Fatty Acid Production. *Microbiol. Spectr.* **2023**, *11*, e03517-22. [CrossRef] [PubMed]
42. Eto, M.; Yahara, T.; Kuroiwa, A.; Shioya, K.; Flores, G.E.; Hamamura, N. Dynamics of rumen microbiome in sika deer (*Cervus nippon yakushimae*) from unique subtropical ecosystem in Yakushima Island, Japan. *Sci. Rep.* **2022**, *12*, 21623. [CrossRef]
43. Wu, Y.; Yao, Y.; Dong, M.; Xia, T.; Li, D.; Xie, M.; Wu, J.; Wen, A.; Wang, Q.; Zhu, G.; et al. Characterisation of the gut microbial community of rhesus macaques in high-altitude environments. *BMC Microbiol.* **2020**, *20*, 68. [CrossRef]
44. Jami, E.; White, B.A.; Mizrahi, I. Potential role of the bovine rumen microbiome in modulating milk composition and feed efficiency. *PLoS ONE* **2014**, *9*, e85423. [CrossRef]
45. Alves-Santos, A.M.; Sugizaki, C.S.A.; Lima, G.C.; Naves, M.M.V. Prebiotic effect of dietary polyphenols: A systematic review. *J. Funct. Foods* **2020**, *74*, 104169. [CrossRef]
46. Sutoyo, D.A.; Atmaka, D.R.; Sidabutar, L. Dietary factors affecting firmicutes and bacteroidetes ratio in solving obesity problem: A literature review. *Media Gizi Indones* **2020**, *15*, 94–109. [CrossRef]
47. Si, H.; Liu, H.; Nan, W.; Li, G.; Li, Z.; Lou, Y. Effects of arginine supplementation on serum metabolites and the rumen bacterial community of sika deer (*Cervus nippon*). *Front. Vet. Sci.* **2021**, *8*, 630686. [CrossRef] [PubMed]
48. Graf, J. The Family Rikenellaceae. In *The Prokaryotes*; Rosenberg, E., DeLong, E., Lory, S., Stackebrandt, E., Thompson, F., Eds.; Springer: Berlin/Heidelberg, Germany, 2014.
49. Cai, Y.; Luo, Y.; Dai, N.; Yang, Y.; He, Y.; Chen, H.; Xing, Z. Functional metagenomic and metabolomics analysis of gut dysbiosis induced by hyperoxia. *Front. Microbiol.* **2023**, *14*, 1197970. [CrossRef] [PubMed]
50. Lagkouvardos, I.; Lesker, T.R.; Hitch, T.C.A.; Gálvez, E.J.C.; Smit, N.; Neuhaus, K.; Wang, J.; Baines, J.F.; Abt, B.; Stecher, B.; et al. Sequence and cultivation study of Muribaculaceae reveals novel species, host preference, and functional potential of this yet undescribed family. *Microbiome* **2019**, *7*, 28. [CrossRef] [PubMed]
51. Du, X.; Cheng, X.; Dong, Q.; Zhou, J.; Degen, A.A.; Jiao, D.; Ji, K.; Liang, Y.; Wu, X.; Yang, G. Dietary Supplementation of Fruit from Nitraria tangutorum Improved Immunity and Abundance of Beneficial Ruminal Bacteria in Hu Sheep. *Animals* **2022**, *12*, 3211. [CrossRef] [PubMed]
52. Filho, P.S.L.; Baseggio, A.M.; Vuolo, M.M.; Reguengo, L.M.; Biasoto, A.C.T.; Correa, L.C.; Junior, S.B.; Cagnon, V.H.A.; Cazarin, C.B.B.; Júnior, M.R.M. Gut microbiota modulation by jabuticaba peel and its effect on glucose metabolism via inflammatory signaling. *Curr. Res. Food Sci.* **2022**, *5*, 382–391. [CrossRef]

53. Chen, M.; Xie, W.; Zhou, S.; Ma, N.; Wang, Y.; Huang, J.; Shen, X.; Chang, G. A high-concentrate diet induces colonic inflammation and barrier damage in Hu sheep. *J. Dairy Sci.* **2023**, *106*, 9644–9662. [CrossRef] [PubMed]
54. Beaumont, M.; Portune, K.J.; Steuer, N.; Lan, A.; Cerrudo, V.; Audebert, M.; Dumont, F.; Mancano, G.; Khodorova, N.; Andriamihaja, M.; et al. Quantity and source of dietary protein influence metabolite production by gut microbiota and rectal mucosa gene expression: A randomized, parallel, double-blind trial in overweight humans. *Am. J. Clin. Nutr.* **2017**, *106*, 1005–1019. [CrossRef]
55. Couch, C.E.; Arnold, H.K.; Crowhurst, R.S.; Jolles, A.E.; Sharpton, T.J.; Witczak, M.F.; Epps, C.W.; Beechler, B.R. Bighorn sheep gut microbiomes associate with genetic and spatial structure across a metapopulation. *Sci. Rep.* **2020**, *10*, 6582. [CrossRef]
56. Waters, J.L.; Ley, R.E. The human gut bacteria Christensenellaceae are widespread, heritable, and associated with health. *BMC Biol.* **2019**, *17*, 83. [CrossRef] [PubMed]
57. Huang, C.; Ge, F.; Yao, X.; Guo, X.; Bao, P.; Ma, X.; Wu, X.; Chu, M.; Yan, P.; Liang, C. Microbiome and metabolomics reveal the effects of different feeding systems on the growth and ruminal development of yaks. *Front. Microbiol.* **2021**, *12*, 682989. [CrossRef] [PubMed]
58. Huang, G.; Guo, L.; Chang, X.; Liu, K.; Tang, W.; Zheng, N.; Zhao, S.; Zhang, Y.; Wang, J. Effect of whole or ground flaxseed supplementation on fatty acid profile, fermentation, and bacterial composition in rumen of dairy cows. *Front. Microbiol.* **2021**, *12*, 760528. [CrossRef]
59. Kong, Y.; Teather, R.; Forster, R. Composition, spatial distribution, and diversity of the bacterial communities in the rumen of cows fed different forages. *FEMS Microbiol. Ecol.* **2010**, *74*, 612–622. [CrossRef]
60. Xu, H.; Zhang, Q.; Wang, L.; Zhang, C.; Li, Y.; Zhang, Y. Growth performance, digestibility, blood metabolites, ruminal fermentation, and bacterial communities in response to the inclusion of gallic acid in the starter feed of preweaning dairy calves. *J. Dairy Sci.* **2022**, *105*, 3078–3089. [CrossRef]
61. Asanuma, N.; Hino, T. Regulation of fermentation in a ruminal bacterium, Streptococcus bovis, with special reference to rumen acidosis. *Anim. Sci. J.* **2002**, *73*, 313–325. [CrossRef]
62. Russell, J.B.; Dombrowski, D. Effect of pH on the efficiency of growth by pure cultures of rumen bacteria in continuous culture. *Appl. Environ. Microbiol.* **1980**, *39*, 604–610. [CrossRef]
63. Kabelitz, T.; Aubry, E.; Van Vorst, K.; Amon, T.; Fulde, M. The Role of *Streptococcus* spp. in Bovine Mastitis. *Microorganisms* **2021**, *9*, 1497. [CrossRef] [PubMed]
64. Makarewicz, M.; Drożdż, I.; Tarko, T.; Duda-Chodak, A. The interactions between polyphenols and microorganisms, especially gut microbiota. *Antioxidants* **2021**, *10*, 188. [CrossRef]
65. Quatrin, A.; Rampelotto, C.; Pauletto, R.; Maurer, L.H.; Nichelle, S.M.; Klein, B.; Rodrigues, R.F.; Junior, M.R.M.; Fonseca, B.d.S.; de Menezes, C.R.; et al. Bioaccessibility and catabolism of phenolic compounds from jaboticaba (*Myrciaria trunciflora*) fruit peel during in vitro gastrointestinal digestion and colonic fermentation. *J. Funct. Foods* **2020**, *65*, 103714. [CrossRef]
66. Pang, D.; Yang, H.; Cao, B.; Wu, T.; Wang, J. The beneficial effect of Enterococcus faecium on the in vitro ruminal fermentation rate and extent of three typical total mixed rations in northern China. *Livest. Sci.* **2014**, *167*, 154–160. [CrossRef]

Disclaimer/Publisher's Note: The statements, opinions and data contained in all publications are solely those of the individual author(s) and contributor(s) and not of MDPI and/or the editor(s). MDPI and/or the editor(s) disclaim responsibility for any injury to people or property resulting from any ideas, methods, instructions or products referred to in the content.

animals

MDPI

Article

Effect of Guanidinoacetic Acid on Production Performance, Serum Biochemistry, Meat Quality and Rumen Fermentation in Hu Sheep

Huayun Jin, Zhijian Du, Xiaoyu Fan, Liwen Qin, Weiwei Liu, Yan Zhang, Jingnan Ren, Changchuan Ye * and Qinghua Liu *

Department of Animal Science, College of Animal Science, Fujian Agriculture and Forestry University, Fuzhou 350002, China; jhy001x@163.com (H.J.); 17607848230@163.com (Z.D.); 17852396033@163.com (X.F.); 18706011090@163.com (L.Q.); lww990609@163.com (W.L.); zhangyan2990@163.com (Y.Z.); 15392547223@163.com (J.R.)

* Correspondence: yecc@fafu.edu.cn (C.Y.); 83793089@163.com (Q.L.)

Simple Summary: Recently, the escalating prices of feedstuff resources have posed significant challenges to animal husbandry, necessitating the creation of innovative approaches to mitigate its effects. GAA (also known as N-amidinoglycine) serves as the sole precursor for creatine synthesis. GAA could enhance growth performance and regulate fat deposition to improve meat quality. Current studies on GAA mainly center on pigs and chickens, with limited reports on its effects on sheep. In our study, we used weaned Hu sheep as subjects. We aimed to explore the impact of dietary GAA supplementation on production performance, serum biochemistry, meat quality and rumen fermentation. This study provides a theoretical foundation for utilizing GAA in sheep meat production. Our findings indicate that GAA enhances growth performance, rumen fermentation and fat accumulation. In addition, GAA might enhance the quality of sheep meat, which presents a promising prospect for meat sheep breeding.

Abstract: Guanidinoacetic acid (GAA) can effectively improve the metabolism of energy and proteins by stimulating creatine biosynthesis. We present a study exploring the impact of GAA on production performance, serum biochemistry, meat quality and rumen fermentation in Hu sheep. A total of 144 weaned male Hu sheep (body weight 16.91 ± 3.1 kg) were randomly assigned to four groups with three replicates of twelve sheep in each group. The diets were supplemented with 0 (CON), 500 (GAA−1), 750 (GAA−2) and 1000 mg/kg (GAA−3) of GAA (weight of feed), respectively. After a comprehensive 90-day experimental period, we discovered that the supplementation of GAA had a remarkable impact on various muscle parameters. Specifically, it significantly enhanced the average daily growth (ADG) of the animals and improved the shear force and fiber diameter of the muscle, while also reducing the drip loss and muscle fiber density. Furthermore, the addition of GAA to the feed notably elevated the serum concentrations of high-density lipoprotein cholesterol (HDL−C), total protein (TP) and globulin (GLB), as well as the enzyme activity of superoxide dismutase (SOD) and glutathione peroxidase (GSH−Px). Concurrently, there was a decrease in the levels of triglycerides (TG) and malondialdehyde (MDA) in the serum. In addition, GAA decreased the pH and the acetate-to-propionate ratio and increased the total volatile fatty acids (TVFA) and ammoniacal nitrogen (NH_3-N) levels of rumen fluid. Additionally, GAA upregulated acetyl-CoA carboxylase (ACC) gene expression in the Hu sheep's muscles. In conclusion, our findings suggest that GAA supplementation not only enhances muscle quality but also positively affects serum biochemistry and ruminal metabolism, making it a potential candidate for improving the overall health and performance of Hu sheep.

Keywords: Hu sheep; guanidinoacetic acid; production performance; rumen fermentation; fatty acid metabolism

Citation: Jin, H.; Du, Z.; Fan, X.; Qin, L.; Liu, W.; Zhang, Y.; Ren, J.; Ye, C.; Liu, Q. Effect of Guanidinoacetic Acid on Production Performance, Serum Biochemistry, Meat Quality and Rumen Fermentation in Hu Sheep. *Animals* **2024**, *14*, 2052. https://doi.org/10.3390/ani14142052

Academic Editors: Giuseppe Pulina and Gonzalo Hervás

Received: 30 April 2024
Revised: 8 July 2024
Accepted: 10 July 2024
Published: 12 July 2024

Copyright: © 2024 by the authors. Licensee MDPI, Basel, Switzerland. This article is an open access article distributed under the terms and conditions of the Creative Commons Attribution (CC BY) license (https://creativecommons.org/licenses/by/4.0/).

1. Introduction

Recently, the escalating prices of feedstuff resources have significantly affected animal husbandry, particularly intensive sheep farming in China. This trend has posed significant challenges to the industry, necessitating the creation of innovative approaches to mitigate its effects. Guanidinoacetic acid (GAA) is a metabolic intermediate synthesized from arginine and glycine [1]. It has been reported that GAA could effectively enhance the metabolism of energy and proteins by stimulating creatine biosynthesis, thereby contributing to improved nutritional utilization and animal performance [2,3]. In monogastric animals, dietary supplement of GAA enhances feed intake and efficiency, growth performance, antioxidant status and meat quality [2,4,5]. Studies have found that adding GAA to the diet could have a regulatory effect on the growth performance, rumen fermentation and fat deposition of ruminants [6,7].

GAA is also known as N-amidinoglycine, which serves as the only precursor for creatine synthesis. As the primary stored form of phosphocreatine within skeletal muscle, creatine serves as a crucial component for recycling ATP and supporting various cellular functions. GAA and creatine are both amino acid derivatives derived from animals. Since plant proteins lack GAA and creatine, ruminants acquire creatine only through de novo synthesis [8]. In living organisms, endogenous creatine biosynthesis occurs via the kidney–liver axis. In this pathway, methionine, glycine and arginine are the substrates for the biosynthesis of creatine [9]. The high-energy phosphate bonds stored in muscles are crucial for muscle activity. Due to the considerable requirement of arginine for protein synthesis, endogenous creatine synthesis places a heavy metabolic burden on arginine and methionine [10]. Ardalan et al. found that post-ruminal GAA supplementation improved creatine availability in cattle [11]. Dietary supplementation with GAA in livestock (especially in broilers) can benefit animal growth performance, enhance muscle development and improve animal health [12–14]. A previous study suggested that GAA supplementation in broilers can enhance carcass yield and lower the levels of abdominal fat [15]. Li et al. found that dietary supplementation with GAA in finishing pigs raised pH45min and reduced drip loss and shear force [16]. Mendoza et al. revealed that GAA supplementation for gestating and lactating sows enhanced litter weight and weaning numbers without compromising their reproductive performance [17]. Recently, researchers have increasingly focused on the effects of GAA in ruminants. Studies conducted on bulls have revealed that GAA supplementation within the diet enhanced daily gain, nutrient digestion, blood creatine level and liver protein synthesis gene expression [18,19]. Another study found that dietary supplementation with GAA in lambs raised the concentrations of phosphocreatine and ATP in the muscle [20]. These findings indicated that GAA may act a pivotal role in promoting ruminant growth and overall health.

With the proposed conditions of use, GAA effectively improves the production performance of poultry and pigs without affecting consumer safety [21,22]. Currently, research on GAA in ruminants is scarce, and no safety concerns have been reported. In addition, few studies have been conducted on the use of GAA in sheep with a diet made primarily from plant-based proteins. Due to there being limited information on the effects of dietary GAA supplementation on sheep's growth performance and fat deposition, there was a gap left in our understanding of its potential benefits and interactions within this nutritional context. Therefore, we conducted an experiment to explore the possibility of GAA application in lamb meat production. In this study, we aimed to explore the impacts of GAA on Hu sheep's production, rumen fermentation and fatty acid metabolism. We believe that our result will provide insights into the potential benefits and optimal dosage of GAA in meat sheep production. Furthermore, this research will provide valuable insights into improving the overall welfare and productivity of livestock.

2. Materials and Methods

2.1. Moral Statement

The study was approved by the Fujian Agriculture and Forestry University Animal Care and Use Committee (Approval ID: PZCASFAFU23012).

2.2. Experimental Design and Feeding Management

Weaned Hu sheep were selected as experimental animals in this experiment. The experiments were conducted at the sheep farm of Longlin Husbandry Co., Ltd., Ningde City, China. This experiment was performed from March to June in 2023. A total of 144 weaned male Hu sheep (average initial body weight: 16.91 ± 3.10 kg; average age: 3 months) were used in a 90-day performance trial. The sheep were assigned to four treatments in a completely randomized block design. Within each block, the twelve sheep were randomly assigned to one of the four treatment groups using a random number generator to ensure unbiased allocation. Body weight was used as a block factor to control its potential impact on the experimental results. Similar body weight distribution was ensured within each block and sheep were randomly assigned to the groups. Each treatment was replicated three times across the different blocks, totaling 36 sheep per treatment (including control group) and 144 sheep in the entire study. The sheep were fed with a basal diet supplemented with 0 (CON group), 500 (GAA−1 group), 750 (GAA−2 group) and 1000 (GAA−3 group) $mg \cdot kg^{-1}$ GAA product (weight of feed), respectively. Table 1 showed the feed composition and nutrient levels for the experimental ration.

Table 1. Experimental diet composition and nutrient level (Dry matter basis, %). The nutritional standards refer to the nutritional requirements for sheep according to NRC (2007) [23].

Dietary Ingredient	Contents
Bamboo shoot shells silage	60
Corn	17
Bran	12
Soybean meal	6
NaCl	1
$CaHCO_3$	1.4
$NaHCO_3$	1.2
Premix (1)	1.4
Total:	100
Nutrient levels (2)	
Metabolizable energy (MJ/Kg)	9.68
Crude protein	11.55
Neutral detergent fiber	58.46
Acid detergent fiber	36.32
Calcium	0.75
Phosphorus	0.50

(1) The premix supplied Cu 20.0 mg, Fe 75.0 mg, Mn 30.0 mg, Zn 80.0 mg, I 1 mg, Se 0.50 mg, Vitamin A 20,000 IU, Vitamin D 5000 IU, and Vitamin E 50.0 mg per kg of diet. (2) Metabolizable energy (ME) were calculated values. The values of CP (crude protein), NDF (neutral detergent fiber), ADF (acid detergent fiber), Ca (calcium) and P (phosphorus) were measured and recorded. CP was determined by the Kjeldahl method; NDF and ADF were measured using the filter bag technique; Ca was determined by the ethylenediaminetetraacetic acid (EDTA) complexometric titration method; and P was measured by spectrophotometry at a wavelength of 400 nm with the molybdenum blue reaction.

The sheep were fed at 08:00 and 16:00 daily. The GAA product labeled "Saint-Lo" (with GAA content over 50% with the remaining comprising a diluent) was acquired from Beijing Gendone Technology Co., Ltd. (Beijing, China). It was pre-mixed into the concentrate for feeding as detailed. Referring to a previous study [20] and our preliminary experiment, the sheep were given a seven-day adaptation period for the diet. The sheep were provided with feed and water ad libitum. Feed intake was tracked on a daily basis. Residual feed from the previous day of each treatment group was collected and weighed before feeding every morning. Disinfection and immunization were conducted according to routine procedures of the farm. As for foot-and-mouth disease, sheep receive their first immunization at 28–35 days of age, followed by a booster shot one month later and then booster immunizations every 4–6 months. As for brucellosis, immunization is conducted at 3–4 months of age. As for coenurosis, sheep receive their first immunization at 3–4 months of age, followed by a booster shot one month later and then annual booster immunizations.

2.3. Growth Performance

During each replication, sheep were weighed on the 90th day, and the total feed intake of each replicate was recorded. Average daily feed intake (ADFI), average daily gain (ADG) and feed/gain ratio (F/G) were calculated for each sheep.

2.4. Sample Collection

Three sheep from each group were randomly selected for fasting and weighing. Blood samples from the jugular vein were stored at 4 °C for 2 h, then centrifuged at 3000× g for 15 min to separate serum. The collected serum samples were stored at −20 °C for the determination of serum biochemical indexes and antioxidant indexes. Then, these sheep were slaughtered, sampled and analyzed. The longissimus dorsi was selected as the muscle that would be sampled for slicing and analysis. Except for the muscle tissue samples used for the slicing and the serum samples, all samples were refrigerated at −80 °C before detection. Measurement indexes included live weight, carcass weight, dressing percentage and eye muscle area. Dressing percentage is calculated as (carcass weight/live weight) × 100%.

2.5. Meat Quality and Histological Analyses

Muscle pH was measured at three spots on the longissimus dorsi muscle using a hand-held pH meter (PHSJ−3F, Shanghai, China). Drip loss was determined by hanging a loin section from a 5 cm × 3 cm × 2 cm longissimus dorsi sample in a sealed plastic bag for 24 h at 4 °C [16]. Shear force was determined by cutting a 3 cm × 1 cm × 1 cm piece of dorsal longissimus muscle and testing it with a digital muscle tenderness tester (C-LM3B, Tenovo, Beijing, China). Intramuscular fat was measured by extraction with petroleum ether.

Muscle samples were fixed, embedded in paraffin, sectioned and mounted on slides. Then, the sections were deparaffinized, rehydrated and stained with hematoxylin and eosin. After glass coverslip mounting, the H&E-stained cross-sections were imaged using a microscope (Nikon Eclipse Ci-L, Tokyo, Japan). The histological analyses of muscle were in accordance with the method of Deng [24]. The muscle fiber diameter was measured by using Image-Pro Plus 6.0 analysis software after imaging, with millimeter as the standard unit, and the diameters of 5 muscle fibers from each slice were measured (mm). The total number of muscle fibers was measured by calculating the total number of muscle fibers in each image. The total muscle fiber area was measured by quantifying the total area (in mm^2) of muscle fibers in each image. The formula for calculating muscle fiber density is Muscle Fiber Density (fibers/mm^2) = Total Number of Muscle Fibers (fibers)/Total Area of Muscle Fibers (mm^2).

2.6. Serum Biochemical Indexes and Antioxidant Capacity

The serum biochemical parameters were determined using a fully automatic biochemical analysis system (BS−200, Mindray Biomedical Electronics Co., Ltd., Shenzhen, China). The corresponding kits were purchased from Hunan Yonghe Sunshine Technology Co., Ltd., Changsha, China. A comprehensive set of biochemical parameters were measured, including total protein (TP), triglyceride (TG), albumin (ALB), globulin (GLB), high-density lipoprotein cholesterol (HDL−C), total cholesterol (TC) and low-density lipoprotein cholesterol (LDL−C).

Serum superoxide dismutase (SOD), serum glutathione peroxidase (GSH−PX) activity and serum malondialdehyde (MDA) content were measured by using the corresponding test kit (Shanghai Liquid Quality Testing Technology Co., Ltd., Shanghai, China).

2.7. Rumen Fermentation

The rumen fluid was collected immediately after slaughter (less than 20 min). We measured the pH of ruminal fluid samples immediately using a pH meter (PHS−25C; Hangzhou Qiwei Instrument Co., Ltd., Hangzhou, China). Then, the rumen fluid was collected in a 10 mL centrifuge tube after being filtered through four layers of sterile gauze. Then, the fluid was stored at −80 °C.

The contents of volatile fatty acids (VFA), including isobutyric acid, acetic acid, isovaleric acid, butyric acid and propionic acid, were measured using a gas chromatograph (Shimadzu GC 2010-FID, Tokyo, Japan) [25,26]. The gas chromatograph system was equipped with a column (KB-FFAP, Kromat, Delran, NJ, USA), and a flame ionization detector was operated at 250 °C. A 0.81 mL/min mobile phase using N_2 was applied to the column. The column was operated at 80 °C.

Ammoniacal nitrogen (NH_3-N) content was measured using phenol-sodium hypochlorite colorimetry, following the Broderick and Kang method [27].

2.8. Quantitative Real-Time PCR for Gene Expression

After slaughter, longissimus dorsi muscle samples were quickly removed from the Hu sheep. These samples were put into 2 mL sterilization centrifuge tubes and frozen in liquid nitrogen. These samples were stored at −80 °C for subsequent analysis.

Muscular total RNAs were isolated using the NucleoZol reagent (Gene, Düren, Germany) and cDNA was synthesized using reverse transcription reagent kits (Servicebio, G3330−100, Wuhan, China). Quantitative real-time PCR was performed with ABI 7300 Real-Time PCR System (Applied Biosystems, Foster City, CA, USA). Primer sequences for hormone-sensitive lipase (HSL), fatty acid synthase (FAS), acetyl-CoA carboxylase (ACC) and the reference gene β-Actin (ACTB) are listed in Table 2. Selected samples from each process were analyzed in triplicate, utilizing ACTB as the reference gene. RNA expression was quantified using the $2^{-\Delta\Delta CT}$ method [28].

Table 2. Nucleotide sequences of primers for quantitative real-time PCR assay.

Items	Primer Sequences (5′−3′)	Genebank Number	Product Length (bp)
ACC	F: CCTGTCCGCCATTGACAT R: TAGGCGATATAAGCCCTTCG	NM_001009256.1	171
FAS	F: CAGTCGGTTGGATCGAGCAT R: AGAAGGAGGGTGGCTTTTGG	XM_015098375.1	151
HSL	F: TACAAACGCAACGAGACTGG R: ACGATAGCACCTGGATCTCG	NM_001128154.1	172
ACTB	F: CCCTGGAGAAGAGCTACGAG R: GGTAGTTTCGTGAATGCCGC	NM_001009784.3	131

2.9. Statistics Analysis

Statistical analyses were undertaken using SPSS 27.0 (SPSS, Inc., Chicago, IL, USA). Data analysis was conducted using one-way analysis of variance (ANOVA). The CON group (without GAA treatment) was set as control. When the ANOVA indicated significant differences among groups, Tukey's multiple range tests were further employed to determine which specific pairs of groups were significantly different. The results are presented as the mean ± standard deviation (SD) and significance was declared at $p \leq 0.05$.

3. Results

3.1. Effects of GAA Supplementation on Growth Performance, Carcass Traits and Meat Quality

According to Table 3, dietary supplementation with 750 and 1000 mg·kg^{-1} GAA (GAA−2 and GAA−3 groups, respectively) significantly increased the ADG of Hu sheep compared to the CON group ($p < 0.05$). However, the F/G and ADFI among these groups did not show significantly differences ($p > 0.05$).

Table 3. Growth performance and carcass traits of Hu sheep following GAA supplementation.

Items	Groups				*p*-Value
	CON	GAA−1	GAA−2	GAA−3	
Growth performance					
Initial BW (kg)	17.45 ± 2.53	17.66 ± 3.28	17.99 ± 3.04	17.71 ± 2.79	0.960
Final BW (kg)	28.05 ± 2.60	28.75 ± 3.01	29.45 ± 3.65	29.42 ± 3.20	0.872
ADFI (kg)	1.17 ± 0.17	1.17 ± 0.22	1.26 ± 0.21	1.24 ± 0.23	0.934
ADG (g)	117.85 ± 1.57 [b]	123.31 ± 3.76 [ab]	127.43 ± 3.24 [a]	130.12 ± 2.39 [a]	0.021
F/G	9.71 ± 1.41	9.51 ± 1.81	10.15 ± 1.71	9.95 ± 1.85	0.969
Carcass traits					
Live weight (kg)	28.50 ± 0.50	28.50 ± 0.50	28.83 ± 0.29	29.00 ± 0.50	0.482
Carcass weight (kg)	12.35 ± 0.10 [b]	12.35 ± 0.30 [b]	12.45 ± 0.26 [b]	13.17 ± 0.40 [a]	0.023
Dressing percentage (%)	43.34 ± 1.11	43.34 ± 1.26	43.19 ± 1.25	45.40 ± 1.14	0.147
Eye muscle area (cm^2)	11.67 ± 1.38	11.47 ± 1.06	13.25 ± 0.90	12.92 ± 1.76	0.318

BW, body weight; ADFI, average daily feed intake; ADG, average daily gain; F/G, feed/gain ratio. The results are presented as the mean ± SD. Different letters indicate significant differences ($p < 0.05$).

The carcass weight of the GAA−3 group was higher compared to the CON group ($p < 0.05$). There were no significant differences among these groups in terms of slaughter weight or dressing percentage (Table 3).

According to Figure 1 and Table 4, dietary supplementation with 1000 mg·kg^{-1} GAA (GAA−3 group) reduced the muscle fiber density and the drip loss of meat, while the shear force and muscle fiber diameter increased ($p < 0.05$). No significant differences were observed in pH_{24h}, crude fat ratio, total muscle fibers and total muscle fiber area among all groups ($p > 0.05$).

Table 4. Meat quality characteristics of Hu sheep following GAA supplementation.

Items	Groups				*p*-Value
	CON	GAA−1	GAA−2	GAA−3	
pH_{24h}	5.22 ± 0.21	5.27 ± 0.19	5.36 ± 0.17	5.38 ± 0.40	0.890
Drip loss (%)	7.15 ± 2.85 [a]	6.48 ± 1.52 [ab]	5.01 ± 0.93 [ab]	3.41 ± 0.55 [b]	0.027
Shear force (N)	17.21 ± 0.9 [b]	21.97 ± 1.55 [ab]	23.49 ± 2.21 [a]	25.71 ± 1.66 [a]	0.034
Intramuscular fat (%)	5.42 ± 0.66	6.00 ± 0.75	6.33 ± 0.75	6.58 ± 0.82	0.328
Muscle fiber diameter (um)	27.77 ± 3.81 [b]	28.08 ± 2.73 [ab]	30.46 ± 3.32 [a]	30.42 ± 3.50 [a]	0.048
Total muscle fibers (N)	225.44 ± 46.75	202.56 ± 81.15	171.11 ± 25.77	213 ± 25.75	0.145
Total muscle fiber area (mm^2)	0.15 ± 0.01	0.14 ± 0.04	0.16 ± 0.03	0.15 ± 0.01	0.362
Muscle fiber density (N/mm^2)	1550.10 ± 298.57 [a]	1375.38 ± 298.47 [a]	1468.34 ± 161.05 [a]	1089.78 ± 2 69.89 [b]	0.005

The results are presented as the mean ± SD. Different letters indicate significant differences ($p < 0.05$).

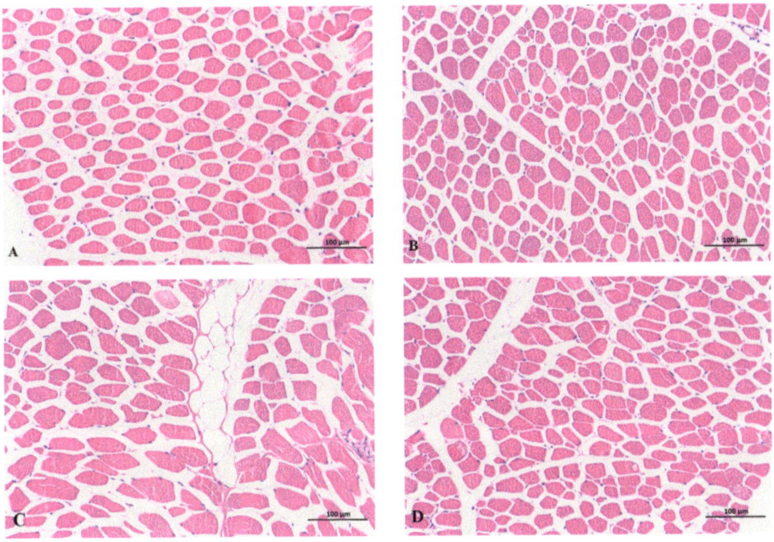

Figure 1. Microscopic images of the longissimus dorsi muscle showing cross-sections from the CON group (**A**), GAA−1 group (**B**), GAA−2 group (**C**), and GAA−3 group (**D**). The diets were supplemented with 0 (CON), 500 (GAA−1), 750 (GAA−2) and 1000 mg/kg (GAA−3) of GAA (weight of feed), respectively. The histological analyses of muscle were shown in Table 4.

3.2. GAA Supplementation Alters Serum Lipid Profile and Protein Levels

As shown in Figure 2, the dietary supplementation with GAA significantly reduced the level of TG ($p < 0.05$) in the sheep's serum. The dietary supplementation with 1000 mg·kg^{-1} GAA (GAA−3 group) significantly increased the levels of HDL−C, TP and GLB ($p < 0.05$) in the serum. However, there were no significant differences in the contents of TC, LDL−C and ALB in the serum among these groups ($p > 0.05$).

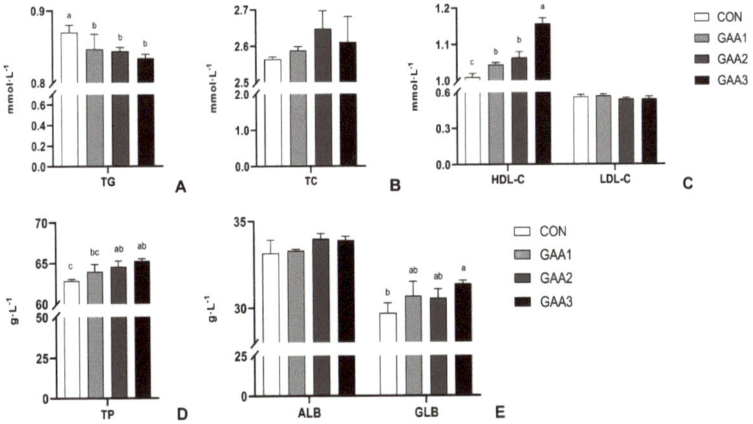

Figure 2. Serum biochemical parameters of Hu Sheep following GAA supplementation (**A**: the level of TG on serum; **B**: the level of TC on serum; **C**: the level of HDL−C and LDL−C on serum; **D**: the level of TP on serum; **E**: the level of ALB and GLB on serum). The error bar indicates standard deviation (data are from Table S1). Different letters indicate significant differences ($p < 0.05$). TG: Triglyceride; TC: total cholesterol; HDL−C: high-density lipoprotein cholesterol; LDL−C: low-density lipoprotein cholesterol; TP: total protein; ALB: albumin; GLB: globulin.

3.3. *GAA Supplementation Enhanced the Antioxidant Capacity on Serum*

According to Figure 3, the addition of GAA to feed would significantly enhance SOD activity in serum ($p < 0.05$). Furthermore, the dietary supplementation with 750 and 1000 mg·kg^{-1} GAA (GAA−2 and GAA−3 groups, respectively) significantly elevated the enzyme activity of GSH−Px ($p < 0.05$). Notably, the level of MDA significantly decreased compared to the CON group ($p < 0.05$), indicating a positive effect on antioxidant status.

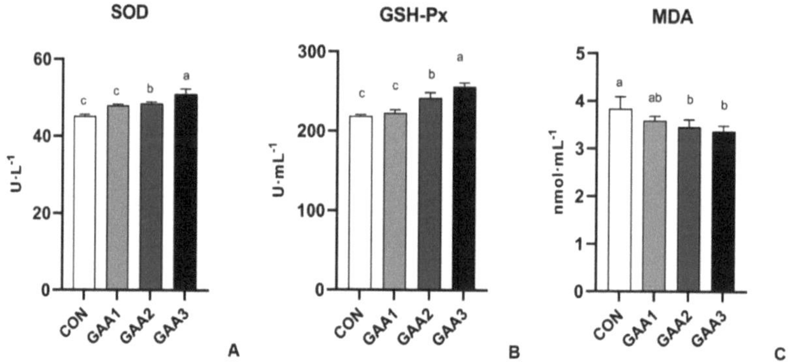

Figure 3. Serum antioxidant capacity of Hu sheep following GAA supplementation (**A**: the enzyme activity of SOD on serum; **B**: the enzyme activity of GSH−Px on serum; **C**: the level of MDA on serum). The error bar indicates standard deviation (data are from Table S2). Different letters indicate significant differences ($p < 0.05$). SOD: superoxide dismutase; GSH−PX: glutathione peroxidase; MDA: malondialdehyde.

3.4. *Modulation of Rumen pH, VFA Profile, and NH_3−N Content by GAA Supplementation*

As shown in Figure 4, ruminal pH exhibits a linear decrease with increasing GAA supplementation. The GAA−3 group showed a significantly lower pH compared to the CON group ($p < 0.05$). The total VFA concentration was significantly higher ($p < 0.05$) in the GAA−3 group compared to other groups. The valerate molar proportion increased ($p < 0.05$) in the GAA−3 group compared to the CON group. While the acetate molar proportion remained unaffected, the propionate molar proportion was greater ($p < 0.05$) in the GAA−3 group compared to the CON group. As a result, the acetate-to-propionate ratio decreased ($p < 0.05$) in the GAA−3 group compared to the CON group. As the supply of GAA increased, the molar proportions of valerate and the NH_3−N content increased ($p < 0.05$), but the levels of butyrate, isobutyrate and isovalerate were unchanged.

3.5. *GAA Supplementation Affects Fatty Acid Metabolism in Longissimus Dorsi Muscle*

As illustrated in Figure 5A, the expression level of ACC mRNA in the longissimus dorsi muscle varied with increasing dietary GAA levels. Specifically, the GAA−2 group exhibited the highest expression, whereas the CON group had the lowest, with a significant difference ($p < 0.05$). According to Figure 5B,C, the expression levels of FAS and HSL mRNA in the longissimus dorsi muscle did not differ significantly ($p > 0.05$).

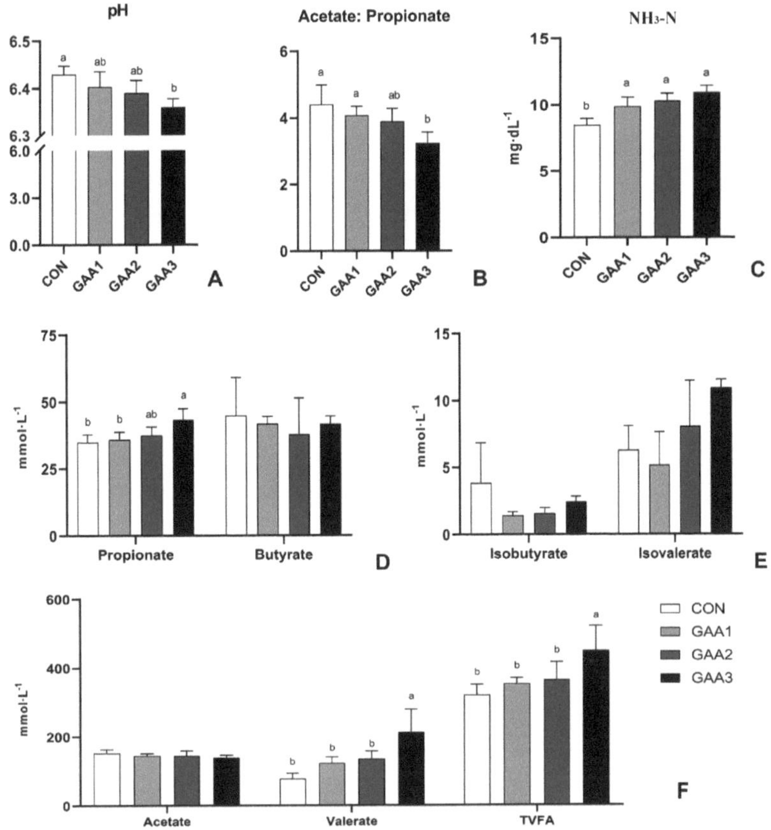

Figure 4. Rumen fermentation of Hu sheep following GAA supplementation (**A**: pH of rumen fluid; **B**: the ratio of acetate to propionate in rumen fluid; **C**: NH_3-N content in rumen fluid; **D–F**: different VFA and TVFA contents in rumen fluid). The error bar indicates standard deviation (data are from Table S3). Different letters indicate significant differences ($p < 0.05$). NH_3-N: Ammoniacal nitrogen; TVFA: Total volatile fatty acids.

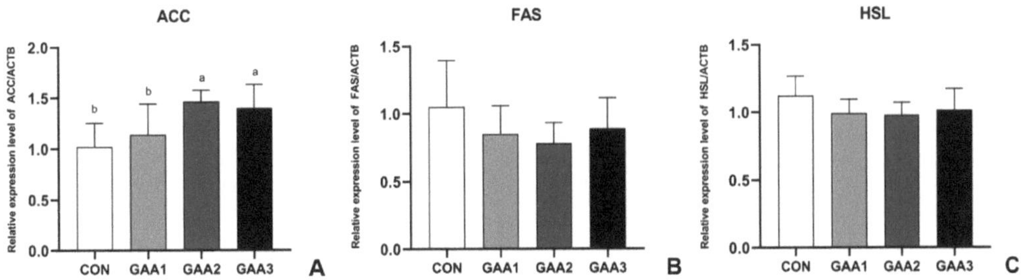

Figure 5. The expression of fatty acid metabolism-related genes (**A**: ACC, **B**: FAS, **C**: HSL) in the longissimus dorsi muscle of Hu sheep following GAA supplementation. ACTB (β-Actin) was set as reference gene. The error bar indicates standard deviation (data are from Table S4). Different letters indicate significant differences ($p < 0.05$). ACC: acetyl-CoA carboxylase; FAS: fatty acid synthase; HSL: sensitive lipase.

4. Discussion

Growth performance serves as a crucial metric for assessing the economic efficiency of livestock farming. Thus, enhancing animal growth performance holds great significance for driving the advancement of the farming industry. As a nutritional additive, GAA enhances energy metabolism, feed conversion efficiency and myoblast growth [10,29]. Jayaraman et al. discovered that dietary supplementation with GAA significantly increased the average daily weight gain and lean meat yield of fattening pigs [4]. Research indicates that GAA can effectively enhance the growth performance of livestock, especially by increasing daily weight gain and decreasing the feed-to-meat ratio [30]. Similarly, dietary supplementation with GAA has effectively improved growth performance and enhanced the growth of muscle and carcass weight in sheep [10]. Notably, dietary supplementation of GAA could promote skeletal muscle development [31]. Our study revealed that the addition of GAA to the sheep's diet not only increased the average daily gain of Hu sheep, but also increased carcass weight. In addition, we discovered that providing sheep with a dietary GAA supplementation of 1000 mg·kg^{-1} increased the muscle fiber diameter of the longissimus dorsi muscle while decreasing muscle fiber density (Figure 1 and Table 4). These results indicated that a dietary GAA supplement might affect sheep muscle fiber traits. Since the number of primary and secondary myofibrils is fixed prenatally [32], the enhanced carcass weight of the GAA-fed sheep might result from modifications in myofibril size and muscle mass accumulation.

Muscle pH can reflect the changes in muscle glycolysis after slaughter, which affect meat drip loss, shear force and meat color [3,33]. It has been reported that the addition of 0.2% GAA had a significant effect on the reduction in the drip loss of the longest dorsal muscle of Jinjiang Bulls [34]. Kim et al. concluded that GAA supplementation delays pH decline, resulting in less protein denaturation and ultimately reduced drip loss [35]. Our results also indicated that the addition of 1000 mg·kg^{-1} GAA to the diet could significantly improve the drip loss of the longest dorsal muscle, which provides further data in support of the application of GAA in ruminants, particularly in sheep.

Skeletal muscle is primarily composed of muscle fibers, with a smaller percentage of connective tissue and adipose tissue. Thus, changes in muscle fibers affect indicators such as muscle shear force. In a previous study, it was found that muscle shear force and muscle fiber diameter showed a positive correlation [36]. Diet composition and nutritional levels could affect the diameter and density of muscle fibers [37]. Li et al. found that the GAA supplementation raised the final body weight, enhanced muscle mass and varied myofiber size distribution [10]. Changes in muscle fiber characteristics can to some extent explain changes in meat quality, for example, the size of myofibers can influence fiber bundle size and muscle growth prospects [38]. In our study, we found that the addition of GAA to the diet significantly increased the shear force of the longest dorsal muscle. The increase in shear force might be a potential adverse effect of GAA application, which might indicate tougher meat. We speculate that GAA enhances the protein utilization efficiency of the Hu sheep, which enlarges the muscle fiber diameter and affects the meat's shear force. In addition, the muscle fiber characteristics of the Hu sheep were altered by the addition of GAA, with an increase in myofibril diameter and a decrease in myofibril density. These results suggested that the change in myofibril characteristics may be one of the reasons for the change in meat quality of the Hu sheep. The hardness of meat may also be related to intramuscular fat content, which did not significantly change in this study. Meanwhile, we need to consider both meat quality and the economic benefits of feeding. Thus, we still need to carry out further research to investigate the adverse effects of GAA on meat quality.

Serum biochemical indexes can intuitively reflect the metabolic level of animal bodies [39], which could also reflect the deposition of nutrients in the animal body. The levels of TC and TG in serum reflect the status of lipid metabolism in the body and are important indicators of adipose tissue development and deposition [40]. In this study, the serum concentrations of TC, TG, LDL−C and HDL−C were examined to ascertain the normality of lipid metabolism in the animals. HDL transports cholesterol from extrahepatic tissues to the liver for metabolism, while LDL carries cholesterol to peripheral tissues. Furthermore, serum LDL−C content is positively correlated with disease incidence [40]. In this study, supplementing with 1000 mg·kg^{-1} GAA significantly reduced the levels of TG in the serum of Hu sheep and increased the levels of HDL−C (Figure 2A,C). In addition, there were significant differences in the TC and LDL−C in serum among these groups (Figure 2B,C). This indicates that supplementing their diet with GAA can improve the serum lipid metabolism status of Hu sheep. Serum TP and ALB are closely related to protein metabolism in the body. They both perform crucial roles in maintaining vascular osmotic pressure, regulating pH levels and facilitating the transportation of nutritional metabolites [7]. Serum TP is an important indicator of liver function, and many enzymes in the body are proteins synthesized in the liver. A decrease in TP content indicates that the body's metabolic capacity may be impaired [41]. Serum ALB is the main carrier for transporting nutrients. In this study, the levels of TP and GLB were notably elevated in the groups supplemented with GAA compared to the control group. However, ALB levels did not exhibit significant variations among the groups (Figure 2D,E). This result matches the outcomes of previous study conducted on Jinjiang Bulls [34]. Based on the enhancement of production performance and changes in muscle fibers, we believe that supplementing with GAA can improve the utilization of protein and other nutrients in Hu sheep.

The antioxidant enzymes SOD and GSH−Px play important roles in the in vivo antioxidant system, whereas MDA contents reflect the rate and intensity of lipid peroxidation in the body [42]. In our study, the addition of GAA led to elevated levels of SOD and GSH−Px and reduced MDA concentrations in the blood of Hu sheep (Figure 3), indicating that GAA enhanced the antioxidant capacity of sheep. It has been reported that the addition of GAA had a significant effect on the antioxidant indexes (SOD and GSH−Px) of Jinjiang Bulls [34]. Zhao et al. reported that supplementation GAA can decrease MDA contents of broilers [43]. GAA-related metabolites (creatine and arginine) have the potential to neutralize free radicals upon GAA consumption [44]. Creatine, which is one of the GAA products, has been shown to effectively eliminate O^{2-} [45]. Therefore, we suggest that GAA may stimulate sheep growth by enhancing their antioxidant potential.

The environment in the rumen can reflect bodily health, influencing digestion, absorption and nutrient utilization in meat-producing sheep. Our results indicate that GAA significantly impacts rumen pH, showing a decreasing trend and negatively correlated with total volatile fatty acids (TVFA) concentration. This is supported by the significant elevation in the molar ratio of propionic acid and valeric acid, when compared to the control group (Figure 4D,F). The changes in rumen pH are associated with increased or decreased concentrations of VFA [46]. VFA accumulation can lead to a decrease in rumen pH, with acetate and propionate being the main influences [47]. Acetic acid can be converted into Acetyl-CoA and is further involved in energy metabolism, fatty acid synthesis and other processes. Propionic acid is linked to carbohydrate metabolism and would inhibit cholesterol synthesis. Butyric acid is converted into β-Hydroxybutyric acid participates in the metabolic activities of tissues such as the liver and muscles to provide the energy needed by the body [48,49]. Our study found that the concentration of propionic acid in the rumen fluid showed an upward trend, which is consistent with the research reports of Li [18]. A significant increase in the rumen propionic acid molar ratio resulted in a corresponding decrease in the acetate-to-propionic acid ratio, indicating that GAA induced a shift in the rumen fermentation pattern towards increased propionic acid production. The increase in TVFA suggests that GAA may promote the synthesis of microbial proteins within the ruminal environment of Hu sheep. VFA is the main energy source for ruminants, providing more than 70% of the total energy needs of ruminants, and is also the main source of carbohydrates for rumen microorganisms [50]. We believe that supplemental feeding of GAA can elevate the VFA content in the rumen of Hu sheep, thereby enhancing energy availability to the body and ultimately resulting in improved weight gain and carcass quality. NH_3−N is an apparent indicator of rumen nitrogen degradation in ruminants. It is both the end product of nitrogenous substance degradation in the diet and the raw material for the microbial synthesis of bacterial proteins [51]. In the present study, the levels of NH_3-N in the experimental groups were significantly higher than that in the CON group, and the GAA−3 group exhibited the highest NH_3-N content (Figure 4C). We hypothesized that the dietary supplementation of GAA in Hu sheep would enhance protein utilization and increase the accumulation of rumen decomposition products. This might lead to an increase in levels of NH_3-N. Based on the above results, we believe that GAA supplementation can improve rumen fermentation by increasing TVFA, propionate and other metabolites, as well as enhancing specific serum biochemical indicators (including enzyme activity of SOD and GSH−Px and the level of MDA) in Hu sheep. This suggests an improved energy utilization efficiency, which likely contributes to changes in muscle fiber characteristics and ultimately improved production performance.

Intramuscular fat (IMF) is deposited in the longest muscle of the back, infraspinatus, biceps femoris and other muscle parts, and it directly affects the level of meat quality. IMF improves tenderness, color, juiciness, texture and flavor. These indicators are crucial for assessing meat quality [52]. Research has shown that GAA may reduce fat deposition by affecting enzyme activities and regulating lipid metabolism [53]. When considering the influence of diet on IMF content, ruminants exhibit a lesser degree of sensitivity compared to monogastric animals. In our study, we selected the longissimus dorsi muscle to investigate the effect of related gene expression on IMF deposition. Acetyl-CoA serves as a crucial intermediate metabolite in the intricate process of fatty acid metabolism. Acetyl-CoA carboxylase (ACC) serves as the initial rate-limiting enzyme in de novo fatty acid synthesis. ACC localizes in the cytosol and catalyzes the conversion of acetyl-CoA to malonyl-CoA, which is subsequently utilized by fatty acid synthase (FAS) to generate long-chain saturated fatty acids [54]. FAS is closely related to the content of fatty acids in muscle and fat. The increase in FAS expression is often accompanied by the increase in triglyceride content in the body, leading to the accumulation of adipose tissue [55,56]. The hormone-sensitive lipase (HSL) gene is an enzyme related to fatty acid hydrolysis. It can hydrolyze triglycerides into non-esterified free fatty acids and glycerol, which plays a central role in regulating fat deposition [57]. We found that the expression level of ACC in the longissimus dorsi muscle of Hu sheep was upregulated in the experimental group (Figure 5A). However, we found no significant differences in FAS and HSL among the experimental groups and the CON group (Figure 5B,C). These findings suggested that GAA stimulated the synthesis of fatty acids in the muscle tissue of Hu sheep via the enhanced expression of ACC in the muscle, thereby promoting the development of intramuscular fat. Based on the observed changes in serum TG and HDL−C levels, we hypothesize that supplementing with GAA may have increased the expression of ACC in the longissimus dorsi muscle. This potential increase in ACC expression could be a contributing factor to the decreased level of triglycerides in serum, leading to greater storage of triglycerides in adipose tissue.

5. Conclusions

This study demonstrates the potential of dietary GAA supplementation to enhance the production performance, serum biochemistry, meat quality and rumen fermentation of Hu sheep. By enhancing the energy supply efficiency and antioxidant potential of Hu sheep, the supplementation of GAA significantly improved their production performance and muscle fiber characteristics. This was possibly associated with the increase in VFA as it plays a major part of providing energy in ruminants. A potential adverse effect of GAA application was found in this study in the form of an increase in shear force, which might indicate tougher meat. Meanwhile, GAA was able to increase the expression of ACC and might enhance the quality of sheep meat. Our results indicate that supplementing the diet with 1000 $mg \cdot kg^{-1}$ of GAA was the most effective in improving the production performance of fattened Hu sheep.

Supplementary Materials: The following supporting information can be downloaded at: https://www.mdpi.com/article/10.3390/ani14142052/s1. Table S1: Effect of GAA on serum biochemistry of Hu sheep; Table S2: Effect of GAA on serum antioxidant capacity of Hu sheep; Table S3: Effect of GAA on rumen fermentation of Hu sheep; Table S4: Effect of GAA on fatty acid metabolism-related genes in the longissimus dorsi.

Author Contributions: Conceptualization, Q.L.; Data curation, H.J. and Z.D.; Investigation, Z.D., X.F., L.Q., W.L., Y.Z., and J.R.; Methodology, Q.L.; Project administration, Q.L.; Resources, X.F.; Supervision, C.Y. and Q.L.; Visualization, H.J. and C.Y.; Writing—original draft, H.J. and C.Y.; Writing—review and editing, H.J., Z.D., X.F., and C.Y. All authors have read and agreed to the published version of the manuscript.

Funding: This work was supported by National Key Research and Development Program of China (2023YFD1000500) and Fujian Province 2022 Local Science and Technology Development Funds "Development and Application of Fungus Grass and Fungus Feed" (2022L3085).

Institutional Review Board Statement: All experimental procedures and animals were approved by the Fujian Agriculture and Forestry University Animal Care and Use Committee (Approval ID: PZCASFAFU23012).

Informed Consent Statement: Not applicable.

Data Availability Statement: Most of the data generated or analyzed in this study are presented in this published article or its supplementary information. Additional data not included here are accessible upon reasonable request to the corresponding author.

Conflicts of Interest: All authors declare no conflict of interest.

References

1. He, D.T.; Gai, X.R.; Yang, L.B.; Li, J.T.; Lai, W.Q.; Sun, X.L.; Zhang, L.Y. Effects of guanidinoacetic acid on growth performance, creatine and energy metabolism, and carcass characteristics in growing-finishing pigs. *J. Anim. Sci.* **2018**, *96*, 3264–3273. [CrossRef] [PubMed]
2. Ostojic, S.M.; Stojanovic, M.D.; Olcina, G. Oxidant-Antioxidant Capacity of Dietary Guanidinoacetic Acid. *Ann. Nutr. Metab.* **2015**, *67*, 243–246. [CrossRef] [PubMed]
3. Wang, L.; Wang, Y.; Xu, D.; He, L.; Zhu, X.; Yin, J. Dietary guanidinoacetic acid supplementation improves water holding capacity and lowers free amino acid concentration of fresh meat in finishing pigs fed with various dietary protein levels. *Anim. Nutr.* **2022**, *11*, 112–120. [CrossRef] [PubMed]
4. Jayaraman, B.; La, K.V.; La, H.; Doan, V.; Carpena, E.M.; Rademacher, M.; Channarayapatna, G. Supplementation of guanidinoacetic acid to pig diets: Effects on performance, carcass characteristics, and meat quality. *J. Anim. Sci.* **2018**, *96*, 2332–2341. [CrossRef] [PubMed]
5. Robinson, J.L.; McBreairty, L.E.; Ryan, R.A.; Randunu, R.; Walsh, C.J.; Martin, G.M.; Brunton, J.A.; Bertolo, R.F. Effects of supplemental creatine and guanidinoacetic acid on spatial memory and the brain of weaned Yucatan miniature pigs. *PLoS ONE* **2020**, *15*, e0226806. [CrossRef] [PubMed]
6. Zhang, J.H.; Li, H.H.; Zhang, G.J.; Zhang, Y.H.; Liu, B.; Huang, S.; Guyader, J.; Zhong, R.Z. Supplementation of guanidinoacetic acid and rumen-protected methionine increased growth performance and meat quality of Tan lambs. *Anim. Biosci.* **2022**, *35*, 1556–1565. [CrossRef] [PubMed]
7. Yi, S.; Hu, S.; Wang, J.; Abudukelimu, A.; Wang, Y.; Li, X.; Wu, H.; Meng, Q.; Zhou, Z. Effect of Guanidinoacetic Acid Supplementation on Growth Performance, Rumen Fermentation, Blood Indices, Nutrient Digestion, and Nitrogen Metabolism in Angus Steers. *Animals* **2024**, *14*, 401. [CrossRef]
8. Brosnan, J.T.; Brosnan, M.E. Creatine metabolism and the urea cycle. *Mol. Genet. Metab.* **2010**, *100*, S49–S52. [CrossRef]
9. Brosnan, J.T.; Wijekoon, E.P.; Warford-Woolgar, L.; Trottier, N.L.; Brosnan, M.E.; Brunton, J.A.; Bertolo, R.F.P. Creatine Synthesis Is a Major Metabolic Process in Neonatal Piglets and Has Important Implications for Amino Acid Metabolism and Methyl Balance. *J. Nutr.* **2009**, *139*, 1292–1297. [CrossRef]
10. Li, X.; Liu, X.; Song, P.; Zhao, J.; Zhang, J.; Zhao, J. Skeletal muscle mass, meat quality and antioxidant status in growing lambs supplemented with guanidinoacetic acid. *Meat Sci.* **2022**, *192*, 108906. [CrossRef]
11. Ardalan, M.; Batista, E.D.; Titgemeyer, E.C. Effect of post-ruminal guanidinoacetic acid supplementation on creatine synthesis and plasma homocysteine concentrations in cattle. *J. Anim. Sci.* **2020**, *98*, skaa072. [CrossRef]
12. Duan, B.B.; Xu, J.W.; Xing, T.; Li, J.L.; Zhang, L.; Gao, F. Creatine nitrate supplementation strengthens energy status and delays glycolysis of broiler muscle via inhibition of LKB1/AMPK pathway. *Poult. Sci.* **2022**, *101*, 101653. [CrossRef]
13. Córdova-Noboa, H.A.; Oviedo-Rondón, E.O.; Sarsour, A.H.; Barnes, J.; Sapcota, D.; López, D.; Gross, L.; Rademacher-Heilshorn, M.; Braun, U. Effect of guanidinoacetic acid supplementation on live performance, meat quality, pectoral myopathies and blood parameters of male broilers fed corn-based diets with or without poultry by-products. *Poult. Sci.* **2018**, *97*, 2494–2505. [CrossRef]

14. Amiri, M.; Ghasemi, H.A.; Hajkhodadadi, I.; Khaltabadi Farahani, A.H. Efficacy of guanidinoacetic acid at different dietary crude protein levels on growth performance, stress indicators, antioxidant status, and intestinal morphology in broiler chickens subjected to cyclic heat stress. *Anim. Feed Sci. Technol.* **2019**, *254*, 114208. [CrossRef]
15. Khalil, S.; Saenbungkhor, N.; Kesnava, K.; Sivapirunthep, P.; Sitthigripong, R.; Jumanee, S.; Chaosap, C. Effects of Guanidinoacetic Acid Supplementation on Productive Performance, Pectoral Myopathies, and Meat Quality of Broiler Chickens. *Animals* **2021**, *11*, 3180. [CrossRef]
16. Li, J.; Zhang, L.; Fu, Y.; Li, Y.; Jiang, Y.; Zhou, G.; Gao, F. Creatine Monohydrate and Guanidinoacetic Acid Supplementation Affects the Growth Performance, Meat Quality, and Creatine Metabolism of Finishing Pigs. *J. Agric. Food Chem.* **2018**, *66*, 9952–9959. [CrossRef] [PubMed]
17. Mendoza, S.M.S.; Bauer, L.; Htoo, J.K.; Asmussen, S.; Richert, S. 282 Effect of dietary supplementation with guanidinoacetic acid, a creatine precursor, in gestating and lactating sows on sow and litter performance. *J. Anim. Sci.* **2024**, *102*, 176–177. [CrossRef]
18. Li, S.Y.; Wang, C.; Wu, Z.Z.; Liu, Q.; Guo, G.; Huo, W.J.; Zhang, J.; Chen, L.; Zhang, Y.L.; Pei, C.X.; et al. Effects of guanidinoacetic acid supplementation on growth performance, nutrient digestion, rumen fermentation and blood metabolites in *Angus bulls*. *Animal* **2020**, *14*, 2535–2542. [CrossRef]
19. Liu, Y.J.; Chen, J.Z.; Wang, D.H.; Wu, M.J.; Zheng, C.; Wu, Z.Z.; Wang, C.; Liu, Q.; Zhang, J.; Guo, G.; et al. Effects of guanidinoacetic acid and coated folic acid supplementation on growth performance, nutrient digestion and hepatic gene expression in *Angus bulls*. *Br. J. Nutr.* **2020**, *126*, 510–517. [CrossRef]
20. Zhang, S.; Zang, C.; Pan, J.; Ma, C.; Wang, C.; Li, X.; Cai, W.; Yang, K. Effects of dietary guanidinoacetic acid on growth performance, guanidinoacetic acid absorption and creatine metabolism of lambs. *PLoS ONE* **2022**, *17*, e0264864. [CrossRef]
21. Safety and efficacy of guanidinoacetic acid for chickens for fattening, breeder hens and roosters, and pigs. *EFSA J.* **2016**, *14*, 4394. [CrossRef]
22. Bampidis, V.; Azimonti, G.; Bastos, M.d.L.; Christensen, H.; Dusemund, B.; Fašmon Durjava, M.; Kouba, M.; López-Alonso, M.; López Puente, S.; Marcon, F.; et al. Safety and efficacy of a feed additive consisting of guanidinoacetic acid for all animal species (Alzchem Trostberg GmbH). *EFSA J.* **2022**, *20*, 7269. [CrossRef]
23. National Research Council. *Nutrient Requirements of Small Ruminants: Sheep, Goats, Cervids, and New World Camelids*; The National Academies Press: Washington, DC, USA, 2007. [CrossRef]
24. Deng, K.; Liu, Z.; Su, Y.; Fan, Y.; Zhang, Y.; Wang, F. Comparison of muscle fiber characteristics and meat quality between newborn and adult Haimen goats. *Meat Sci.* **2024**, *207*, 109361. [CrossRef] [PubMed]
25. Ye, C.C.; Zeng, X.Z.; Zhu, J.L.; Liu, Y.; Ye, Q.H.; Qiao, S.Y.; Zeng, X.F. Dietary N-Carbamylglutamate Supplementation in a Reduced Protein Diet Affects Carcass Traits and the Profile of Muscle Amino Acids and Fatty Acids in Finishing Pigs. *J. Agric. Food Chem.* **2017**, *65*, 5751–5758. [CrossRef] [PubMed]
26. Du Preez, J.C.; Lategan, P.M. Gas chromatographic determination of C2-C5 fatty acids in aqueous media with a Porapak N column. *J. Chromatogr.* **1976**, *124*, 63–65. [CrossRef] [PubMed]
27. GA, B.; JH, K. Automated simultaneous determination of ammonia and total amino acids in ruminal fluid and in vitro media. *J. Dairy Sci* **1980**, *63*, 64–75.
28. Harshitha, R.; Arunraj, D.R. Real-time quantitative PCR: A tool for absolute and relative quantification. *Biochem. Mol. Biol. Educ.* **2021**, *49*, 800–812. [CrossRef] [PubMed]
29. Ceylan, N.; Koca, S.; Golzar Adabi, S.; Kahraman, N.; Bhaya, M.N.; Bozkurt, M.F. Effects of dietary energy level and guanidino acetic acid supplementation on growth performance, carcass quality and intestinal architecture of broilers. *Czech J. Anim. Sci.* **2021**, *66*, 281–291. [CrossRef]
30. Majdeddin, M.; Golian, A.; Kermanshahi, H.; De Smet, S.; Michiels, J. Guanidinoacetic acid supplementation in broiler chickens fed on corn-soybean diets affects performance in the finisher period and energy metabolites in breast muscle independent of diet nutrient density. *Br. Poult. Sci.* **2018**, *59*, 443–451. [CrossRef]
31. Yan, Z.; Yan, Z.; Liu, S.; Yin, Y.; Yang, T.; Chen, Q. Regulative Mechanism of Guanidinoacetic Acid on Skeletal Muscle Development and Its Application Prospects in Animal Husbandry: A Review. *Front. Nutr.* **2021**, *8*, 714567. [CrossRef]
32. Du, M.; Zhao, J.X.; Yan, X.; Huang, Y.; Nicodemus, L.V.; Yue, W.; McCormick, R.J.; Zhu, M.J. Fetal muscle development, mesenchymal multipotent cell differentiation, and associated signaling pathways1,2. *J. Anim. Sci.* **2011**, *89*, 583–590. [CrossRef] [PubMed]
33. Lealiifano, A.K.; Cadogan, D.J.; White, S.; Grigg, H.C.; Brewster, C.J. 68. In-feed guanidinoacetic acid improves meat quality in finisher pigs. *Anim. Sci. Proc.* **2021**, *12*, 230. [CrossRef]
34. Li, Z.; Liang, H.; Xin, J.; Xu, L.; Li, M.; Yu, H.; Zhang, W.; Ge, Y.; Li, Y.; Qu, M. Effects of Dietary Guanidinoacetic Acid on the Feed Efficiency, Blood Measures, and Meat Quality of Jinjiang Bulls. *Front. Vet. Sci.* **2021**, *8*, e684295. [CrossRef]
35. Kim, Y.H.B.; Warner, R.D.; Rosenvold, K. Influence of high pre-rigor temperature and fast pH fall on muscle proteins and meat quality: A review. *Anim. Prod. Sci.* **2014**, *54*, 375–395. [CrossRef]
36. Listrat, A.; Lebret, B.; Louveau, I.; Astruc, T.; Bonnet, M.; Lefaucheur, L.; Picard, B.; Bugeon, J. How Muscle Structure and Composition Influence Meat and Flesh Quality. *Sci. World J.* **2016**, *2016*, 1–14. [CrossRef] [PubMed]
37. Qi, Y.Y.; Zhang, K.Y.; Tian, G.; Bai, S.P.; Ding, X.M.; Wang, J.P.; Peng, H.W.; Lv, L.; Xuan, Y.; Zeng, Q.F. Effects of dietary corn germ meal levels on growth performance, serum biochemical parameters, meat quality, and standardized ileal digestibility of amino acids in Pekin ducks. *Poult. Sci.* **2022**, *101*, 101779. [CrossRef] [PubMed]

38. Kokoszyński, D.; Piwczyński, D.; Arpášová, H.; Hrnčar, C.; Saleh, M.; Wasilewski, R. A comparative study of carcass characteristics and meat quality in genetic resources Pekin ducks and commercial crossbreds. *Asian-Australas. J. Anim. Sci.* **2019**, *32*, 1753–1762. [CrossRef] [PubMed]
39. Niu, Y.; Wan, X.L.; Zhang, L.L.; Wang, C.; He, J.T.; Bai, K.W.; Zhang, X.H.; Zhao, L.G.; Wang, T. Effect of different doses of fermented Ginkgo biloba leaves on serum biochemistry, antioxidant capacity hepatic gene expression in broilers. *Anim. Feed Sci. Technol.* **2019**, *248*, 132–140. [CrossRef]
40. Zhang, B.; Wang, Z.; Huang, C.; Wang, D.; Chang, D.; Shi, X.; Chen, Y.; Chen, H. Positive effects of Mulberry leaf extract on egg quality, lipid metabolism, serum biochemistry, and antioxidant indices of laying hens. *Front. Vet. Sci.* **2022**, *9*, e1005643. [CrossRef]
41. Zhang, Y.; Qi, S.; Fan, S.; Jin, Z.; Bao, Q.; Zhang, Y.; Zhang, Y.; Xu, Q.; Chen, G. Comparison of growth performance, meat quality and blood biochemical indexes of Yangzhou goose under different feeding patterns. *Poult. Sci.* **2023**, *103*, 103349. [CrossRef]
42. Liu, H.N.; Liu, Y.; Hu, L.L.; Suo, Y.L.; Zhang, L.; Jin, F.; Feng, X.A.; Teng, N.; Li, Y. Effects of dietary supplementation of quercetin on performance, egg quality, cecal microflora populations, and antioxidant status in laying hens. *Poult. Sci.* **2014**, *93*, 347–353. [CrossRef]
43. Zhao, W.; Li, J.; Xing, T.; Zhang, L.; Gao, F. Effects of guanidinoacetic acid and complex antioxidant supplementation on growth performance, meat quality, and antioxidant function of broiler chickens. *J. Sci. Food Agric.* **2021**, *101*, 3961–3968. [CrossRef] [PubMed]
44. Ostojic, S.M. Advanced physiological roles of guanidinoacetic acid. *Eur. J. Nutr.* **2015**, *54*, 1211–1215. [CrossRef] [PubMed]
45. Lawler, J.M.; Barnes, W.S.; Wu, G.; Song, W.; Demaree, S. Direct Antioxidant Properties of Creatine. *Biochem. Biophys. Res. Commun.* **2002**, *290*, 47–52. [CrossRef]
46. Liu, Y.; Zhang, J.; Wang, C.; Guo, G.; Huo, W.; Xia, C.; Chen, L.; Zhang, Y.; Pei, C.; Liu, Q. Effects of guanidinoacetic acid supplementation on lactation performance, nutrient digestion and rumen fermentation in Holstein dairy cows. *J. Sci. Food Agric.* **2022**, *103*, 1522–1529. [CrossRef]
47. Dijkstra, J.; Ellis, J.L.; Kebreab, E.; Strathe, A.B.; López, S.; France, J.; Bannink, A. Ruminal pH regulation and nutritional consequences of low pH. *Anim. Feed Sci. Technol.* **2012**, *172*, 22–33. [CrossRef]
48. Chen, Y.J.; Kim, I.H.; Cho, J.H.; Yoo, J.S.; Kim, H.J.; Shin, S.O. Utilization of δ-aminolevulinic acid for livestock: Blood characteristics and immune organ weight in broilers. *J. Anim. Feed. Sci.* **2008**, *17*, 215. [CrossRef]
49. Brunetto, A.L.R.; Giacomelli, C.M.; Favero, J.F.; Bissacotti, B.F.; Copeti, P.M.; Morsch, V.M.; de Oliveira, F.d.C.; Wagner, R.; Alves, R.; Pereira, W.A.B.; et al. Phytogenic blend in the diet of growing Holstein steers: Effects on performance, digestibility, rumen volatile fatty acid profile, and immune and antioxidant responses. *Anim. Feed Sci. Technol.* **2023**, *297*, 115595. [CrossRef]
50. Stepanchenko, N.; Stefenoni, H.; Hennessy, M.; Nagaraju, I.; Wasson, D.E.; Cueva, S.F.; Räisänen, S.E.; Dechow, C.D.; Pitta, D.W.; Hristov, A.N. Microbial composition, rumen fermentation parameters, enteric methane emissions, and lactational performance of phenotypically high and low methane-emitting dairy cows. *J. Dairy Sci.* **2023**, *106*, 6146–6170. [CrossRef]
51. Thao, N.T.; Wanapat, M.; Cherdthong, A.; Kang, S. Effects of Eucalyptus Crude Oils Supplementation on Rumen Fermentation, Microorganism and Nutrient Digestibility in Swamp Buffaloes. *Asian-Australas. J. Anim. Sci.* **2014**, *27*, 46–54. [CrossRef]
52. Frank, D.; Joo, S.-T.; Warner, R. Consumer Acceptability of Intramuscular Fat. *Korean J. Food Sci. Anim. Resour.* **2016**, *36*, 699–708. [CrossRef] [PubMed]
53. Zhao, J.-M.; Li, F.-Q.-Y.; Li, X.-Y.; Jiao, D.-R.; Liu, X.-D.; Lv, X.-Y.; Zhao, J.-X. Guanidinoacetic Acid Attenuates Adipogenesis through Regulation of miR−133a in Sheep. *Animals* **2023**, *13*, 3108. [CrossRef] [PubMed]
54. Menendez, J.A.; Lupu, R. Fatty acid synthase and the lipogenic phenotype in cancer pathogenesis. *Nat. Rev. Cancer* **2007**, *7*, 763–777. [CrossRef] [PubMed]
55. Qiao, Y.; Huang, Z.; Li, Q.; Liu, Z.; Hao, C.; Shi, G.; Dai, R.; Xie, Z. Developmental Changes of the FAS and HSL mRNA Expression and Their Effects on the Content of Intramuscular Fat in Kazak and Xinjiang Sheep. *J. Genet. Genom.* **2007**, *34*, 909–917. [CrossRef] [PubMed]
56. Xu, X.; Wei, X.; Yang, Y.; Niu, W.; Kou, Q.; Wang, X.; Chen, Y. PPARγ, FAS, HSL mRNA and protein expression during Tan sheep fat-tail development. *Electron. J. Biotechnol.* **2015**, *18*, 122–127. [CrossRef]
57. Recazens, E.; Mouisel, E.; Langin, D. Hormone-sensitive lipase: Sixty years later. *Prog. Lipid Res.* **2021**, *82*, 101084. [CrossRef]

Disclaimer/Publisher's Note: The statements, opinions and data contained in all publications are solely those of the individual author(s) and contributor(s) and not of MDPI and/or the editor(s). MDPI and/or the editor(s) disclaim responsibility for any injury to people or property resulting from any ideas, methods, instructions or products referred to in the content.

animals

MDPI

Article

The Impact of Different Dietary Ratios of Soluble Carbohydrate-to-Neutral Detergent Fiber on Rumen Barrier Function and Inflammation in Dumont Lambs

Shufang Li [1], Tian Ma [1], Yawen An [1], Yu Zhang [1], Xiaodong Yang [1], Aiwu Gao [2] and Hairong Wang [1,*]

[1] Animal Nutrition and Feed Science, Inner Mongolia Agricultural University, Hohhot 010018, China; lishufang0325@163.com (S.L.)

[2] Food Science, Inner Mongolia Agricultural University, Hohhot 010018, China

* Correspondence: wanghairong97@163.com

Simple Summary: The rumen is a digestive organ in ruminants that is affected by soluble carbohydrate (SCHO) levels in the diet. This study confirmed that as the proportion of SCHO to NDF in the diet increased, the rumen pH decreased, the LPS concentration increased, and the papilla height and area increased. When the ratio of SCHO to NDF was 2.0, the tight junctions were disrupted, inflammatory cytokines levels were increased, and NF-κB/MAPK was activated.

Abstract: Appropriate soluble carbohydrate (SCHO)-to-NDF ratios in the diet are essential for rumen health. The effects of different SCHO-to-NDF ratios (1.0, 1.5, and 2.0) on rumen barrier function and inflammation in Dumont lambs (n = 18, 6 replicates per treatment) was investigated. The SCHO:NDF ratio was altered by replacing the forage (*Leynus chinensis*) with corn grain. With an increase in the proportion of SCHO, the final body weight (FBW), average daily gain (ADG), soluble carbohydrate intake (SCHOI), and LPS level increased; and the neutral detergent fiber intake (NDFI), ruminal papillae height, papillae area, and pH decreased ($p < 0.05$, $p_{lin} < 0.05$). The medium CHO:NDF group had increased claudin-1 mRNA ($p < 0.05$, $p_{lin} = 0.005$, $p_{quad} = 0.003$) and protein ($p < 0.05$, $p_{quad} < 0.001$) levels; the high CHO:NDF group had increased occludin mRNA and protein ($p < 0.05$, $p_{lin} = 0.001$) levels. The level of the anti-inflammatory cytokine IL-10 was significantly greater in the medium CHO:NDF group than in the high CHO:NDF group ($p < 0.05$, $p_{quad} < 0.001$). With an increase in the ratio of SCHO, the mRNA level and concentration of the proinflammatory cytokines IL-1β, IL-6, and TNF-α linearly increased ($p < 0.05$, $p_{lin} < 0.05$), and those in the high CHO:NDF group were significantly greater than those in the low CHO:NDF group. The levels of phosphorylated p65 ($p_{lin} = 0.003$), IκB-α ($p_{lin} < 0.001$), and JNK ($p_{lin} = 0.001$) increased linearly, and those in the high CHO:NDF group were significantly greater than those in the other two groups ($p < 0.05$). Therefore, when the SCHO-to-NDF ratio was increased to 1.5, the rumen epithelium was not affected, but when the ratio was increased to 2.0, NF-κB and MAPK were activated in the rumen epithelium, leading to impaired barrier function and inflammation. The suitable NFC:NDF ratio for the short-term fattening of Dumont lambs was found to be 1.50.

Keywords: soluble carbohydrate-to-neutral detergent-fiber ratio; rumen epithelium; tight junctions; inflammation; NF-κB /MAPK; Dumont lambs

Citation: Li, S.; Ma, T.; An, Y.; Zhang, Y.; Yang, X.; Gao, A.; Wang, H. The Impact of Different Dietary Ratios of Soluble Carbohydrate-to-Neutral Detergent Fiber on Rumen Barrier Function and Inflammation in Dumont Lambs. *Animals* **2024**, *14*, 1666. https://doi.org/10.3390/ani14111666

Academic Editors: Dingfu Xiao and Tiantian Meng

Received: 20 April 2024
Revised: 27 May 2024
Accepted: 30 May 2024
Published: 2 June 2024

Copyright: © 2024 by the authors. Licensee MDPI, Basel, Switzerland. This article is an open access article distributed under the terms and conditions of the Creative Commons Attribution (CC BY) license (https://creativecommons.org/licenses/by/4.0/).

1. Introduction

Maximizing animal production efficiency to achieve the greatest economic benefits while ensuring animal health and the quality and safety of livestock products is a key objective in livestock production. Roughage is a source of NDF, which stimulates rumination and gastrointestinal peristalsis [1]. Cereal grains, as the main source of soluble carbohydrates, are used as primary ingredients in concentrates with the aim of increasing energy levels and digestibility in the growing–finishing diets of ruminants [2]. The rumen is the main site of nutrient digestion and metabolism in ruminants, and it is affected by the ratio of concentrate to roughage in the diet. The rumen buffering capacity and volatile fatty acid (VFA) absorption in ruminants are correlated with the carbohydrate fermentation rate, with the pH in the rumen remaining within the normal range of 6.5–7.5 [3]. When ruminants are fed a large amount of concentrate, rumen peristalsis slows, and the rumen epithelium shows cuticle shedding and vacuolization [4], which affects the absorption of VFAs, resulting in elevated levels of VFAs and lactic acid and a decrease in pH [5,6]; these changes can induce subacute gastric acidosis (SARA) in severe cases [7]. Moreover, a diet high in soluble carbohydrate leads to rumen microbiota disorders, increased levels of bacterial endotoxins and biogenic amines, and impaired tight junctions between rumen epithelial cells, which affects rumen epithelial barrier function [8–10]. In addition, when ruminants are fed a high-concentrate diet, LPS produced in the rumen can be transferred into blood or gastrointestinal epithelial cells through the paracellular pathway and transcellular transport and is recognized by epithelial cells or immune cells [11], which activate NF-κB and MAPK via the LPS/TLR4 pathway, leading to rumen or systemic inflammation [12,13].

The Inner Mongolia region is the largest pasture area in China, and herders usually graze their sheep with a small amount of supplemental energy and protein feed, such as corn and soybean meal; this diet pattern is low-energy and low-nitrogen and cannot satisfy the growth demand of hybrid meat sheep. However, in the mainstream commercial goat market, many grain starch-based total mixed rations (TMRs) and pellet feeds are added to the diet to improve productivity, which increases the incidence of nutritional metabolic diseases and can damage the digestive tract of animals [14]. Therefore, special attention should be given to rumen health when shifting from a low-concentrate diet to a high-concentrate diet. Thus, in the present study, Dumont lambs were used to study the effects of low-, medium-, and high-SCHO:NDF-ratio diets on the morphology of the rumen mucosa epithelium tissue, rumen barrier function, and inflammation in lambs. Taken together, these findings reveal the mechanism by which diet structure influences the health of the rumen of lambs and provide a theoretical basis for the healthy fattening of lambs.

2. Materials and Methods

2.1. Animal Feeding and Experimental Design

Eighteen healthy male Dumont lambs (25.73 $\pm$ 2.17 kg BW, 3 mo. age) were used to evaluate the effect of treatment on growth performance, rumen pH, rumen lipopolysaccharide (LPS) content, papillae morphology, and inflammatory modulators. Lambs were individually housed and fed diets with different SCHO:NDF ratios of 1.0 (L group), 1.5 (M group), or 2.0 (H group). The SCHO:NDF ratio was altered by replacing the forage (*Leynus chinensis*) with corn grain (Table 1). According to the NY/T 816-2004 China Mut-ton Sheep feeding standard, the experiment lasted 75 days, with a 15-day adaptation period and a 60-day formal period. These lambs were randomly assigned to a metabolic cage (1.5 $\times$ 1 $\times$ 1 m) in a sheep house and fed twice a day at 08:00 and 18:00, with free access to feed and water.

2.2. Growth Performance and Feed Intake

The crude protein, calcium, phosphorus, ADF, and NDF levels were determined according to the feed analysis and feed detection method described by Zhang [15]; the SCHO level was determined according to the method described by Huang [16]. Each lamb was weighed on an empty stomach on day 1 and day 60 of the official period of the trial, and daily feed intake was recorded. The ADG, AGFI, SCHOI, NDFI, and F/G were subsequently calculated.

Table 1. Composition and nutritional status of the experimental diets (%).

Items	Diets with Different NSC:NDF Levels		
	1.0	1.5	2.0
Ingredients (% of DM)			
Leymus chinensis	70.00	50.00	30.00
Corn	28.80	41.00	53.00
Soybean meal	0.00	7.30	14.9
Calcium phosphate dibasic	0.00	0.10	0.10
Salt	0.70	0.60	0.50
Premix1	0.50	0.50	0.50
Sodium bicarbonate	0.00	0.50	1.00
Total	100.00	100.00	100.00
Nutrient levels (%)			
Crude protein	10.70	12.60	14.40
Soluble carbohydrates	37.50	43.60	49.90
Neutral detergent fiber	36.90	30.10	23.30
Calcium	1.40	1.00	0.70
Phosphorus	0.30	0.30	0.30
Acid detergent fiber	27.90	21.50	15.00
Soluble carbohydrates/NDF ratio	1.02	1.45	2.14
Metabolizable energy (MJ/kg)	9.00	9.70	10.40

Provided per kg of premix: Fe 25 mg; Zn 35 mg; Cu 9 mg; Co 0.1 mg; I 0.9 mg; Se 0.25 mg; Mn 19.5 mg; nicotinic acid 60 mg; vitamin E 15 U; vitamin A 3000 U; vitamin D3 1000 U. Metabolizable energy is a calculated value, and the rest of the nutritional indicators are measured values.

2.3. Sample Collection

At days 50, 51, and 52, rumen fluid was sampled 0 h, 2 h, 4 h, 6 h, 8 h, and 10 h after feeding through a rubber stomach tube with a vacuum pump and filtered through four-layer gauze immediately after pH was measured with a pH meter (OHAUS Corporation, Parsippany, NJ, USA). The rumen was washed with PBS after the animals had been sacrificed once the experiment ended. Approximately 4 cm^2 of rumen ventral sac tissue was collected and fixed using a 4% paraformaldehyde solution for histological analysis. Five grams of rumen epithelial tissue was also collected and stored at −80 °C.

2.4. Determination of LPS Levels in the Rumen Epithelium

The rumen epithelium was centrifuged at 3000× g for 20 min, and the supernatant was carefully collected. The LPS concentration was determined using an ELISA kit according to the manufacturer's instructions (Wuhan Genemy Biotechnology Co., Ltd., Wuhan, China).

2.5. Histological Analysis

Hematoxylin–eosin-stained sections were prepared according to the methods described by Dai et al. [17] and Gui [18], with some modifications. The 4% paraformaldehyde-fixed rumen epithelium was paraffin-embedded after gradual dehydration with different concentrations of ethanol, and 6 μm sections were cut via a paraffin microtome (Leica RM2235; Leica Microsystems, Wetzlar, Germany), spread on slides, and stained with hematoxylin–eosin after hydration. After the sections were sealed with neutral gum, they were observed under an orthostatic microscope (Nikon ECLIPSE Ni; Nikon Corporation, Tokyo, Japan), and images were captured and analyzed using image-acquisition software (Image View 4.1; Century Kexin Scientific Instrument Co., Ltd., Beijing, China). Five fields of view were selected for each section, and the three longest papillae under the field of view were utilized to determine the height, width, and area of the rumen papillae.

2.6. RNA Analysis

Total RNA was extracted from rumen tissues using TRIzol reagent (Takara Corporation, Otsu, Japan). After approximately 0.1 g of rumen epithelium was removed and ground in liquid nitrogen, chloroform fractionation, isopropanol precipitation, and ethanol washing were performed to obtain pure total RNA. The concentration and purity of total RNA were assessed using an Implen P330 NanoPhotometer (Implen, Germany) to ensure that the 260/280 ratio in all the samples was between 1.8 and 2.0, after which, the RNA was transcribed to cDNA. The sequences of primers used are provided in Supplementary Table S1. PCR amplification was performed using a Light Cycler 480 real-time PCR system (Roche, Switzerland) under the following conditions: pre-denaturation at 95 °C for 30 s; 40 cycles of 95 °C for 5 s and 60 °C for 20 s; melting at 95 °C for 5 s, 60 °C for 60 s, and 95 °C for continuous acquisition; and cooling at 50 °C for 30 s. The relative mRNA expression of the target genes was calculated by the $2^{-\Delta\Delta Ct}$ method with β-actin as an internal reference.

2.7. Cytokine Content Analysis

A total of 0.1 g of rumen epithelium was homogenized using a hand-held glass homogenizer, and 0.9 mL of PBS was added. The supernatant was collected carefully after centrifugation at 3000× *g* for 20 min. The concentrations of inflammatory cytokines (IL-1β, IL-6, TNF-α, IL-10, and IFN-γ) were determined using ELISA reagent according to the manufacturer's instructions (Wuhan Genemy Biotechnology Co., Ltd., Wuhan, China).

2.8. Western Blot Analysis

Approximately 0.1 g of rumen epithelium was added to 1 mL of Western protein isolation buffer (Beyotime, Shanghai, China) and fully homogenized using a hand-held glass homogenizer. Protein samples were adjusted to a concentration of 50 µg/20 µL using a BCA protein concentration kit (Beyotime). The amount of protein that was loaded in each gel well was 40 µg, and the proteins were transferred to a polyvinylidene difluoride (PVDF) membrane (Pall Corporation, Washington, NY, USA) after SDS–PAGE by a TE22 Mighty Small Transfer (Hoefer Inc., Holliston, MA, USA). Subsequently, the PVDF membrane was blocked in a non-protein-blocking solution (Shanghai Sangon Biotech Co. Ltd., Shanghai, China) for 1 h at room temperature, the primary antibody was applied, and the membrane was incubated at 4 °C overnight. The PVDF membrane was washed three times with 1× Tris-buffered saline containing Tween 20 (TBST) for 5 min each time. Then, the sections were incubated with the fluorescent secondary antibody for 1 h at room temperature in the dark and washed with TBST three times for 10 min each. Information on the antibody types and dilution ratios is listed in Supplementary Table S2. The signals were scanned on an Odyssey DLx imaging system (LI-COR Inc., Lincoln, NE, USA), and the imaging results were analyzed using Image Studio Version 5.2 software (LI-COR Inc., USA).

2.9. Statistical Analysis

The data were analyzed using one-way ANOVA in SPSS 23 (IBM Corp., New York, NY, USA), followed by Duncan's post hoc test and linear and quadratic trend fitting. The results are presented as the mean ± SEM, with significance levels set at $p < 0.05$.

3. Results

3.1. Growth Performance and Feed Intake

As shown in Table 2, the medium CHO:NDF group had an increased AGFI ($p < 0.05$, $p_{quad} = 0.002$). Polynomial contrast analysis revealed that with an increase in the NFC:NDF ratio in the diet, the FBW ($p < 0.05$, $p_{lin} < 0.010$, $p_{quad} = 0.018$), ADG ($p < 0.05$, $p_{lin} < 0.050$, $p_{quad} = 0.012$), and SCHOI ($p < 0.05$, $p_{lin} = 0.001$, $p_{quad} = 0.002$) showed linear and quadratic increases, with the medium and high CHO:NDF groups showing greater increases than the low CHO:NDF group ($p < 0.05$); the NDFI ($p < 0.05$, $p_{lin} < 0.001$, $p_{quad} = 0.006$) showed linear and quadratic decreases, with that of the high CHO:NDF group being lower than that of the low and medium CHO:NDF groups ($p < 0.05$).

Table 2. Effects of different soluble carbohydrate-to-neutral detergent fiber ratios in the diet on the growth performance and feed intake of lambs.

Item	Diets			SEM	*p* Value	P_{linear}	$P_{quadratic}$
	L Group	M Group	H Group				
FBW (kg)	31.16 [b]	37.51 [a]	35.51 [a]	0.87	0.004	0.010	0.018
ADG (g)	124.33 [b]	190.50 [a]	164.58 [a]	10.29	0.009	0.050	0.012
ADFI (kg/d)	1.02 [b]	1.19 [a]	1.00 [b]	0.003	0.009	0.669	0.003
SCHOI (g/d)	382.99 [b]	518.31 [a]	499.76 [a]	20.34	<0.001	0.001	0.002
NDFI (g/d)	376.87 [a]	357.83 [a]	233.36 [b]	22.67	<0.001	<0.001	0.006
F/G ratio	8.41	6.42	6.34	0.50	0.177	0.093	0.354

FBW, final body weight; ADG, average daily gain; ADFI, average daily feed intake; SCHOI, soluble carbohydrate intake; NDFI, neutral detergent fiber intake; F/G, feed to gain ratio; data were expressed as mean ± SEM. In the same row, values with different letter superscripts mean significant difference ($p < 0.05$). L group = SCHO-to-NDF ratio was 1.0; M group = SCHO-to-NDF ratio was 1.5; H group = SCHO-to-NDF ratio was 2.0.

3.2. Changes in pH and LPS Concentration in the Rumen

As the proportion of SCHO to NDF in the diet increased, the ruminal pH linearly decreased ($p_{lin} < 0.05$), and in the high-SCHO:NDF group, the rumen fluid pH significantly decreased compared to that in both the low- and medium-SCHO:NDF groups 2 h and 4 h after morning feeding ($p < 0.05$, Figure 1A). The high-SCHO:NDF group also had significantly greater LPS concentrations than the low-SCHO:NDF group did, and the LPS levels increased linearly as the SCHO:NDF ratio increased ($p_{lin} < 0.05$, Figure 1A).

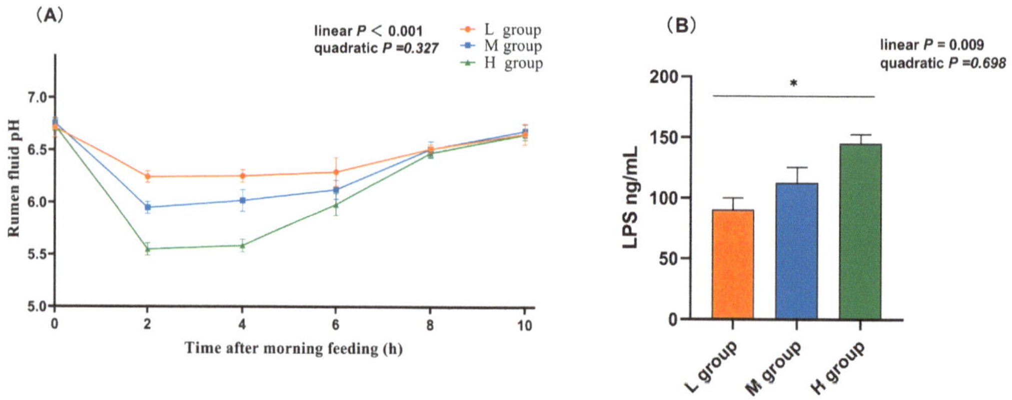

Figure 1. Effect of different soluble carbohydrate-to-neutral detergent fiber ratios in diets on pH (**A**) of rumen fluid and LPS (**B**) concentration of rumen epithelium. Data were expressed as mean ± SEM. * $p < 0.05$ mean significant differences. L group = SCHO-to-NDF ratio was 1.0; M group = SCHO-to-NDF ratio was 1.5; H group = SCHO-to-NDF ratio was 2.0.

3.3. Morphological Analysis of the Rumen Epithelium

We found that in the low-SCHO:NDF group, the surface of the rumen papillae was smooth without noticeable shedding (Figure 2A). The thickness of the rumen papilla cuticle slightly increased in the medium-SCHO:NDF group (Figure 2B). However, the rumen epithelial cuticle thickness increased in the high-SCHO:NDF group, and rumen papillae were notably enlarged and irregular in shape (Figure 2C). Furthermore, polynomial analyses revealed that papillae height followed a linear ($p_{lin} < 0.001$) and quadratic ($p_{quad} < 0.001$) pattern of increase with increasing levels of dietary SCHO:NDF, and papillae area followed a linear ($p_{lin} < 0.001$) pattern of increase, with the high-SCHO:NDF group having significantly greater papillae height and area than the low- and medium-SCHO:NDF groups ($p < 0.05$). The papillae width did not significantly differ among the three groups ($p > 0.05$) (Figure 2D, Table S3).

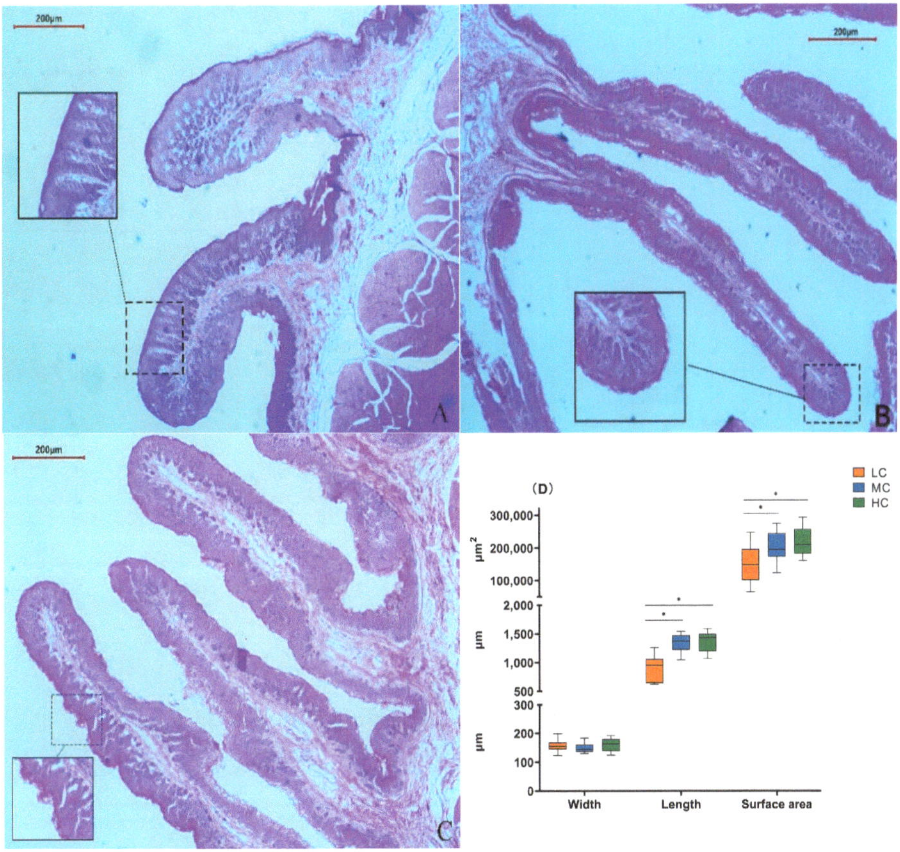

Figure 2. Effect of different soluble carbohydrate-to-neutral detergent fiber ratios in diets on rumen epithelial morphology of lambs. Error lines represent maximum and minimum values. (**A**) Morphology of rumen papillae in the L group; (**B**) morphology of rumen papillae in the M group; (**C**) morphology of rumen papillae in the H group. (**D**) Morphological statistics of rumen papillae. * $p < 0.05$ mean significant difference. L group = SCHO-to-NDF ratio was 1.0; M group = SCHO-to-NDF ratio was 1.5; H group = SCHO-to-NDF ratio was 2.0.

3.4. Rumen Epithelium Tight Junction mRNA and Protein Expression

As the proportion of SCHO to NDF in the diet increased, the occludin mRNA and protein expression levels linearly decreased (p_{lin} = 0.001) and were significantly lower in the high-SCHO:NDF group than in the low-SCHO:NDF group ($p < 0.05$, Table 3 and Figure 3C). The claudin-1 mRNA level increased and then decreased in a linear (p_{lin} = 0.021) and quadratic (p_{quad} = 0.003) manner, and the claudin-1 mRNA expression in the medium-SCHO:NDF group was significantly greater than that in the low-SCHO:NDF group ($p < 0.05$). The protein abundance was also significantly greater in the medium group than that in the low- and high-SCHO:NDF groups ($p < 0.05$) (Figure 3, Table 3).

Table 3. Effect of different soluble carbohydrate-to-neutral detergent fiber ratios in diets on rumen epithelium tight junction mRNA expression in lambs.

Item	Diets			SEM	*p* Value	P_{linear}	$P_{quadratic}$
	L Group	M Group	H Group				
occludin	1.00 [a]	0.82 [ab]	0.60 [b]	0.06	0.004	0.001	0.820
claudin-1	1.00 [b]	1.93 [a]	1.53 [ab]	0.13	0.005	0.021	0.003
ZO-1	1.00	1.04	1.02	0.02	0.756	0.816	0.486

In the same row, values with different letter superscripts mean significant difference ($p < 0.05$).

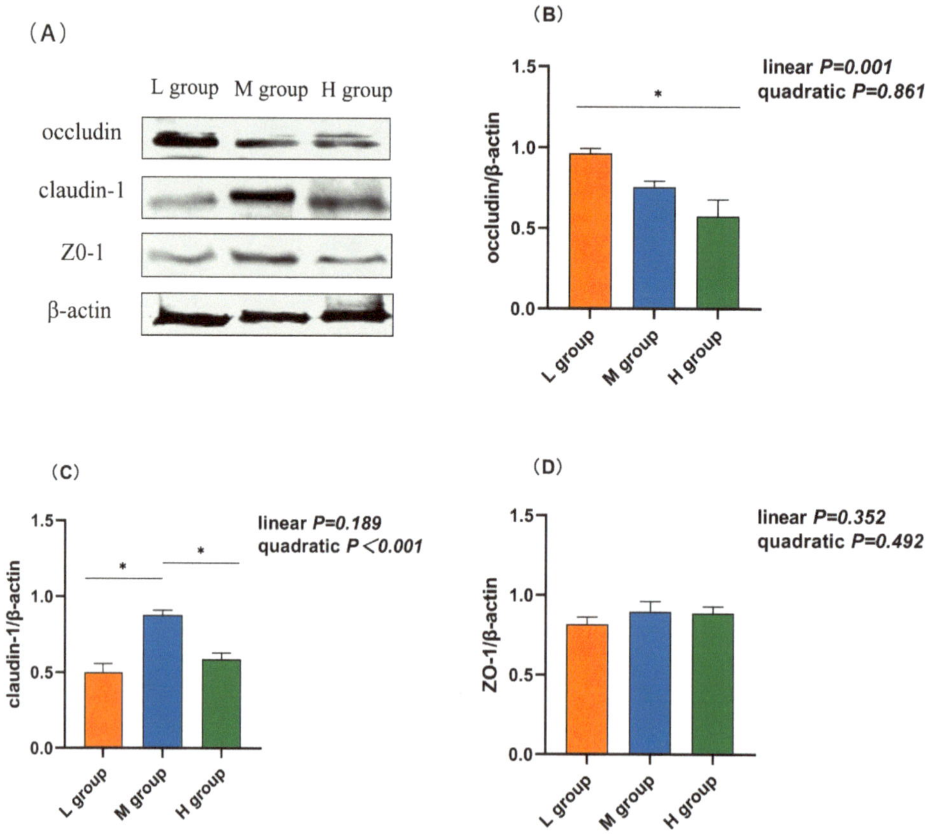

Figure 3. Effect different soluble carbohydrate-to-neutral detergent fiber ratios in diets on the protein abundance of rumen epithelium (**B**) occludin, (**C**) cloudin-1, and (**D**) ZO-1. (**A**) Representative bands from Western blot analysis. Data were expressed as mean ± SEM. In the same row, values with different letter superscripts mean significant difference ($p < 0.05$). * $p < 0.05$ mean significant differences. L group = SCHO-to-NDF ratio was 1.0; M group = SCHO-to-NDF ratio was 1.5; H group = SCHO-to-NDF ratio was 2.0.

3.5. Rumen Epithelium Inflammatory Factor mRNA and Protein Expression

Fluorescence quantitative PCR was used to detect changes in the mRNA expression of inflammatory factors in the rumens of lambs fed diets supplemented with different soluble carbohydrate-to-neutral detergent fiber ratios. As presented in Table 4, polynomial analyses revealed that the IL-10 mRNA level increased and then decreased quadratically ($p_{quad} < 0.001$) with increasing SCHO:NDF ratios in the diet, and the IL-10 level in the medium-SCHO:NDF group was significantly greater than that in the low- and high-SCHO:NDF groups ($p < 0.05$). The mRNA expression of IL-1β ($p_{lin} = 0.008$), IL-6 ($p_{lin} = 0.011$), and TNF-α ($p_{lin} = 0.003$) increased linearly, while the expression of IL-1β and TNF-α in the high-SCHO:NDF group was significantly greater than that in the low-SCHO:NDF group ($p < 0.05$); the expression of IL-6 in the medium- and high-SCHO:NDF groups was significantly greater than that in the low-SCHO:NDF group ($p < 0.05$).

Table 4. Effect of different soluble carbohydrate-to-neutral detergent fiber ratios in the diet on the mRNA expression level of inflammatory factors in the rumen epithelium.

Item	Diets			SEM	p Value	P_{linear}	$P_{quadratic}$
	L Group	M Group	H Group				
IL-10	1.00 ^{b}	1.50 ^{a}	0.84 ^{b}	0.09	0.001	0.263	<0.001
IL-1β	1.00 ^{b}	1.16 ^{ab}	1.44 ^{a}	0.07	0.025	0.008	0.647
IL-6	1.00 ^{b}	1.82 ^{a}	2.08 ^{a}	0.17	0.016	0.011	0.344
TNF-α	1.00 ^{b}	1.14 ^{b}	1.60 ^{a}	0.09	0.008	0.003	0.286
IFN-γ	1.00	0.03	1.08	0.03	0.509	0.648	0.293

In the same row, values with different letter superscripts mean significant difference ($p < 0.05$). L group = SCHO-to-NDF ratio was 1.0; M group = SCHO-to-NDF ratio was 1.5; H group = SCHO-to-NDF ratio was 2.0.

3.6. Inflammatory Factor Concentrations in the Rumen Epithelium

The rumen IL-10 concentration in the medium-SCHO:NDF (195.00 pg/mL) group was significantly greater than that in the high-SCHO:NDF (137.41 pg/mL) group ($p < 0.05$, Figure 4A). Polynomial analyses revealed that the level of IL-10 increased and then decreased quadratically ($p_{quad} = 0.005$) and that the levels of IL-1β ($p_{lin} = 0.017$), IL-6 ($p_{lin} < 0.01$) and TNF-α ($p_{lin} = 0.006$) increased linearly with increasing SCHO:NDF levels in the diet. The IL-1β concentration was significantly greater in the high-SCHO:NDF (111.01 pg/mL) group than in the low-SCHO:NDF (86.98 pg/mL) group ($p < 0.05$, Figure 3B). The IL-6 concentration in the high-SCHO:NDF (56.30 pg/mL) group was significantly greater than that in the low-SCHO:NDF (34.41 pg/mL) group, and that in the medium-SCHO:NDF (46.69 pg/mL) group was also significantly lower than that in the low-SCHO:NDF (34.41 pg/mL) group ($p < 0.05$, Figure 4C). Furthermore, the concentration of TNF-α was significantly greater in the high-SCHO:NDF (147.76 pg/mL) group than in the low- and medium-SCHO:NDF (122.00 and 125.96 pg/mL) groups ($p < 0.05$, Figure 4D).

3.7. Abundance of Proteins in the NF-κB/MAPK Pathway

The NF-κB and MAPK pathways are vital inflammatory pathways. Western blotting revealed that the abundance of phosphorylated p65 ($p_{lin} = 0.003$), IκB-α ($p_{lin} < 0.001$), and JNK ($p_{lin} = 0.001$) increased linearly with an increase in the ratio of SCHO to NDF in the diets, and the levels of phosphorylated p65, IκB-α, and JNK in the group with a high SCHO:NDF ratio were significantly greater than those in the low- and medium-SCHO:NDF groups ($p < 0.05$, Figure 5C,D,F). There was no significant difference in the phosphorylation levels of proteins between the low- and medium-SCHO:NDF groups ($p > 0.05$, Figure 5E,G).

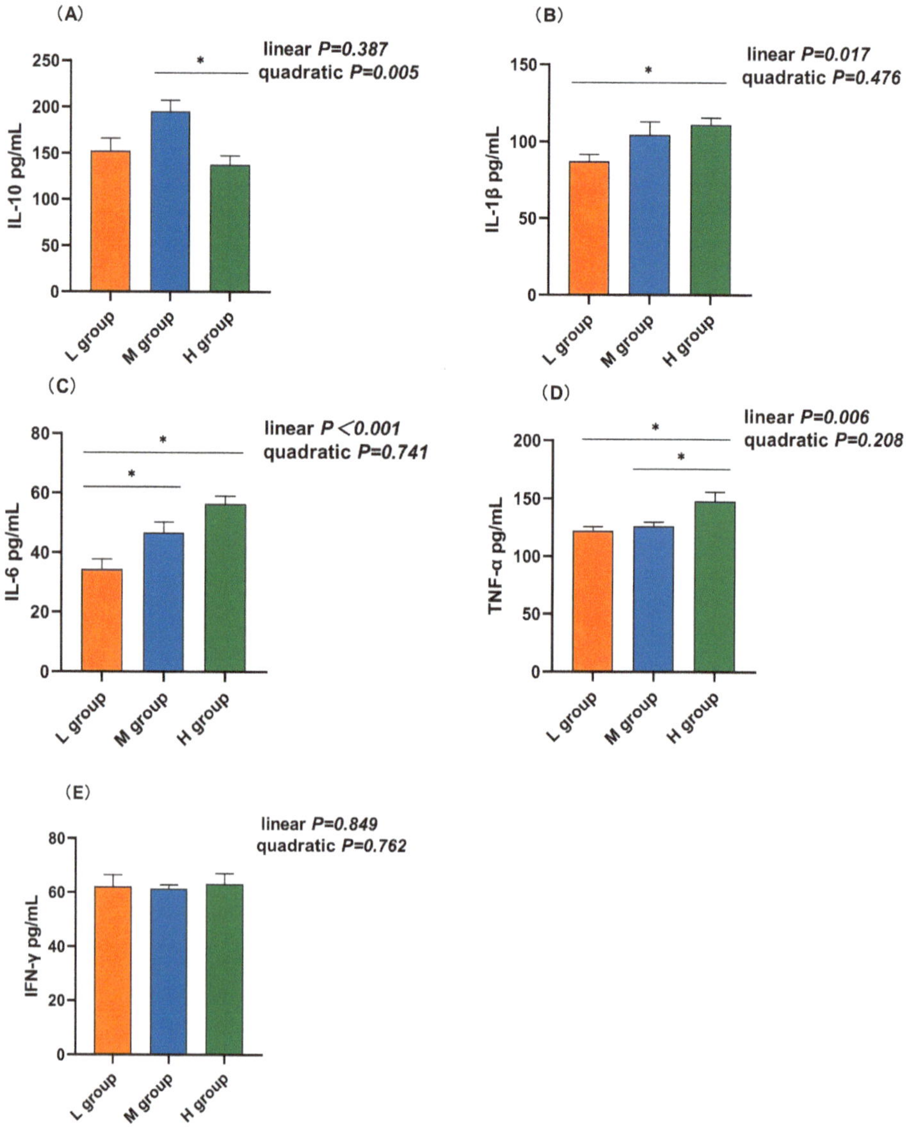

Figure 4. Effect of different soluble carbohydrate-to-neutral detergent fiber ratios in diets on rumen epithelium inflammatory factors: (**A**) IL-10, (**B**) IL-1β, (**C**) IL-6, (**D**) TNF-α, and (**E**) IFN-γ concentrations. Data were expressed as mean ± SEM. * $p < 0.05$ mean significant differences. L group = SCHO-to-NDF ratio was 1.0; M group = SCHO-to-NDF ratio was 1.5; H group = SCHO-to-NDF ratio was 2.0.

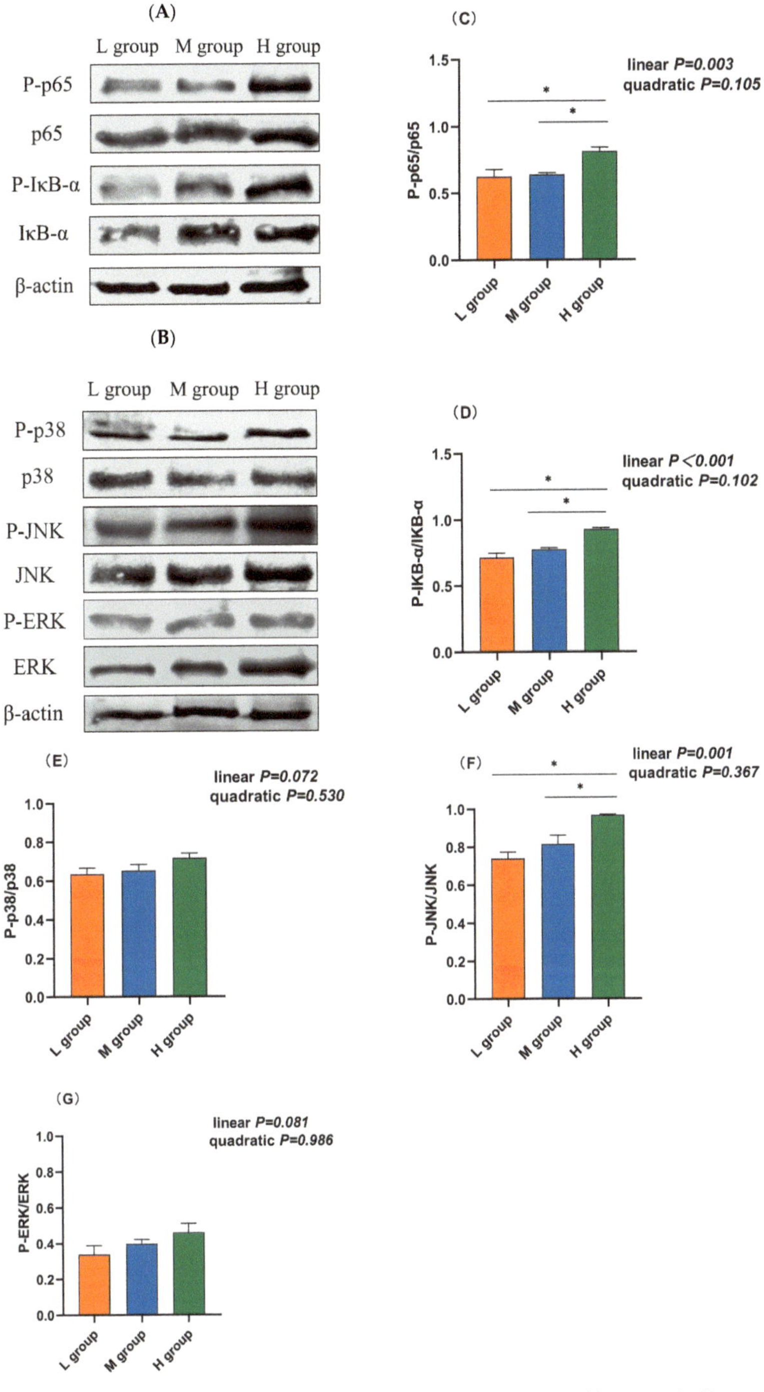

Figure 5. Effect of different soluble carbohydrate-to-neutral detergent fiber ratios in diets on phosphorylation levels of (**C**) p65 and (**D**) IκB-α, (**E**) p38, (**F**) JNK, and (**G**) ERK in the rumen epithelium. (**A**,**B**) Representative bands from Western blot analysis. Data were expressed as mean ± SEM. * $p < 0.05$ mean significant differences. L group = SCHO-to-NDF ratio was 1.0; M group = SCHO-to-NDF ratio was 1.5; H group = SCHO-to-NDF ratio was 2.0.

4. Discussion

Growth performance is an important quantitative parameter that is used to measure animal development and nutrition. In the present study, we showed that the growth performance of lambs increased significantly with an increase in the dietary SCHO:NDF ratio, but a high SCHO:NDF ratio did not result in greater weight gain, which may have been a result of reduced feed intake in the high-SCHO:NDF group compared to the medium-SCHO:NDF group. Rumen pH plays a crucial role in maintaining homeostasis of the internal environment [3], and stability is closely related to the nutritional level of feed. When the fiber level in the ration is too low or the starch content is too high, the rumen rapidly ferments to produce a large amount of VFA and lactic acid, and the pH decreases [5,19,20]. We found that after morning feeding, the rumen pH tended to decrease and then increase, and the rate of decrease in the rumen pH increased with an increasing SCHO:NDF ratio. This result was consistent with the findings of Kennelly et al. [21] and Han et al. [22]. The rumen pH of lambs in the high-SCHO:NDF group was considerably lower than that in the low-SCHO:NDF group at 2 and 4 h after morning feeding. However, this value did not reach the standard for SARA, which is usually defined as a rumen pH < 5.6 lasting at least 3 h [23–25]. This result could have been due to the oral sampling method, which caused saliva to mix into the rumen fluid and increased the pH.

A normal rumen morphology is a fundamental prerequisite for ensuring the proper digestion and absorption of nutrients. The composition of diets could influence both the structural and functional aspects of the rumen epithelium [4,26]. The rational composition of diets, along with high-quality roughage, could mitigate symptoms such as the aggregation and keratinization of rumen papillae and reduce the incidence of ruminal digestive disorders attributed to feed [27]. In the present study, we revealed that the height and area of the rumen papillae increased linearly with increasing SCHO:NDF ratios. In addition, irregularities and shedding of rumen papillae were observed in the high-SCHO:NDF group. This study is consistent with previous research conducted using Hu sheep [28], goats [29], and calves [30] fed diets with different SCHO-to-NDF ratios. This consistency suggested that a high SCHO-to-NDF ratio in the diet affects rumen morphology and disrupts barrier function.

Tight junctions (TJs) are intercellular connections that are located in the apical part of the epithelium, and they are vital in the formation of the mucosal barrier [31,32]. Occludin and claudins are two essential transmembrane proteins that make up TJs, together with the ZO protein family, and form scaffolds [32]. The loss or enrichment of tight junction proteins is closely associated with various pathological conditions [33–35]. Claudin-1 regulates paracellular permeability, and its overexpression improves barrier function [36]. Occludin, which was identified as the first transmembrane protein of TJs, plays a pivotal role in maintaining the integrity of TJ barriers. Our findings indicated that the medium-SCHO:NDF group exhibited significantly elevated claudin-1 levels in the rumen epithelium. Conversely, occludin expression was significantly lower in the high-SCHO:NDF group than in the low-SCHO:NDF group, consistent with the results of previous studies of Zhang et al. [37], Ma et al. [29], and Liu et al. [38]. In addition, previous studies have shown that a high SCHO-to-NDF ratio in the diet reduces occludin expression in the jejunum [39] and colon [40].

Previous studies have shown that a high SCHO-to-NDF ratio in the diet influences tight junctions in several ways. On the one hand, a high SCHO-to-NDF ratio in the diet decreases the rumen pH, thereby increasing the LPS concentration [17]. LPS induces excessive NO production and modifies TJ protein expression and subcellular localization [41]. On the other hand, a high SCHO-to-NDF ratio in the diet can reduce the expression of rumen epithelial tight junction proteins through the MAPK pathway [37]. Furthermore, previous studies have confirmed that inflammatory cytokines regulate tight junction protein expression and distribution [42]. TNF-α increases the paracellular permeability of the intestinal epithelium, reduces epithelium transepithelial resistance, and disrupts tight junction distribution by interfering with assembly and disassembly mechanisms. This experiment illustrated that a medium SCHO-to-NDF ratio in the diet enhanced rumen barrier function by promoting claudin-1 expression. In contrast, a high SCHO-to-NDF ratio in the diet disrupted the morphology and integrity of the gastrointestinal epithelium in ruminants and impaired gastrointestinal epithelium barrier function.

The amount of SCHO in the diet has been found to be positively correlated with the amount of LPS in the rumen fluid and the BA content [9,17,20,43]. LPS or other microbe-related molecular patterns can bind to specific pattern-recognition receptors in the gastrointestinal epithelium (such as TLR2 and TLR4) to induce local inflammation [44,45]. A high SCHO-to-NDF ratio in the diet resulted in elevated levels of rumen LPS and increased the mRNA and protein expression of IL-1β and IL-8 [17]. In the present study, we found that a high SCHO-to-NDF ratio in the diet resulted in inflammation in the rumen epithelium, as indicated by increased mRNA levels and concentrations of IL-1β, IL-6, and TNF-α. The results of this study align with the previous findings of Zhang et al. [46] on the rumen of dairy cows fed a high-concentrate diet. Additionally, our previous study revealed that long-term feeding of a high SCHO-to-NDF ratio in the diet increases the concentrations of TNF-α and IL-6 in the blood, leading to an inflammatory response [47]. This result shows that a medium SCHO-to-NDF ratio in the diet did not have a discernible impact on rumen health. However, a high SCHO-to-NDF ratio in the diet can lead to an increase in proinflammatory cytokines and induce rumen inflammation.

The NF-κB and MAPK pathways are the primary inflammatory pathways that mediate LPS stimulation. When LPS binds to the TLR4 complex, IκB phosphorylation leads to the release of NF-κB and a cascade reaction to activate MAPKs, thus promoting inflammatory cytokine expression [13,48,49]. A high SCHO-to-NDF ratio in the diet induces significant changes in the gastrointestinal microbiota of ruminants, resulting in increased numbers of Gram-negative bacteria, increased concentrations of LPS and histamine, and the activation of the NF-κB and MAPK pathway, leading to inflammatory damage in the rumen epithelium [45,50,51]. In this study, we found that feeding a high SCHO-to-NDF ratio diet resulted in a lower pH, a higher LPS concentration, and the increased phosphorylation of p65, IκB, and JNK proteins in the rumen epithelium compared to those in the low-SCHO:NDF group. Under normal physiological conditions, NF-κB and IκB bind in the cytoplasm in an inactivated state. When stimulated by LPS, the phosphorylation and degradation of IκB permit the translocation of NF-κB to the nucleus. This translocation triggers the release of inflammatory cytokines and mediators, leading to inflammatory responses [11,13]. This finding is consistent with the findings of Jiang et al. [52] and Zhao et al. [12,53] in the bovine rumen epithelium. Zhao et al. [12] discovered that SARA caused an increase in LPS levels in the rumen epithelia of dairy cows, thereby overactivating the inflammatory NF-κB and MAPK pathways, which caused the rumen epithelia to release more TNF-α, IL-6, and IL-1β. This result indicates that feeding a high concentration of nutrients may lead to inflammatory reactions in the rumens of lambs by activating the NF-κB and MAPK pathways. In addition, some recent studies have shown that the MAPK pathway can influence changes in tight junctions [54,55]. Zhang et al. [37] reported that a high SCHO-to-NDF ratio in the diet reduced tight junction protein expression and impaired rumen epithelium function through the MAPK pathway. The downregulation of occludin expression in the high-SCHO:NDF group in this study may also be related to the activation of JNK.

5. Conclusions

In conclusion, when the SCHO-to-NDF ratio was increased to 1.5, the rumen epithelium was not affected, but when the ratio was increased to 2.0, the NF-kB and MAPK pathways were activated in the rumen epithelium, leading to impaired barrier function and inflammation. The suitable NFC:NDF ratio for the short-term fattening of Dumont lambs was found to be 1.50.

Supplementary Materials: The following supporting information can be downloaded at https://www.mdpi.com/article/10.3390/ani14111666/s1: Table S1: PCR primer sequences of the target genes; Table S2: Antibody types and dilution ratios; Table S3: Effects of different soluble carbohydrate-to-neutral detergent fiber ratios in the diet on rumen epithelial morphology of lambs.

Author Contributions: Conceptualization, S.L., A.G., and H.W.; validation, S.L., T.M., Y.Z., and X.Y.; data curation, H.W.; writing—original draft preparation, S.L.; writing—review and editing, S.L., Y.A., and H.W.; project administration, H.W.; funding acquisition, H.W. All authors have read and agreed to the published version of the manuscript.

Funding: This research was supported by the Research Fund for Universities in Inner Mongolia Autonomous Region (project No. BR231520), the National Natural Science Foundation of China (NSFC) (31860658), and the Natural Science Foundation of Inner Mongolia Autonomous Region (2022MS03041).

Institutional Review Board Statement: All animal experimentation protocols were approved by the Specialized Committee on Scientific Research and Academic Ethics and Morality of Inner Mongolia Agricultural University ([2019] 034) and were based on the guidelines for the management of experimental animals at IMAU.

Informed Consent Statement: Not applicable.

Data Availability Statement: The data presented in this study are available on request from the corresponding author.

Conflicts of Interest: The authors declare no conflicts of interest. The funders had no role in the design of the study; in the collection, analyses, or interpretation of data; in the writing of the manuscript; or in the decision to publish the results.

References

1. Richeson, J.T.; Samuelson, K.L.; Tomczak, D.J. Beef species-ruminant nutrition cactus beef symposium: Energy and roughage levels in cattle receiving diets and impacts on health, performance, and immune responses1. *J. Anim. Sci.* **2019**, *97*, 3596–3604. [CrossRef]
2. Jiang, Y.; Dai, P.; Dai, Q.; Ma, J.; Wang, Z.; Hu, R.; Zou, H.; Peng, Q.; Wang, L.; Xue, B. Effects of the higher concentrate ratio on the production performance, ruminal fermentation, and morphological structure in male cattle-yaks. *Vet. Med. Sci.* **2022**, *8*, 771–780. [CrossRef]
3. Plaizier, J.C.; Mulligan, F.J.; Neville, E.W.; Guan, L.L.; Steele, M.A.; Penner, G.B. Invited review: Effect of subacute ruminal acidosis on gut health of dairy cows. *J. Dairy Sci.* **2022**, *105*, 7141–7160. [CrossRef] [PubMed]
4. Steele, M.A.; Greenwood, S.L.; Croom, J.; Mcbride, B.W. An increase in dietary non-structural carbohydrates alters the structure and metabolism of the rumen epithelium in lambs. *Can. J. Plant Sci.* **2012**, *92*, 123–130. [CrossRef]
5. Wang, L.; Li, Y.; Zhang, Y.; Wang, L. The Effects of Different Concentrate-to-Forage Ratio Diets on Rumen Bacterial Microbiota and the Structures of Holstein Cows During the Feeding Cycle. *Animals* **2020**, *10*, 957. [CrossRef]
6. Ogata, T.; Makino, H.; Ishizuka, N.; Iwamoto, E.; Masaki, T.; Ikuta, K.; Kim, Y.H.; Sato, S. Long-term high-grain diet altered the ruminal pH, fermentation, and composition and functions of the rumen bacterial community, leading to enhanced lactic acid production in Japanese Black beef cattle during fattening. *PLoS ONE* **2019**, *14*, e0225448. [CrossRef] [PubMed]
7. Gozho, G.N.; Plaizier, J.C.; Krause, D.O.; Kennedy, A.D.; Wittenberg, K.M. Subacute ruminal acidosis induces ruminal lipopolysaccharide endotoxin release and triggers an inflammatory response. *J. Dairy Sci.* **2005**, *88*, 1399–1403. [CrossRef] [PubMed]
8. Guo, P.; Liu, D.; Zhao, P.; Gao, M.; Hu, H. Effects of Increasing the Ratio of Non-fiber Carbohydrate to Neutral Detergent Fiber on Bacteria Flora, Endotoxin and Histamine Content in Rumen and Plasma of Dairy Goats. *Acta Vet. Zootech. Sin.* **2015**, *46*, 96–103.
9. Ma, Y.; Wang, C.; Zhang, H.; Yu, L.; Dong, L.; Gong, D.; Yao, J.; Wang, H. Illumina Sequencing and Metabolomics Analysis Reveal Thiamine Modulation of Ruminal Microbiota and Metabolome Characteristics in Goats Fed a High-Concentrate Diet. *Front. Microbiol.* **2021**, *12*, 653283. [CrossRef]
10. Fu, Y.; He, Y.; Xiang, K.; Zhao, C.; He, Z.; Qiu, M.; Hu, X.; Zhang, N. The Role of Rumen Microbiota and Its Metabolites in Subacute Ruminal Acidosis (SARA)-Induced Inflammatory Diseases of Ruminants. *Microorganisms* **2022**, *10*, 1495. [CrossRef]
11. Park, B.S.; Lee, J.O. Recognition of lipopolysaccharide pattern by TLR4 complexes. *Exp. Mol. Med.* **2013**, *45*, e66. [CrossRef] [PubMed]
12. Zhao, C.; Liu, G.; Li, X.; Guan, Y.; Wang, Y.; Yuan, X.; Sun, G.; Wang, Z.; Li, X. Inflammatory mechanism of Rumenitis in dairy cows with subacute ruminal acidosis. *BMC Vet. Res.* **2018**, *14*, 135. [CrossRef] [PubMed]
13. Sen, R.; Smale, S.T. Selectivity of the NF-{kappa}B response. *Cold Spring Harb. Perspect. Biol.* **2010**, *2*, a000257. [CrossRef] [PubMed]
14. Wang, H. Research Advances in System-Nutritional Regulation on Mechanism of Carbohydrates Metabolism and Gastrointestin Health of Ruminants. *Chin. J. Anim. Nutr.* **2020**, *32*, 4686–4696.

15. Zhang, L. *Feed Analysis and Feed Quality Detection Technology*, 3rd ed.; China Agricultural University Press: Beijing, China, 2007.
16. Wang, X.; Huang, J. *Principles and Techniques of Plant Physiological and Biochemical Experiments*, 3rd ed.; Higher Education Press: Beijing, China, 2019.
17. Dai, H.; Liu, X.; Yan, J.; Aabdin, Z.U.; Bilal, M.S.; Shen, X. Sodium Butyrate Ameliorates High-Concentrate Diet-Induced Inflammation in the Rumen Epithelium of Dairy Goats. *J. Agric. Food Chem.* **2017**, *65*, 596–604. [CrossRef] [PubMed]
18. Gui, H. Effect of Dietary Concentarte Level on RumenEpithelium Growth and Its Underlying Mechanism. Ph.D. Thesis, Nanjing Agricultural University, Nanjing, China, 2020.
19. Belanche, A.; Doreau, M.; Edwards, J.E.; Moorby, J.M.; Pinloche, E.; Newbold, C.J. Shifts in the rumen microbiota due to the type of carbohydrate and level of protein ingested by dairy cattle are associated with changes in rumen fermentation. *J. Nutr.* **2012**, *142*, 1684–1692. [CrossRef] [PubMed]
20. Plaizier, J.C.; Danesh Mesgaran, M.; Derakhshani, H.; Golder, H.; Khafipour, E.; Kleen, J.L.; Lean, I.; Loor, J.; Penner, G.; Zebeli, Q. Review: Enhancing gastrointestinal health in dairy cows. *Anim. Int. J. Anim. Biosci.* **2018**, *12*, s399–s418. [CrossRef] [PubMed]
21. Kennelly, J.J.; Robinson, B.; Khorasani, G.R. Influence of carbohydrate source and buffer on rumen fermentation characteristics, milk yield, and milk composition in early-lactation Holstein cows. *J. Dairy Sci.* **1999**, *82*, 2486–2496. [CrossRef] [PubMed]
22. Han, H.; Liu, D.; Gao, M.; Hu, H.; Xie, C.; Deng, W.; Liu, Y.; Wang, P.; Lu, D. Effects of Different Dietary NFC/NDF Ratios on Ruminal Microorganisms and pH in Dairy Goats. *Chin. J. Anim. Nutr.* **2011**, *23*, 597–603.
23. Kleen, J.L.; Hooijer, G.A.; Rehage, J.; Noordhuizen, J.P. Subacute ruminal acidosis (SARA): A review. *J. Vet. Med. A Physiol. Pathol. Clin. Med.* **2003**, *50*, 406–414. [CrossRef]
24. Penner, G.B.; Beauchemin, K.A.; Mutsvangwa, T. Severity of ruminal acidosis in primiparous holstein cows during the periparturient period. *J. Dairy Sci.* **2007**, *90*, 365–375. [CrossRef] [PubMed]
25. Plaizier, J.C.; Krause, D.O.; Gozho, G.N.; McBride, B.W. Subacute ruminal acidosis in dairy cows: The physiological causes, incidence and consequences. *Vet. J.* **2008**, *176*, 21–31. [CrossRef] [PubMed]
26. Steele, M.A.; AlZahal, O.; Hook, S.E.; Croom, J.; McBride, B.W. Ruminal acidosis and the rapid onset of ruminal parakeratosis in a mature dairy cow: A case report. *Acta Vet. Scand.* **2009**, *51*, 39. [CrossRef] [PubMed]
27. Suárez, B.J.; Van Reenen, C.G.; Beldman, G.; van Delen, J.; Dijkstra, J.; Gerrits, W.J. Effects of supplementing concentrates differing in carbohydrate composition in veal calf diets: I. Animal performance and rumen fermentation characteristics. *J. Dairy Sci.* **2006**, *89*, 4365–4375. [CrossRef]
28. Bo Trabi, E.; Seddik, H.-E.; Xie, F.; Wang, X.; Liu, J.; Mao, S. Effect of pelleted high-grain total mixed ration on rumen morphology, epithelium-associated microbiota and gene expression of proinflammatory cytokines and tight junction proteins in Hu sheep. *Anim. Feed. Sci. Technol.* **2020**, *263*, 114453. [CrossRef]
29. Ma, Y.; Wang, C.; Elmhadi, M.; Zhang, H.; Han, Y.; Shen, B.; He, B.L.; Liu, X.Y.; Wang, H.R. Thiamine ameliorates metabolic disorders induced by a long-term high-concentrate diet and promotes rumen epithelial development in goats. *J. Dairy Sci.* **2021**, *104*, 11522–11536. [CrossRef] [PubMed]
30. Kong, L. The Comparative Study of Calf Stomach Anatomy and Histologicalstructure Development under the Different Feeding Condition. Master's Thesis, Gansu Agricultural University, Lanzhou, China, 2012.
31. Otani, T.; Furuse, M. Tight Junction Structure and Function Revisited. *Trends Cell Biol.* **2020**, *30*, 805–817. [CrossRef] [PubMed]
32. Hartsock, A.; Nelson, W.J. Adherens and tight junctions: Structure, function and connections to the actin cytoskeleton. *Biochim. Biophys. Acta* **2008**, *1778*, 660–669. [CrossRef] [PubMed]
33. Sawada, N. Tight junction-related human diseases. *Pathol. Int.* **2013**, *63*, 1–12. [CrossRef]
34. Zheng, X.; Ren, B.; Gao, Y. Tight junction proteins related to blood-brain barrier and their regulatory signaling pathways in ischemic stroke. *Biomed. Pharmacother.* **2023**, *165*, 115272. [CrossRef]
35. Wei, W.; Li, W.; Yang, L.; Weeramantry, S.; Ma, L.; Fu, P.; Zhao, Y. Tight junctions and acute kidney injury. *J. Cell. Physiol.* **2023**, *238*, 727–741. [CrossRef] [PubMed]
36. Markov, A.G.; Aschenbach, J.R.; Amasheh, S. Claudin clusters as determinants of epithelial barrier function. *IUBMB Life* **2015**, *67*, 29–35. [CrossRef] [PubMed]
37. Zhang, K.; Meng, M.; Gao, L.; Tu, Y.; Bai, Y. Rumen-derived lipopolysaccharide induced ruminal epithelium barrier damage in goats fed a high-concentrate diet. *Microb. Pathog.* **2019**, *131*, 81–86. [CrossRef] [PubMed]
38. Liu, J.H.; Xu, T.T.; Liu, Y.J.; Zhu, W.Y.; Mao, S.Y. A high-grain diet causes massive disruption of ruminal epithelial tight junctions in goats. *Am. J. Physiol. Regul. Integr. Comp. Physiol.* **2013**, *305*, R232–R241. [CrossRef] [PubMed]
39. Lai, Z.; Lin, L.; Zhang, J.; Mao, S. Effects of high-grain diet feeding on mucosa-associated bacterial community and gene expression of tight junction proteins and inflammatory cytokines in the small intestine of dairy cattle. *J. Dairy Sci.* **2022**, *105*, 6601–6615. [CrossRef]
40. Chen, M.; Xie, W.; Zhou, S.; Ma, N.; Wang, Y.; Huang, J.; Shen, X.; Chang, G. A high-concentrate diet induces colonic inflammation and barrier damage in Hu sheep. *J. Dairy Sci.* **2023**, *106*, 9644–9662. [CrossRef]
41. Han, X.; Fink, M.P.; Yang, R.; Delude, R.L. Increased iNOS activity is essential for intestinal epithelial tight junction dysfunction in endotoxemic mice. *Shock* **2004**, *21*, 261–270. [CrossRef] [PubMed]
42. Gitter, A.H.; Bendfeldt, K.; Schmitz, H.; Schulzke, J.D.; Bentzel, C.J.; Fromm, M. Epithelial barrier defects in HT-29/B6 colonic cell monolayers induced by tumor necrosis factor-alpha. *Ann. N. Y. Acad. Sci.* **2000**, *915*, 193–203. [CrossRef] [PubMed]

43. Li, S.; Khafipour, E.; Krause, D.O.; Kroeker, A.; Rodriguez-Lecompte, J.C.; Gozho, G.N.; Plaizier, J.C. Effects of subacute ruminal acidosis challenges on fermentation and endotoxins in the rumen and hindgut of dairy cows. *J. Dairy Sci.* **2012**, *95*, 294–303. [CrossRef]
44. Kent-Dennis, C.; Aschenbach, J.R.; Griebel, P.J.; Penner, G.B. Effects of lipopolysaccharide exposure in primary bovine ruminal epithelial cells. *J. Dairy Sci.* **2020**, *103*, 9587–9603. [CrossRef]
45. Sun, X.; Yuan, X.; Chen, L.; Wang, T.; Wang, Z.; Sun, G.; Li, X.; Li, X.; Liu, G. Histamine Induces Bovine Rumen Epithelial Cell Inflammatory Response via NF-κB Pathway. *Cell Physiol. Biochem.* **2017**, *42*, 1109–1119. [CrossRef] [PubMed]
46. Zhang, R.; Zhu, W.; Mao, S. High-concentrate feeding upregulates the expression of inflammation-related genes in the ruminal epithelium of dairy cattle. *J. Anim. Sci. Biotechnol.* **2016**, *7*, 42. [CrossRef] [PubMed]
47. Ma, T.; Gao, Z.; Yang, X.; Guo, S.; Li, S.; Wang, H. Effects of Forage with Varied Concentrate/Roughage Ratios on Growth, Serum Biochemistry, and Immunity of Dumeng Lambs. *Fujian J. Agric. Sci.* **2022**, *37*, 1519–1527. [CrossRef]
48. Lawrence, T. The nuclear factor NF-kappaB pathway in inflammation. *Cold Spring Harb. Perspect. Biol.* **2009**, *1*, a001651. [CrossRef] [PubMed]
49. Zhao, C.; Wang, Y.; Yuan, X.; Sun, G.; Shen, B.; Xu, F.; Fan, G.; Jin, M.; Li, X.; Liu, G. Berberine inhibits lipopolysaccharide-induced expression of inflammatory cytokines by suppressing TLR4-mediated NF-κB and MAPK signaling pathways in rumen epithelial cells of Holstein calves. *J. Dairy Res.* **2019**, *86*, 171–176. [CrossRef] [PubMed]
50. Sarmikasoglou, E.; Chu, L.; Yue, F.; Faciola, A.P. Effects of ruminal LPS exposure on primary bovine ruminal epithelial cells. *J. Dairy Sci.* **2023**, *107*, 1244–1262. [CrossRef] [PubMed]
51. Pang, K.; Dai, D.; Yang, Y.; Wang, X.; Liu, S.; Huang, W.; Xue, B.; Chai, S.; Wang, S. Effects of high concentrate rations on ruminal fermentation and microbiota of yaks. *Front. Microbiol.* **2022**, *13*, 957152. [CrossRef]
52. Jiang, M.; Wang, K.; Huang, Y.; Zhang, X.; Yang, T.; Zhan, K.; Zhao, G. Quercetin Alleviates Lipopolysaccharide-Induced Cell Oxidative Stress and Inflammatory Responses via Regulation of the TLR4-NF-κB Signaling Pathway in Bovine Rumen Epithelial Cells. *Toxins* **2023**, *15*, 512. [CrossRef] [PubMed]
53. Zhao, C.; Yi, F.; Wei, B.; Tan, P.; Huang, Y.; Zeng, F.; Wang, Y.; Xu, C.; Wang, J. Sodium Propionate Relieves LPS-Induced Inflammation by Suppressing the NF-κB and MAPK Signaling Pathways in Rumen Epithelial Cells of Holstein Cows. *Toxins* **2023**, *15*, 438. [CrossRef]
54. Naydenov, N.G.; Hopkins, A.M.; Ivanov, A.I. c-Jun N-terminal kinase mediates disassembly of apical junctions in model intestinal epithelia. *Cell Cycle* **2009**, *8*, 2110–2121. [CrossRef]
55. Carrozzino, F.; Pugnale, P.; Féraille, E.; Montesano, R. Inhibition of basal p38 or JNK activity enhances epithelial barrier function through differential modulation of claudin expression. *Am. J. Physiol. Cell Physiol.* **2009**, *297*, C775–C787. [CrossRef] [PubMed]

Disclaimer/Publisher's Note: The statements, opinions and data contained in all publications are solely those of the individual author(s) and contributor(s) and not of MDPI and/or the editor(s). MDPI and/or the editor(s) disclaim responsibility for any injury to people or property resulting from any ideas, methods, instructions or products referred to in the content.

animals

Article

Nutritional Evaluation of Milk Thistle Meal as a Protein Feedstuff for Diets of Dairy Cattle

Modinat Tolani Lambo †, Rui Liu †, Xianglong Zhang, Yonggen Zhang, Yang Li * and Manji Sun *

College of Animal Science and Technology, Northeast Agricultural University, Harbin 150030, China; motolanilambo@neau.edu.cn (M.T.L.); lr110916@163.com (R.L.); zhangxianglong0908@163.com (X.Z.); zhangyonggen@neau.edu.cn (Y.Z.)

* Correspondence: liyang1405053@neau.edu.cn (Y.L.); sunmanji@163.com (M.S.)

† These authors contributed equally to this work.

Simple Summary: This study aimed to assess the feeding value of milk thistle meal for dairy cattle production by comparing it with seven conventional protein feed resources. The nutritional evaluation revealed that milk thistle meal had a higher fiber content and non-degradable carbohydrate levels compared to most other feedstuffs. Despite its poor carbohydrate composition, it exhibited a higher crude protein rumen effective degradation rate, similar to soybean meal, and had a valuable source of methionine, an essential amino acid. However, it showed lower small intestinal rumen undegradable protein digestibility. The study highlights the potential of milk thistle meal as a nutrient-rich feed supplement for dairy cattle, suggesting avenues for further improvement to enhance its application in dairy feeding practices.

Abstract: The objective of this work is to investigate the chemical and nutritional value of milk thistle meal (MTM) in order to improve it and to provide theoretical support for its application in dairy cattle production. MTM was assessed in comparison with seven conventional protein feed sources, namely, soybean meal (SBM), cottonseed meal (CS), canola meal (CN), palm kernel meal (PK), rice bran meal (RB), corn germ meal (CG), and sesame meal (SS). The chemical composition of these feedstuffs was assessed using wet chemical analysis, the Cornell Net Carbohydrate and Protein System was used to evaluate the carbohydrate and protein fractions, and the in situ nylon bag technique and the modified three-step in vitro method were used to assess the rumen degradation and intestinal digestibility. Additionally, Fourier transform infrared technology was used to determine the feedstuff protein spectral molecular structure and its amino acid profile was also assessed. The result showed that MTM acid detergent fiber, lignin, unavailable nitrogen, and non-degradable carbohydrate content were higher than those of the other feedstuffs. It had a 17% and 36% rumen effective degradation rate of neutral detergent fiber and dry matter, respectively, and had the lowest small intestinal rumen undegradable protein digestibility rate. It was low in leucine, histidine, arginine, and proline, but high in methionine. The total area of amide I and amide II in the protein secondary structure was similar to that of CN and CS, and the amide I and II ratio was not different from that of RB. To sum up, MTM has a poor carbohydrate composition and is high in fiber but, in comparison to most other protein feeds, has a higher crude protein rumen effective degradation rate, similar to that of SBM, and it is a good source of methionine, a limiting amino acid. Hence, its nutritional value can be further improved for application in dairy feeding through processes such as microbial or enzymatic fermentation.

Keywords: milk thistle meal; nutritional value; Fourier transform infrared technology; soybean meal; dairy cattle; CNCPS

Citation: Lambo, M.T.; Liu, R.; Zhang, X.; Zhang, Y.; Li, Y.; Sun, M. Nutritional Evaluation of Milk Thistle Meal as a Protein Feedstuff for Diets of Dairy Cattle. *Animals* **2024**, *14*, 1864. https://doi.org/10.3390/ani14131864

Academic Editor: Marian Czauderna

Received: 24 April 2024
Revised: 1 June 2024
Accepted: 19 June 2024
Published: 24 June 2024

Copyright: © 2024 by the authors. Licensee MDPI, Basel, Switzerland. This article is an open access article distributed under the terms and conditions of the Creative Commons Attribution (CC BY) license (https://creativecommons.org/licenses/by/4.0/).

1. Introduction

The livestock feed industry continues to face a global challenge of feed resource shortage due to the rapid expansion in animal production coupled with the growing human population, resulting in severe competition for human-edible feed resources and culminating in an increasing rise in the prices of conventional animal feed ingredients. This impacts the animal industry severely and necessitates the need to explore alternative, less expensive, locally available and unconventional feed resources, i.e., feed ingredients not commonly utilized in feed composition [1]. Studies have shown that protein remains a major and critical nutritional limiting factor for ruminants [2]. However, feedstuffs with high-quality protein sources like soybean meal and canola meal may be difficult to replace, due to their balanced amino acid (AA) ratio and rich nutrient composition. Milk thistle (*Silybum marianum*) belongs to the family *Asteraceae* and is an annual/biennial plant with high economic value and prospects. Milk thistle contains a considerable amount of protein, fat, flavonolignans, and linoleic and oleic acids [3,4]. It is known for its richness in bioactive compounds, such as total phenolics, flavonoids, tannins, tocopherol, and silymarin, a flavonolignan, which is its most abundant bioactive compound [5,6], hence the increasing interest in its use in animal feeding. Studies have reported the benefits of utilizing various milk thistle byproducts, such as its oil, fruits, and seeds, in the feed of different livestock species. It was reported that feeding 2% milk thistle seed powder enhanced the carcass characteristics and the physical and chemical characteristics of lamb, *Longissimus thoracic*, and also improved its growth performance [7]. Supplementing the fattener diet of pigs with milk thistle seed increased the body weight gain and enhanced meat color, texture, water-holding capacity, and oxidative capacity, while it increased polyunsaturated fatty acid (PUFA) [8]. Karaiskou et al. [9] also observed that supplementing milk thistle oil in ewe's diet translated into higher daily milk yield and enhanced its nutrient value by increasing its composition of monounsaturated fatty acids and PUFA and lowering saturated fatty acid content.

Milk thistle meal (MTM) is a byproduct of milk thistle seed oil extraction [10], and it is gaining attention in the animal industry. Milk thistle seed is made up of up to 26% oil [11], and, after extraction, there remains a residual of silymarin [12], which has various biological and functional properties, such as improving lipid metabolism, inflammation, hepato-protection, and anti-oxidation [13,14], and could be a prospective alternative plant protein feedstuff. Using an alternative feedstuff like MTM is beneficial to alleviate the scarcity of feed resources and lower feed prices; however, unconventional feed sometimes has the general drawback of poor nutrient compositions such as low protein content or AA imbalance and possibly high antinutrients [15,16]. Halmemies-Beauchet-Filleau et al. [17] stated that the possibility of adopting new and unconventional feedstuffs for ruminants relies on the feeding value of such feed ingredients, the response of the animal, its impact on productivity, and its economic value in comparison to its conventional counterpart. According to Hashemi Jabali et al. [18], MTM has a protein content of about 25–30%, and its neutral detergent fiber (NDF) could also be as high as 580 g/kg, which often limits its utilization in ruminants or even results in it being discarded. In addition, it is speculated that it has an insufficient limiting AA content. Currently, information is scarce on the nutritional value, digestibility, and degradation characteristics of MTM, which will limit the possibility of developing it by improving its feeding value and maximizing its potential for application in ruminants.

The Cornell Net Carbohydrate and Protein System (CNCPS) is an appropriate nutritional model that can easily and accurately assess the available nutrient resources in feeds [19] and reflect its degradation rate in the rumen, its digestibility, and its nutrient absorption efficiency. The CNCPS model can also be used to assess the nutrient value of the carbohydrate and protein components of the feed resource [20]. The in situ nylon bag technique can predict the actual rumen degradation and digestion of the nutrients of prospective feedstuff and the feeding value of feeds for livestock [21]. Furthermore, with regards to protein feedstuff utilization in ruminants, the quality and nutritional value, degradability and digestibility, availability, and metabolism of protein do not depend solely on the crude protein and AA composition but are also dependent on the feeds' protein intrinsic molecular structures [22]. The adoption of molecular spectroscopy, such as Fourier transform infrared (FTIR) technology, provides a fast, simple, highly sensitive, and noninvasive analysis of the protein quality of various feed samples by evaluating the primary and secondary protein structures with high resolution. The nutrient value of a protein feed can be predicted by the height and area of the amide I and II bands and the α-helix peak [23].

Therefore, this research was carried out to compensate for the gap in information on the feeding value of milk thistle meal for potential utilization as an alternative feed ingredient by comparing it with other conventional protein feedstuffs. The study compared milk thistle meal with seven other conventional protein feedstuffs based on their chemical composition, rumen degradation, small intestinal digestibility, carbohydrate and protein sub-fractions, amino acid profile, and protein secondary structure using the wet chemical method, the in situ nylon bag technique, the modified three-step in vitro procedure, CNCPS, and FTIR spectroscopy.

2. Materials and Methods

2.1. Feed Sampling and Chemical Composition

Soybean meal (SBM), canola meal (CNM), cottonseed meal (CSM), palm kernel meal (PKM), rice bran meal (RBM), corn germ meal (CGM), and sesame meal (SSM) were all purchased from Cargill Purina Company (Harbin, China). The MTM was procured from Daxinganling Liwokan Biotechnology Co., Ltd. (Daxinganling, China). The composition of the dry matter (DM), crude protein (CP), crude fat (EE), and crude ash were determined following the AOAC methods 930.15, 976.05, 920.39, and 942.05, respectively [24]. The starch content, neutral detergent fiber (NDF), acid detergent fiber (ADF), acid detergent lignin (ADL), neutral detergent insoluble protein (NDICP), acid detergent insoluble protein (ADICP), soluble crude protein (SCP), and non-protein nitrogen (NPN) contents were also analyzed following the methods of Van Soest [25] and Licitra [26]. The non-fibrous carbohydrate (NFC) content was calculated as being approximately 50 g per feedstuff, and all samples were oven-dried at 55 °C for 48 h, ground, sieved through a 1-mm mesh, and then stored at 4 °C for evaluating the chemical composition. The samples (n = 6) were collected per feedstuff and used as replicates.

$$NFC(\text{g/kg}DM) = 1000 - NDF(\text{g/kg}DM) - CP(\text{g/kg}DM) - EE(\text{g/kg}DM) - Ash(\text{g/kg}DM) \tag{1}$$

The feedstuff AA profile was evaluated at Heilongjiang Academy of Agricultural Sciences (Harbin, China) using a fully automated amino acid detector (S-433D; Sekam Scientific Instruments Co., Ltd., Beijing, China), and 17 AA were assayed.

2.2. Protein and Carbohydrate Fractions

The nutritional value of the feedstuffs was evaluated using the CNCPS system following the method of Wang et al. [27]. The CNCPS divides the CP in the feed into non-protein nitrogen (PA), rapidly degradable protein (PB1), moderately degradable protein (PB2), slowly degradable protein (PB3), and unavailable protein (PC). As reported by Sniffen [28], the calculations were as follows:

$$\begin{aligned}
&PA(\text{g/kg}CP) = NFN(\text{g/kg}SCP) \times 0.001 \times SCP(\text{g/kg}CP)\\
&PB1(\text{g/kg}/CP) = SCP(\text{g/kg}CP) - PA(\text{g/kg}CP)\\
&PB2(\text{g/kg}CP) = 1000 - PA(\text{g/kg}CP) - PB1(\text{g/kg}CP) - PB3(\text{g/kg}CP) - PC(\text{g/kg}CP)\\
&PB3(\text{g/kg}CP) = NDIP(\text{g/kg}CP) - ADIP(\text{g/kg}CP)\\
&PC(\text{g/kg}CP) = ADIP(\text{g/kg}CP)
\end{aligned} \tag{2}$$

The carbohydrate was also divided into sugars and soluble fraction (CA), starch and pectin (CB1), fraction of available cell wall (CB2), and unavailable cell wall (CC), and was estimated as follows [28]:

$$
\begin{aligned}
&CHO(\text{g/kg}DM) = 1000 - CP(\text{g/kg}DM) - EE(\text{g/kg}DM) - Ash(\text{g/kg}DM)\\
&CC(\text{g/kg}CHO) = 1000 \times [NDF(\text{g/kg}DM) \times 0.001 \times ADL(\text{g/kg}NDF) \times 2.4]/CHO(\text{g/kg}DM)\\
&CB2(\text{g/kg}CHO) = 1000 \times \begin{bmatrix} NDF(\text{g/kg}DM) - NDIP(\text{g/kg}CP) \times 0.001 \times CP(\text{g/kg}DM) \\ -NDF\text{g/kg}DM) \times 0.001 \times ADL(\text{g/kg}NDF) \times 2.4 \end{bmatrix} /CHO(\text{g/kg}DM)\\
&NSC(\text{g/kg}CHO) = 1000 - CB2(\text{g/kg}CHO) - CC(\text{g/kg}CHO)\\
&CB1(\text{g/kg}CHO) = Starch(\text{g/kg}NSC) \times [1000 - CB2(\text{g/kg}CHO) - CC(\text{g/kg}CHO)]\\
&CA(\text{g/kg}CHO) = [1000 - Starch(\text{g/kg}NSC)] \times [1000 - CB2\text{g/kg}CHO) - CC(\text{g/kg}CHO)]/1000
\end{aligned} \tag{3}
$$

2.3. *In Situ Rumen Degradation*

Three fistulated Chinese Holstein cows with a body weight (BW) of 613 ± 10.2 kg (mean ± SD) in good health and body condition were utilized for in situ rumen degradation evaluation of the feedstuffs. The cows were fed daily at 06:00 and 18:00 and were offered fresh water without restriction. Table 1 shows the chemical and nutritional composition of the cows' diets. Seven grams each of the feedstuffs was measured into pre-labelled nylon bags (n = 3) with a 50 μm pore diameter and size of 10 cm × 20 cm (Ankom Technology, Macedon, NY, USA), which were then tightened with rubber bands before being placed in a large mesh bag with a size 36 cm × 42 cm and then set for fermentation in the rumen ventral sac of each cow underneath the particulate mat layer for 0, 2, 4, 8, 12, and 16, 24, 36, and 48 h. After the fermentation process, the nylon bags were carefully washed with clean cold water, oven-dried at 55 °C for 48 h, and weighed. The residue was milled and sieved with a 1-mm sieve to analyze the DM, CP, and NDF. The rumen degradation rate and effective degradability were estimated following Ørskov and McDonald [29], as follows:

$$p = a + b(1 - e^{-ct}) \tag{4}$$

$$ED = a + b[c/(c + kp)] \tag{5}$$

where p is the degradation rate (%) at time t (h), a is the rapidly degradable fraction (%), b is the potentially degradable fraction (%), c is the rate of potentially degradable fraction, ED is the effective degradability, and the kp (outflow velocity) is 0.06 h^{-1} [30].

Table 1. Experimental diet composition and nutrient level (DM basis) %.

Items	Content
Ingredients	
Chinese wildrye	42.69
Corn silage	15.77
Corn	13.27
Wheat bran	3.75
Molasses	0.99
Soybean meal	3.18
Dried distillers' grain	5.38
Cottonseed meal	2.08
Corn fiber feed	7.45
Corn germ meal	4.96
Premix [1]	0.50
Nutrient composition [2]	
NE_L/(Mcal/kg)	1.30
CP	14.87
NDF	50.81
ADF	31.60
Ca	0.62
P	0.41

[1] Each kilogram of premix contains VA 8,000,000 IU, VD 700,000 IU, VE 10,000 IU, Fe 1600 mg, Cu 1500 mg, Zn 10,000 mg, Mn 3500 mg, Se 80 mg, I 120 mg, and Co 50 mg. [2] Net lactation energy was calculated, and the rest was measured.

2.4. Intestinal Digestion

As described by Calsamiglia and Stern [31], the feedstuff small intestinal DM and protein digestibility were determined using the three-step in vitro method. A total of 1 g of pepsin (Sigma Chemical, St. Louis, MO, USA) was dissolved in pH 1.9 hydrochloric acid solution, and 1 g each of the residues from 16 h in situ rumen fermentation was pre-weighed into a nylon bag. The bags were then placed in the solution before being incubated at 39 °C for 1 h. The bags were then washed with cold water and placed in 1L pH 7.75 KH_2PO_4 buffer solution, which had been previously mixed with 50 μg thymol and 3 g trypsin (sigma P-7545, Sigma Chemical, St. Louis, MO, USA). The samples were then incubated again at 39 °C for 24 h before the bags were washed and dried at 55 °C for 48 h for N content determination. Intestinal digestion is calculated as follows:

$$\text{Intestinal digestibility} = [(\text{Nutrient content of the residue after rumen degradation for 16 h} - \text{Nutrient content of the residue after digestion in the small intestine})/\text{Nutrient content of the residue after rumen degradation for 16 h}] \times 100 \quad (6)$$

2.5. Molecular Structural Characteristics of Milk Thistle Protein Spectra

Variations in the protein structure of the feedstuffs were determined by FTIR using a double-beam infrared spectrophotometer (WGH-30 Tianjin Port Dong Technology, Tianjin, China). A total of 2 mg each of the feed samples and 150 mg of kBr were used to prepare the disks, and scanning was carried out 128 times at a 4000~400 cm^{-1} range and 4 cm^{-1} spectral resolution. In total, five spectra were collected repeatedly per sample. The obtained spectra were analyzed with OMNIC 8.2 software (Thermo Nicolet Co., Madison, WI, USA). First, the spectrum was automatically smoothed, and the peak areas of the amide I and II bands were measured. Then, the peak height positions of the amide I and II bands, α-helix, and β-sheet were determined by Fourier automatic deconvolution function.

2.6. Statistical Analysis

All data were sorted and organized in Excel (2016). Rumen degradation and intestinal digestibility data were analyzed on SAS software (version 9.4, SAS Institute Inc., Cary, NC, USA). The NLIN program was used to determine the degradation parameters a, b, and c values before being analyzed with other indicators. All other experimental data were also analyzed on SAS 9.4 software using the general linear model (GLM) procedure to perform variance analysis, and multiple comparisons were performed using the Tukey test, and the result was declared significant when $p < 0.05$.

3. Results and Discussion

3.1. Chemical Composition

Accessing readily available quality protein feed resources—which are not only affordable, but can also meet the nutrient requirement for animals—remains a challenge within the livestock industry. Over the years, feedstuffs like soybean, canola, and cottonseed meal have remained the major protein feedstuffs with great value. However, going by the recent rise in the market prices of conventional feedstuffs, due to high competition for its utilization between humans and animals, it is imperative to develop affordable, locally sourced protein alternatives. To achieve this, evaluating the basic nutritional components of potential alternative feeds is crucial for identifying their deficiencies as a basis for improving them for utilization as a protein feed resource for livestock. Studies have reported the nutritional composition of milk thistle seed cake or oil [32,33]. Still, to our knowledge, this is the first study to assess the nutritional and feeding value of MTM fully. The chemical composition of all of the evaluated feedstuffs is shown in Table 2. By comparison, we found that MTM has a higher NDF content than soybean meal, which had the lowest NDF of all of the feedstuffs (59% vs. 29%). The NDF composition of MTM was, in fact, not significantly different from that of corn germ and palm kernel meal. Its ADF and ADL compositions were also higher than those of the other feedstuffs. Its ash content was not significantly different from SBM, CNM, and CSM, while its starch, EE, and NFC content was lower than those of most other feedstuffs, including SBM.

Table 2. Chemical composition of milk thistle meal and seven conventional protein feedstuffs (DM basis) %.

Items	Soybean Meal	Canola Meal	Cottonseed Meal	Palm Kernel Meal	Rice Bran Meal	Corn Germ Meal	Sesame Meal	Milk Thistle Meal	SEM	*p*-Value
DM	88.77 [d]	90.78 [a]	90.43 [ab]	90.09 [bc]	88.63 [d]	89.60 [c]	88.77 [d]	89.62 [c]	0.148	<0.0001
Ash	7.31 [cd]	6.94 [cd]	6.62 [d]	4.72 [e]	11.27 [b]	1.25 [f]	19.50 [a]	7.97 [c]	0.248	<0.0001
EE	2.29 [cd]	3.29 [b]	2.80 [bcd]	7.86 [a]	2.03 [d]	2.92 [bc]	2.34 [cd]	0.20 [e]	0.179	<0.0001
CP	44.54 [c]	40.28 [d]	48.16 [a]	16.42 [g]	16.72 [g]	20.90 [f]	46.09 [b]	25.57 [e]	0.328	<0.0001
SP	8.40 [b]	9.01 [b]	4.76 [d]	0.39 [e]	3.42 [d]	4.32 [d]	10.65 [a]	6.74 [c]	0.348	<0.0001
NDICP	12.71 [b]	5.62 [e]	5.71 [e]	10.23 [c]	5.57 [e]	8.26 [d]	15.36 [a]	6.13 [e]	0.416	<0.0001
ADICP	0.58 [e]	1.95 [c]	1.02 [d]	1.31 [d]	0.46 [e]	0.37 [e]	4.01 [a]	2.61 [b]	0.0759	<0.0001
NPN	0.59 [d]	4.98 [b]	2.84 [c]	0.00 [d]	2.20 [c]	2.49 [c]	8.61 [a]	4.34 [b]	0.258	<0.0001
NDF	29.19 [b]	33.05 [b]	35.57 [b]	61.64 [a]	31.69 [b]	56.93 [a]	35.72 [b]	59.19 [a]	1.564	<0.0001
ADF	8.90 [f]	20.97 [c]	17.88 [cd]	34.27 [b]	12.75 [e]	15.11 [ed]	17.92 [cd]	43.15 [a]	0.697	<0.0001
ADL	1.65 [e]	8.17 [bc]	7.00 [c]	9.28 [b]	4.53 [d]	1.82 [e]	7.10 [c]	16.83 [a]	0.294	<0.0001
Starch	2.29 [c]	0.23 [e]	0.22 [e]	0.00 [f]	6.51 [b]	6.73 [a]	0.00 [f]	0.82 [d]	0.0186	<0.0001
NFC	16.67 [b]	16.46 [b]	6.84 [cd]	9.36 [c]	38.29 [a]	18.00 [b]	2.47 [d]	7.08 [cd]	1.225	<0.0001

DM, dry matter; EE, ether extract; CP, crude protein; SCP, soluble crude protein; NPN, non-protein nitrogen; NDICP, neutral detergent insoluble crude protein; ADICP, acid detergent insoluble crude protein; NDF, neutral detergent fiber; ADF, acid detergent fiber; ADL, acid detergent lignin; NFC, non-fiber carbohydrate. [a–g] Values in the same line with different capital letter superscripts mean samples have significant differences ($p < 0.001$); SEM, standard error of the mean.

NDF is vital for the digestive process of dairy cows as it has a significant impact on their feed intake, rumen fermentation, and nutrient utilization, and it is essential for maintaining the rumen environment and sustaining milk production and fat content [34,35]. When formulating the diets for high-yielding dairy cows, it is necessary to ensure an appropriate concentration of NDF while meeting energy density requirements, as this can maximize milk production and fat content [34]. According to the National Research Council (NRC) recommendations, they should have an NDF level of about 30–40% in their diet [36]. The result of this study revealed that MTM has a very high fiber content. Shi [35] reported that, during peak lactation, cows fed a high NDF diet had lower milk production, particularly the total milk yield, milk lactose and protein yield, the percentage of milk protein, and nitrogen efficiency. Moreover, even though ruminants need to consume sufficient NDF to ensure various physiological functions, they tend to obtain it from fresh and tender forage. The high lignin content in MTM, reaching up to 16.83%, may also negatively impact the digestion rate of the feed. NDF includes cellulose, hemicellulose, and lignin. For ruminant animals, the plant cellulose and hemicellulose can be fermented by rumen microorganisms and maximized as energy sources. However, lignin is less digestible and can impact the rumen digestion rate because it is not degraded in the rumen anaerobic environment but only aerobically; therefore, its level in feed has a negative correlation to DM digestibility [37,38]. NFC includes starch, sugars, and pectin, and it has been noted that increasing their levels in the diet can enhance milk production performance and benefit the use of energy and nitrogen in dairy cows [39]. From this study, MTM had an NFC composition of about 7%, which was relatively lower than that of soybean meal (16.67%); therefore, caution is needed when replacing it in ruminant ration.

Furthermore, the CP content of MTM was lower (25%) compared to that of SBM, CNM, CSM, and SSM, but it was still higher than that of PKM, RBM, and CGM, making it a potential protein feed candidate with prospect. Its NDICP content was also lesser than that of other feedstuff, except for CNM, CSM, and RBM, while its ADICP content was higher than other feedstuffs, except for SSM. SCP and NPN are high-quality sources of rumen-degradable protein (RDP). MTM has relatively high levels of SCP and NPN, similar to SBM and RSM, indicating a higher proportion of rapidly degradable components in the rumen. However, it also has a high level of ADICP, which is less available to the rumen as an AA, reducing its feeding value [40]. The literature has revealed that, after pressing or extracting oilseed meals, only a small amount of fat is retained, yet there could be variations due to differences in extraction processes. In this study, PKM retained a higher fat content, and MTM had the lowest fat content among the feedstuff, at only 0.20%, indicating that it could be limited in essential fatty acids, such as linoleic and linolenic acid. A low fat content could also reduce its palatability and, consequently, the feed intake and growth performance of the animal.

3.2. Amino Acid Profile

The AA profile of MTM and other feedstuff is presented in Table 3. MTM has a poor AA composition, with leucine, histidine, arginine, and proline being all significantly lower than in the other feedstuffs. In addition, the proportions of aspartic acid, glycine, isoleucine, tyrosine, and total AA were lower than the other feedstuffs, except CGM. Its lysine proportion was also lower than that of SBM, CNM, and CSM, although still higher than that found in PKM, RBM, CGM, and SSM. Nevertheless, it is noteworthy that it has high methionine content, exceeded by only SSM, and it has high serine and cystine contents, exceeded by only SBM. The main limiting AAs for dairy cows are methionine, histidine, and lysine [41]. Even though MTM has a higher methionine content than SBM, it is lower in lysine and histidine composition; therefore, when used as a substitute for soybean meal, careful consideration is needed. Zhu previously reported on the protein and AA composition of *Silybum marianum* meal and observed a 47% CP, which differs greatly from this study. This variation could result from plant variety, origin, and processing techniques, which will, in turn, influence the difference in the AA composition as well. Generally, SBM has a high total AA content and a balanced AA proportion, and MTM has a considerable gap in comparison. As a result, when reducing the amount of SBM in the dairy cow ration, MTM may not have an advantage in terms of AA content, more so because its CP content is slightly lower. However, considering the price advantage, it is still an unconventional feed that is worth developing.

Table 3. Amino acid contents of milk thistle meal and seven conventional protein feedstuffs.

Items	Soybean Meal	Canola Meal	Cottonseed Meal	Palm Kernel Meal	Rice Bran Meal	Corn Germ Meal	Sesame Meal	Milk Thistle Meal	SEM	*p*-Value
Aspartate	5.45 ^{a}	2.49 ^{d}	4.36 ^{b}	1.22 ^{g}	1.33 ^{f}	1.35 ^{f}	2.59 ^{c}	1.79 ^{e}	0.017	<0.0001
Threonine	1.68 ^{a}	0.70b ^{bc}	1.01 ^{b}	0.32 ^{c}	0.39 ^{c}	0.56 ^{bc}	0.68 ^{bc}	0.78 ^{bc}	0.120	<0.0001
Serine	2.59 ^{a}	0.56 ^{d}	0.83 ^{c}	0.28 ^{f}	0.28 ^{f}	0.42 ^{e}	0.40 ^{e}	1.12 ^{b}	0.006	<0.0001
Glutamic acid	8.84 ^{b}	6.66 ^{d}	10.35 ^{a}	3.21 ^{f}	2.00 ^{h}	2.90 ^{g}	7.08 ^{c}	3.93 ^{e}	0.013	<0.0001
Glycine	2.07 ^{a}	1.92 ^{c}	2.02 ^{b}	0.71 ^{g}	0.80 ^{f}	1.07 ^{e}	1.87 ^{d}	1.10 ^{e}	0.008	<0.0001
Alanine	2.05 ^{a}	1.61 ^{d}	1.80 ^{c}	0.63 ^{h}	0.89 ^{f}	1.21 ^{e}	1.87 ^{b}	0.82 ^{g}	0.008	<0.0001
Cysteine	0.68 ^{a}	0.31 ^{d}	0.55 ^{c}	0.11 ^{g}	0.20 ^{e}	0.14 ^{f}	0.30 ^{d}	0.61 ^{b}	0.005	<0.0001
Valine	2.00 ^{c}	2.01 ^{c}	2.23 ^{b}	1.02 ^{f}	0.96 ^{f}	1.35 ^{d}	2.57 ^{a}	1.22 ^{e}	0.012	<0.0001
Methionine	0.62 ^{d}	0.67 ^{c}	0.61 ^{d}	0.30 ^{f}	0.15 ^{g}	0.39 ^{e}	1.71 ^{a}	0.77 ^{b}	0.005	<0.0001
Isoleucine	1.81 ^{a}	1.50 ^{c}	1.55 ^{b}	0.59 ^{g}	0.58 ^{g}	0.75 ^{f}	1.45 ^{d}	0.93 ^{e}	0.007	<0.0001
Leucine	2.92 ^{a}	2.62 ^{a}	2.80 ^{a}	1.05 ^{c}	1.07 ^{c}	1.71 ^{b}	2.58 ^{a}	1.39 ^{b}	0.007	<0.0001
Tyrosine	1.43 ^{a}	0.97 ^{d}	1.20 ^{c}	0.37 ^{h}	0.43 ^{g}	0.56 ^{f}	1.39 ^{b}	0.72 ^{e}	0.007	<0.0001
Phenylalanine	2.23 ^{b}	1.53 ^{c}	2.67 ^{a}	0.77 ^{e}	0.77 ^{e}	0.93 ^{d}	2.23 ^{b}	0.90 ^{d}	0.007	<0.0001
Lysine	2.66 ^{a}	2.09 ^{b}	2.03 ^{c}	0.44 ^{h}	0.72 ^{f}	0.85 ^{e}	0.59 ^{g}	0.97 ^{d}	0.008	<0.0001
Histidine	1.09 ^{b}	1.01 ^{c}	1.36 ^{a}	0.28 ^{h}	0.38 ^{g}	0.66 ^{e}	0.83 ^{d}	0.44 ^{f}	0.005	<0.0001
Arginine	3.18 ^{b}	2.15 ^{d}	5.71 ^{a}	1.74 ^{e}	1.02 ^{h}	1.22 ^{g}	2.51 ^{c}	1.65 ^{f}	0.009	<0.0001

Table 3. *Cont.*

Items	Soybean Meal	Canola Meal	Cottonseed Meal	Palm Kernel Meal	Rice Bran Meal	Corn Germ Meal	Sesame Meal	Milk Thistle Meal	SEM	*p*-Value
Proline	2.41 [a]	2.18 [b]	1.80 [d]	0.80 [h]	0.92 [g]	1.34 [e]	1.99 [c]	1.03 [f]	0.006	<0.0001
Total Amino Acid	43.69 [a]	30.98 [d]	42.88 [b]	13.84 [g]	12.89 [h]	17.41 [f]	32.64 [c]	20.16 [e]	0.155	<0.0001

Values are expressed as a percentage of crude protein (%CP). [a–h] Values in the same line with different letter superscripts mean samples have significant differences ($p < 0.001$); SEM, standard error of the mean.

3.3. Protein and Carbohydrate Fraction

Table 4 shows the CNCPS components of all of the feedstuffs. Regarding the protein fractions, MTM had a higher PA component than SBM, CNM, CSM, PKM, RBM, and CGM. The PA component is mainly related to NPN. Based on this study, MTM contains 4.34% NPN (Table 2), resulting in a PA component of 16.83%. NPN is made up specifically of the nitrogen present in the nucleic acids and ammonia, and it can be rapidly utilized by the rumen microorganisms to synthesize microbial protein [42], which provides most of the protein available in ruminants' small intestine and constitutes about 50–80% of the total protein that is absorbed [43]. The PB component represents the true protein portion of the feed. Our results revealed that the PB1 component of MTM was similar to that of SBM, CNM, CSM, RBM, CGM, and SSM; moreover, its PB2 component was also not different ($p > 0.05$) from SBM, RB, and SS, and its PB3 component was lower than that of SBM, PKM, RBM, CGM, and SSM. PB1 and PB2 are rapidly and moderately degradable, respectively. The PB2 component is soluble in a neutral detergent but insoluble in a buffer. Accelerated degradation of feed protein could result in peptides and AA accumulation two hours after feeding [20]. It could be lost as microbial protein or result in increased ammonia production, because true protein is initially degraded into peptides and AA and later deaminated into ammonia nitrogen [44]. Therefore, regulating the feed content of PA and PB feeds can better control the rate of microbial protein, reduce nitrogen loss, and promote nitrogen utilization in feeds. MTM has optimum levels of these two components, similar to soybean meal, indicating that it has good protein quality. However, this study revealed that it had the highest PC component of about 10%, which may affect the evaluation of its protein composition because the PC represents protein bound to substances like lignin and tannins that rumen microorganisms cannot utilize [19].

Regarding the carbohydrate components, MTM has a significantly lower CA component than that of SBM, CNM, CSM, RBM, and SSM; moreover, its CB1 component was lower than that of SBM, RBM, and CGM. Its CB2 component was lower than that of CSM, PKM, and CGM, and not different from SBM, CNM, RBM, and SSM. It had the highest CC component, reaching about 61%. On a dry matter basis, its CHO content was significantly lower than PKM, RBM, and CGM, but higher than that of SBM, CNM, CSM, and SSM, and the NSC content in its CHO was lower than that of SBM. The CA component is the water-soluble fraction that is rapidly fermentable and is composed mainly of sugars and some proportions of oligosaccharides and organic acids [20], and the CB1 component is the fraction composed of starch and pectin. The MTM fractions of CA and CB1 are much lower than those of most protein feedstuffs. The CC component, which is the indigestible fraction composed of unavailable cell wall, accounts for 60.94% of the entire carbon component, which could lead to a lower digestibility of the carbon component and, therefore, limit the application of MTM. In short, MTM has a good proportion of protein components but a poor proportion of carbon components. To further develop this feedstuff for a better feeding performance, the digestibility of the carbon components needs to be improved in order to maximize its feeding value.

Table 4. CNCPS components of milk thistle meal and seven conventional protein feedstuffs.

Items	Soybean Meal	Canola Meal	Cottonseed Meal	Palm Kernel Meal	Rice Bran Meal	Corn Germ Meal	Sesame Meal	Milk Thistle Meal	SEM	*p*-Value
PA (%CP)	7.64 ^{c}	12.35 ^{b}	5.92 ^{c}	0.00 ^{d}	13.14 ^{b}	11.96 ^{b}	18.71 ^{a}	16.83 ^{a}	0.79	<0.0001
PB_1 (%CP)	11.56 ^{a}	10.02 ^{ab}	3.95 ^{ab}	2.34 ^{b}	7.34 ^{ab}	9.02 ^{ab}	4.43 ^{ab}	9.79 ^{ab}	1.803	0.0071
PB_2 (%CP)	54.54 ^{bc}	63.72 ^{b}	78.29 ^{a}	35.35 ^{e}	43.84 ^{de}	39.46 ^{de}	43.52 ^{de}	49.28 ^{cd}	2.276	<0.0001
PB_3 (%CP)	23.97 ^{d}	9.07 ^{e}	9.72 ^{e}	54.32 ^{a}	32.90 ^{c}	37.77 ^{b}	24.63 ^{d}	13.75 ^{e}	1.060	<0.0001
PC (%CP)	2.30 ^{d}	4.84 ^{c}	2.13 ^{d}	8.00 ^{b}	2.78 ^{d}	1.80 ^{d}	8.71 ^{b}	10.37 ^{a}	0.265	<0.0001
CA (%CHO)	59.05 ^{a}	44.10 ^{abc}	29.04 ^{cd}	27.60 ^{cd}	53.39 ^{ab}	26.05 ^{d}	36.19 ^{bcd}	18.68 ^{d}	3.859	<0.0001
CB_1 (%CHO)	5.00 ^{c}	0.46 ^{e}	0.51 ^{e}	0.00 ^{f}	9.30 ^{a}	8.98 ^{b}	0.00 ^{f}	1.23 ^{d}	0.0390	<0.0001
CB_2 (%CHO)	27.33 ^{cd}	15.77 ^{de}	30.84 ^{bc}	41.04 ^{b}	21.76 ^{cde}	59.14 ^{a}	13.39 ^{e}	19.14 ^{cde}	2.739	<0.0001
CC (%CHO)	8.62 ^{ef}	39.66 ^{c}	39.61 ^{c}	31.35 ^{d}	15.55 ^{e}	5.83 ^{f}	53.08 ^{b}	60.94 ^{a}	1.589	<0.0001
CHO (%DM)	45.86 ^{e}	49.50 ^{d}	42.41 ^{f}	70.99 ^{b}	69.98 ^{b}	74.93 ^{a}	32.08 ^{g}	66.27 ^{c}	0.398	<0.0001
NSC (%CHO)	64.05 ^{a}	44.56 ^{b}	29.55 ^{bc}	27.60 ^{bc}	62.69 ^{a}	35.03 ^{bc}	36.19 ^{bc}	19.92 ^{c}	3.854	<0.0001

PA, non-protein nitrogen; PB1, rapidly degraded protein; PB2, intermediately degraded protein; PB3, slowly degraded protein; PC, unavailable protein; CHO, carbohydrate; NSC, non-structural carbohydrate; CA, sugars and soluble fraction; CB1, starch and pectin; CB2, fraction of available cell wall; CC, unavailable cell wall. $^{a–g}$ Values in the same line with different letter superscripts mean samples have significant differences ($p < 0.001$); SEM, standard error of the mean.

3.4. In Situ Rumen Degradation and Small Intestinal Digestibility

The in situ DM, CP, and NDF rumen degradation is shown in Table 5. With respect to the rumen degradation kinetics of DM, the *a*-value of MTM was lower than that of SBM, CNM, RBM, and SSM, and the *b*-value was lower than that of the other feedstuffs. Similarly, it had the least effective degradability, the same as PKM. Among the CP rumen degradation kinetics, MTM had the highest *a*-value, and its *b*-value did not differ significantly from the others, except SSM, while its *c*-value was the lowest. Its CP effective degradation was not different from SBM and CNM and was higher than the others, while it had a rumen bypass protein lower than that of PKM, RBM, CGM, and SSM. Furthermore, it had the lowest NDF rapid degradation fraction, a slow degradation fraction lower than SBM and PKM, and an NDF effective degradation rate lower than CNM, CSM, PKM, CGM, and SSM.

The MTM rumen DM rapid, slow, and effective degradation rates were lower than those of most protein feeds determined in this experiment. This may be due to the excessively high NDF level and the relatively low rumen NDF effective degradation rate observed. The NDF content of MTM reached up to 59%, but the rumen NDF rapid, slow, and effective degradation rates were 0.17%, 30.41%, and 17.78%, respectively. The MTM low NDF degradation is also reflected and associated with the low DM degradation rate and relatively high CC component predicted by CNCPS. We speculate that this likely results from the excessively high ADL content in MTM, accounting for a large proportion of NDF, as ADL is negatively correlated with the rumen fiber digestion rate. Lignin is cross-linked with arabinoxylan, hindering the physical contact of hydrolytic microbial enzymes [45]. When MTM is used in large amounts in the ruminant diet, the low rumen digestion rate could cause rumen expansion or bloating by filling up a large space, hence reducing feed intake and making it challenging to meet the nutritional needs for high milk production [34]. Nevertheless, because of the good rumen CP effective degradation rate of MTM, which is similar to that of soybean meal, the nutritive value of the feedstuff can be further developed with a focus on lowering NDF and increasing the rumen DM and NDF degradation rate via processes such as microbial fermentation.

Table 5. Rumen degradation parameters of milk thistle meal and seven conventional protein feedstuffs.

Items	Soybean Meal	Canola Meal	Cottonseed Meal	Palm Kernel Meal	Rice Bran Meal	Corn Germ Meal	Sesame Meal	Milk Thistle Meal	SEM	p-Value
In situ dry matter (DM) rumen degradation kinetics										
a (g/kg)	357 ^{a}	237 ^{bc}	137 ^{cd}	114 ^{d}	321 ^{ab}	136 ^{cd}	281 ^{ab}	132 ^{d}	21.21	<0.0001
b (g/kg)	643 ^{c}	682 ^{bc}	636 ^{cd}	886 ^{a}	544 ^{de}	764 ^{b}	376 ^{f}	464 ^{ef}	19.03	<0.0001
c (g/kg h^{-1})	28.0 ^{de}	37 ^{cd}	41.1 ^{cd}	16.2 ^{e}	43 ^{bc}	55 ^{b}	89 ^{a}	39 ^{c}	3.06	<0.0001
ED (g/kg)	622 ^{a}	571 ^{bc}	510 ^{d}	369 ^{e}	602 ^{ab}	577 ^{bc}	541 ^{cd}	364 ^{e}	8.9	<0.0001
In situ crude protein (CP) rumen degradation kinetics										
a (g/kg)	202 ^{b}	168 ^{bc}	94.7 ^{d}	119 ^{d}	167 ^{bc}	85 ^{d}	212 ^{b}	292 ^{a}	13.01	<0.0001
b (g/kg)	798 ^{a}	832 ^{a}	852 ^{a}	687 ^{ab}	833 ^{a}	902 ^{a}	406 ^{b}	708 ^{ab}	66.20	0.0019
c (g/kg h^{-1})	26.9 ^{b}	38.3 ^{b}	40.5 ^{b}	19.9 ^{b}	23.6 ^{b}	33.2 ^{b}	95.4 ^{a}	27.0 ^{b}	7.01	<0.0001
ED (g/kg)	526 ^{abc}	571 ^{ab}	516 ^{abc}	282 ^{d}	473 ^{c}	493 ^{c}	496 ^{bc}	577 ^{a}	15.90	<0.0001
RUP (g/kg)	474 ^{bcd}	429 ^{cd}	484 ^{bcd}	718 ^{a}	527 ^{b}	507 ^{b}	504 ^{bc}	423 ^{d}	15.94	<0.0001
In situ neutral detergent fiber (NDF) rumen degradation kinetics										
a (g/kg)	249 ^{a}	74.7 ^{cd}	34.8 ^{b}	5.16 ^{cd}	6.83 ^{cd}	7.5 ^{cd}	8.86 ^{c}	1.70 ^{d}	18.3	<0.0001
b (g/kg)	657 ^{a}	361 ^{bc}	633 ^{ab}	995 ^{a}	85.10 ^{bc}	311 ^{bc}	55.21 ^{c}	304 ^{bc}	11.8	0.0005
c (g/kg h^{-1})	38.7 ^{ab}	10.6 ^{bc}	4.0 ^{ab}	3.90 ^{c}	30.8 ^{abc}	26.7 ^{abc}	39.8 ^{ab}	56.2 ^{a}	0.07	0.0005
ED (g/kg)	562 ^{a}	33.13 ^{f}	53.043 ^{e}	80.12 ^{d}	26.34 ^{f}	123 ^{c}	35.61 ^{f}	177.8 ^{b}	2.81	<0.0001

a, rapidly degradable fraction in rumen degradation; b, slowly degradable fraction in rumen degradation; a + b, potentially degradable fraction in rumen degradation; c, the degradation rate of the slowly degradable fraction; ED, effective degradability of the incubated samples; RUP, rumen undegradable protein. $^{a–f}$ Values in the same line with different letter superscripts mean samples have significant differences ($p < 0.001$); SEM, standard error of the mean.

A large portion of macronutrients, including lipids, escape starch, and proteins, are broken down and absorbed in the small intestine [2]. The small intestinal DM and RUP digestibility of MTM was lower than that of most of the other feedstuffs and far lower than that of SBM (Table 6). The high NDF composition of MTM can account for this, and the nutrient might have been bound to lignin and difficult to digest. The implication for dairy cows is less efficient nutrient uptake, which could impact their growth performance and milk production. Typically, a greater portion of the protein in a high-quality protein feedstuff should be available to reach the small intestine, where it is digested and absorbed [46]; therefore, MTM is not a good source of RUP. Nevertheless, its digestibility can be improved by the fermentation process, the addition of exogenous enzymes, or both [47].

Table 6. Small intestine digestibility of milk thistle meal and seven conventional protein feedstuffs.

Items	Small Intestinal Dry Matter Digestibility (g/kg)	Small Intestinal Protein Degradation Rate (g/kg)
Soybean meal	656 ^{a}	895 ^{a}
Canola meal	248 ^{d}	483 ^{c}
Cottonseed meal	366 ^{b}	702 ^{b}
Palm kernel meal	16.46 ^{e}	559 ^{c}
Rice bran meal	327 ^{bc}	551 ^{c}
Corn germ meal	95.5 ^{f}	136 ^{e}
Sesame meal	299 ^{bc}	302 ^{d}
Milk thistle meal	276 ^{cd}	367 ^{d}
SEM	14.88	23.15
p-Value	<0.0001	<0.0001

$^{a–f}$ Values in the same line with different letter superscripts mean samples have significant differences ($p < 0.001$); SEM, standard error of the mean.

3.5. Spectral Molecular Structure Characteristics of Milk Thistle Meal and Seven Conventional Protein Feedstuffs

The results in Table 7 reveal that the total area of amide I and II regions of MTM was significantly lower that of than other feedstuffs, except for RSM and CSM, while its peak area of amide II was lower than that of SBM and CSM but not different from the other feedstuff. In addition, its peak area ratio of amide I and II regions was higher than that of SBM, CNM, CSM, PKM, and CGM, while the peak height was higher than that of the other feedstuff. Additionally, the MTM protein secondary structure was significantly lower when compared to the other feedstuff, but the ratio of the α-helix to the β-sheet was not different. The peak height and area intensity of the amide I and II bands are related to the number of protein functional groups, and differences in their ratios signify variations in protein structural composition [48]. The amide I band was mainly composed of C=O stretching vibrations and a small amount of C-N stretching vibrations, while the amide II band was mainly composed of N-H bending vibrations and a small amount of C-N stretching vibrations [49]. In this study, SBM, CSM, CNM, and SSM exhibited higher values in the peak area and peak height of the amide I and amide II bands, possibly because of their higher total protein content. Conversely, PKM, RBM, CGM, and MTM showed lower values, due to their relatively lower protein content.

Table 7. Protein spectral molecular structure characteristics of milk thistle meal and seven conventional protein feedstuffs.

Items	Soybean Meal	Canola Meal	Cottonseed Meal	Palm Kernel Meal	Rice Bran Meal	Corn Germ Meal	Sesame Meal	Milk Thistle Meal	SEM	*p*-Value
Amide I										
A_Amide I and Amide II	14.27 ^{b}	10.04 ^{c}	9.29 ^{c}	1.41 ^{d}	3.50 ^{d}	2.24 ^{d}	19.45 ^{a}	8.33 ^{c}	0.858	<0.0001
A_Amide I	6.74 ^{b}	5.36 ^{b}	4.98 ^{b}	0.35 ^{d}	2.66 ^{c}	1.77 ^{cd}	10.55 ^{a}	4.88 ^{b}	0.461	<0.0001
A_Amide II	2.00 ^{a}	1.36 ^{ab}	1.94 ^{a}	0.77 ^{bc}	0.50 ^{c}	0.51 ^{c}	1.17 ^{bc}	0.93 ^{bc}	0.164	<0.0001
A_Amide I/Amide II ratio	3.62 ^{c}	4.06 ^{c}	2.59 ^{d}	0.46 ^{e}	5.33 ^{b}	3.62 ^{c}	9.04 ^{a}	5.27 ^{b}	0.200	<0.0001
Amide I and II										
H_Amide I	0.079 ^{b}	0.075 ^{b}	0.068 ^{b}	0.019 ^{d}	0.036 ^{cd}	0.024 ^{d}	0.11 ^{a}	0.057 ^{bc}	0.006	<0.0001
H_Amide II	0.036 ^{a}	0.043 ^{a}	0.043 ^{a}	0.016 ^{b}	0.012 ^{b}	0.011 ^{b}	0.049 ^{a}	0.018 ^{b}	0.003	<0.0001
H_Amide I/Amide II ratio	2.31 ^{b}	1.74 ^{cd}	1.62 ^{de}	1.20 ^{e}	3.13 ^{a}	2.26 ^{b}	2.18 ^{bc}	3.10 ^{a}	0.103	<0.0001
Protein secondary structure spectral structural characteristics										
H_α-helix	0.072 ^{b}	0.072 ^{b}	0.064 ^{b}	0.018 ^{d}	0.034 ^{cd}	0.022 ^{d}	0.011 ^{a}	0.052 ^{bc}	0.006	<0.0001
H_β-sheet	0.077 ^{ab}	0.070 ^{bc}	0.066 ^{bc}	0.015 ^{e}	0.031 ^{de}	0.023 ^{de}	0.103 ^{a}	0.049 ^{cd}	0.006	<0.0001
H_α-helix/β-sheet ratio	0.95 ^{b}	1.03 ^{b}	0.97 ^{b}	1.30 ^{a}	1.08 ^{ab}	0.97 ^{b}	1.03 ^{b}	1.08 ^{ab}	0.075	0.0526

A_Amide I and Amide II: peak area intensity of total amide I and amide II; A_Amide I: peak area intensity of Amide I; A_Amide I/Amide II ratio: peak area ratio of amide I to amide II; H_Amide I: peak height intensity of amide I; H_Amide II: peak height intensity of amide II; H_Amide I/Amide II ratio: peak height ratio of amide I to amide II; H_α-helix: peak height intensity of α-helix; H_β-sheet: peak height intensity of β-sheet; H_α-helix/β-sheet ratio: peak height ratio of α-helix to β-sheet. $^{a–e}$ Values in the same line with different letter superscripts mean samples have significant differences ($p < 0.001$); SEM, standard error of the mean.

The amide I band includes the secondary protein structures of α-helix and β-sheet spatial structures, both of which are related to protein digestion in animals [50]. The amide I to amide II band ratio and the α-helix to β-sheet ratio are often related to feed protein solubility, rumen degradation rate, and small intestinal digestibility [51]. MTM and RBM show no difference in the amide I to II bands ratio and α-helix to β-sheet height ratio, even though their rumen degradation rate and small intestinal digestibility do not exhibit similarity. Studies have shown that the ratio of the amide I band to the amide II band is closely influenced by deoiling methods [52] and feed types [53]. Although there is a correlation in the protein molecular structure between MTM and RBM, there are significant differences in their nutritional composition. Therefore, a spectral structure may not fully reflect the digestive characteristics of the feeds' proteins.

4. Conclusions

Based on the outcome revealed in this study, milk thistle meal, in comparison to most other conventional feedstuffs, has a beneficial nutritional quality. Although it has a lower crude protein, it exhibits moderate levels of essential amino acids, especially methionine, has a significant fraction of indigestible protein and slowly digestible carbohydrates, and has high crude protein effective degradability and rumen undegradable protein values, making it a potential protein feed resource. However, caution is required when including milk thistle meal in the ruminant diet, because of its high fiber content, low effective degradability of dry matter, and neutral detergent fiber, coupled with a very low rumen-effective degradation rate of NDF. There is, therefore, a need to investigate further ways of improving its nutritional value and developing it as an unconventional feed suitable for dairy cattle.

Author Contributions: Conceptualization, Y.L. and M.S.; data curation, R.L., X.Z. and Y.L.; funding acquisition, M.T.L. and Y.L.; investigation, R.L. and X.Z.; resources, Y.Z.; writing—original draft preparation, M.T.L. and Y.L.; writing—review and editing, M.T.L. and Y.Z. All authors have read and agreed to the published version of the manuscript.

Funding: This research was funded by the National Key Research and Development Program of China (2023YFD2000701), the Key Research and Development Program of Heilongjiang Province of China (GZ20230028), the Natural Science Foundation of Heilongjiang Province (YQ2023C011), the China Agriculture Research System (CARS-36), and the China Postdoctoral Research fund.

Institutional Review Board Statement: The animal study protocol was approved by the Northeast Agricultural University Animal Science and Technology College Animal Care and Use Committee (Protocol number: NEAUEC20181007).

Informed Consent Statement: Not applicable.

Data Availability Statement: Data are contained within the article.

Conflicts of Interest: The authors declare no conflicts of interest.

References

1. Duguma, B.; Janssens, G.P.J. Assessment of Feed Resources, Feeding Practices and Coping Strategies to Feed Scarcity by Smallholder Urban Dairy Producers in Jimma Town, Ethiopia. *SpringerPlus* **2016**, *5*, 717. [CrossRef] [PubMed]
2. Harmon, D.L.; Swanson, K.C. Review: Nutritional Regulation of Intestinal Starch and Protein Assimilation in Ruminants. *Animal* **2020**, *14*, s17–s28. [CrossRef] [PubMed]
3. Bijak, M. Silybin, a Major Bioactive Component of Milk Thistle (*Silybum marianum* L. Gaernt.)—Chemistry, Bioavailability, and Metabolism. *Molecules* **2017**, *22*, 1942. [CrossRef] [PubMed]
4. Marceddu, R.; Dinolfo, L.; Carrubba, A.; Sarno, M.; Di Miceli, G. Milk Thistle (*Silybum marianum* L.) as a Novel Multipurpose Crop for Agriculture in Marginal Environments: A Review. *Agronomy* **2022**, *12*, 729. [CrossRef]
5. Elhassaneen, Y.A.; Nasef, A.Z.; Arafa, R.S.; Bayomi, A.I. Bioactive Compounds and Antioxidant Activities of Milk Thistle (*Silybum marianum*) Extract and Their Potential Roles in the Prevention of Diet-Induced Obesity Complications. *AJFST* **2023**, *11*, 70–85. [CrossRef]
6. Abenavoli, L.; Capasso, R.; Milic, N.; Capasso, F. Milk Thistle in Liver Diseases: Past, Present, Future. *Phytother. Res.* **2010**, *24*, 1423–1432. [CrossRef]
7. Rayni, P.M.; Dayani, O.; Shahrbabak, M.S.; Hosseini, M.M.S. Growth Performance, Carcass Traits, and Physicochemical Characteristics of Longissimus Thoracis of Fattening Lambs Feeding Milk Thistle (*Silybum mariamum*) Seed Powder. *Small Rumin. Res.* **2023**, *227*, 107082. [CrossRef]

8. Grela, E.R.; Świątkiewicz, M.; Florek, M.; Wojtaszewska, I. Impact of Milk Thistle (*Silybum marianum* L.) Seeds in Fattener Diets on Pig Performance and Carcass Traits and Fatty Acid Profile and Cholesterol of Meat, Backfat and Liver. *Livest. Sci.* **2020**, *239*, 104180. [CrossRef]
9. Karaiskou, C.; Kasapidou, E.; Michailidis, G.; Markantonatos, X.; Basdagianni, Z. Effect of Dietary Milk Thistle (*Silybum marianum* L.) Oil Supplementation on Animal Performance and Milk Fatty Acid Composition in Dairy Ewes. *Small Rumin. Res.* **2021**, *203*, 106493. [CrossRef]
10. Gholamalian, R.; Mahdavi, A.H.; Riasi, A. Hepatic Fatty Acids Profile, Oxidative Stability and Egg Quality Traits Ameliorated by Supplementation of Alternative Lipid Sources and Milk Thistle Meal. *Anim. Physiol. Nutr.* **2022**, *106*, 860–871. [CrossRef]
11. Fathi-Achachlouei, B.; Azadmard-Damirchi, S. Milk Thistle Seed Oil Constituents from Different Varieties Grown in Iran. *J. Am. Oil Chem. Soc.* **2009**, *86*, 643–649. [CrossRef]
12. Abascal, K.; Yarnell, E. The Many Faces of *Silybum marianum* (Milk Thistle): Part 2—Clinical Uses, Safety, and Types of Preparations. *Altern. Complement. Ther.* **2003**, *9*, 251–256. [CrossRef]
13. Murray, M.T. *Silybum marianum* (Milk Thistle). In *Textbook of Natural Medicine*; Elsevier: Amsterdam, The Netherlands, 2020; pp. 851–855.e1, ISBN 978-0-323-52342-4.
14. Porwal, O.; Mohammed Ameen, M.S.; Anwer, E.T.; Uthirapathy, S.; Ahamad, J.; Tahsin, A. *Silybum marianum* (Milk Thistle): Review on Its Chemistry, Morphology, Ethno Medical Uses, Phytochemistry and Pharmacological Activities. *J. Drug Deliv. Ther.* **2019**, *9*, 199–206. [CrossRef]
15. Aristides, L.G.A.; Venancio, E.J.; Alfieri, A.A.; Otonel, R.A.A.; Frank, W.J.; Oba, A. Carcass Characteristics and Meat Quality of Broilers Fed with Different Levels of *Saccharomyces cerevisiae* Fermentation Product. *Poult. Sci.* **2018**, *97*, 3337–3342. [CrossRef] [PubMed]
16. Lu, Z.; Zeng, N.; Jiang, S.; Wang, X.; Yan, H.; Gao, C. Dietary Replacement of Soybean Meal by Fermented Feedstuffs for Aged Laying Hens: Effects on Laying Performance, Egg Quality, Nutrient Digestibility, Intestinal Health, Follicle Development, and Biological Parameters in a Long-Term Feeding Period. *Poult. Sci.* **2023**, *102*, 102478. [CrossRef] [PubMed]
17. Halmemies-Beauchet-Filleau, A.; Rinne, M.; Lamminen, M.; Mapato, C.; Ampapon, T.; Wanapat, M.; Vanhatalo, A. Review: Alternative and Novel Feeds for Ruminants: Nutritive Value, Product Quality and Environmental Aspects. *Animal* **2018**, *12*, s295–s309. [CrossRef] [PubMed]
18. Hashemi Jabali, N.S.; Mahdavi, A.H.; Ansari Mahyari, S.; Sedghi, M.; Akbari Moghaddam Kakhki, R. Effects of Milk Thistle Meal on Performance, Ileal Bacterial Enumeration, Jejunal Morphology and Blood Lipid Peroxidation in Laying Hens Fed Diets with Different Levels of Metabolizable Energy. *Anim. Physiol. Nutr.* **2018**, *102*, 410–420. [CrossRef] [PubMed]
19. Higgs, R.J.; Chase, L.E.; Ross, D.A.; Van Amburgh, M.E. Updating the Cornell Net Carbohydrate and Protein System Feed Library and Analyzing Model Sensitivity to Feed Inputs. *J. Dairy Sci.* **2015**, *98*, 6340–6360. [CrossRef]
20. Fox, D.G.; Barry, M.C.; Pitt, R.E.; Roseler, D.K.; Stone, W.C. Application of the Cornell Net Carbohydrate and Protein Model for Cattle Consuming Forages. *J. Anim. Sci.* **1995**, *73*, 267. [CrossRef]
21. Wang, E.; Wang, J.; Lv, J.; Sun, X.; Kong, F.; Wang, S.; Wang, Y.; Yang, H.; Cao, Z.; Li, S.; et al. Comparison of Ruminal Degradability, Indigestible Neutral Detergent Fiber, and Total-Tract Digestibility of Three Main Crop Straws with Alfalfa Hay and Corn Silage. *Animals* **2021**, *11*, 3218. [CrossRef] [PubMed]
22. Theodoridou, K.; Yu, P. Application Potential of ATR-FT/IR Molecular Spectroscopy in Animal Nutrition: Revelation of Protein Molecular Structures of Canola Meal and Presscake, As Affected by Heat-Processing Methods, in Relationship with Their Protein Digestive Behavior and Utilization for Dairy Cattle. *J. Agric. Food Chem.* **2013**, *61*, 5449–5458. [CrossRef] [PubMed]
23. Li, L.; Zhang, Q.; Yuan, X.; Yang, H.; Qin, S.; Hong, L.; Pu, L.; Li, L.; Zhang, P.; Zhang, J. Study of the Molecular Structure of Proteins in Fermented Maize-Soybean Meal-Based Rations Based on FTIR Spectroscopy. *Food Chem.* **2024**, *441*, 138310. [CrossRef] [PubMed]
24. AOAC. *Official Methods of Analysts*, 17th ed.; AOAC International: Arlington, VA, USA, 2000; Volume 1.
25. Van Soest, P.J.; Robertson, J.B.; Lewis, B.A. Methods for Dietary Fiber, Neutral Detergent Fiber, and Nonstarch Polysaccharides in Relation to Animal Nutrition. *J. Dairy Sci.* **1991**, *74*, 3583–3597. [CrossRef] [PubMed]
26. Licitra, G.; Hernandez, T.M.; Van Soest, P.J. Standardization of Procedures for Nitrogen Fractionation of Ruminant Feeds. *Anim. Feed Sci. Technol.* **1996**, *57*, 347–358. [CrossRef]
27. Wang, Y.; Gao, J.; Cheng, C.; Lv, J.; Lambo, M.T.; Zhang, G.; Li, Y.; Zhang, Y. Nutritional Values of Industrial Hemp Byproducts for Dairy Cattle. *Animals* **2022**, *12*, 3488. [CrossRef] [PubMed]
28. Sniffen, C.J.; O'Connor, J.D.; Van Soest, P.J.; Fox, D.G.; Russell, J.B. A Net Carbohydrate and Protein System for Evaluating Cattle Diets: II. Carbohydrate and Protein Availability. *J. Anim. Sci.* **1992**, *70*, 3562–3577. [CrossRef]
29. Ørskov, E.R.; McDonald, I. The Estimation of Protein Degradability in the Rumen from Incubation Measurements Weighted According to Rate of Passage. *J. Agric. Sci.* **1979**, *92*, 499–503. [CrossRef]
30. Nuez-Ortín, W.G.; Yu, P. Estimation of Ruminal and Intestinal Digestion Profiles, Hourly Effective Degradation Ratio and Potential N to Energy Synchronization of Co-Products from Bioethanol Processing: Ruminal and Intestinal Digestion Profiles of Co-Products of Bioethanol Production. *J. Sci. Food Agric.* **2010**, *90*, 2058–2067. [CrossRef]
31. Calsamiglia, S.; Stern, M.D. A Three-Step in Vitro Procedure for Estimating Intestinal Digestion of Protein in Ruminants. *J. Anim. Sci.* **1995**, *73*, 1459–1465. [CrossRef]

32. Stastnik, O.; Pavlata, L.; Mrkvicova, E. The Milk Thistle Seed Cakes and Hempseed Cakes Are Potential Feed for Poultry. *Animals* **2020**, *10*, 1384. [CrossRef]
33. Šťastní k, O.; Jůzl, M.; Karásek, F.; Štenclová, H.; Nedomová, Š.; Pavlata, L.; Mrkvicová, E.; Doležal, P.; Jarošová, A. The Effect of Feeding Milk Thistle Seed Cakes on Quality Indicators of Broiler Chickens Meat. *Potravin. Slovak J. Food Sci.* **2016**, *10*, 248–254. [CrossRef] [PubMed]
34. Beauchemin, K.A. Using ADF and NDF in Dairy Cattle Diet Formulation—A Western Canadian Perspective. *Anim. Feed Sci. Technol.* **1996**, *58*, 101–111. [CrossRef]
35. Shi, R.; Dong, S.; Mao, J.; Wang, J.; Cao, Z.; Wang, Y.; Li, S.; Zhao, G. Dietary Neutral Detergent Fiber Levels Impacting Dairy Cows' Feeding Behavior, Rumen Fermentation, and Production Performance during the Period of Peak-Lactation. *Animals* **2023**, *13*, 2876. [CrossRef] [PubMed]
36. NRC. *Nutrient Requirements of Dairy Cattle: Eighth Revised Edition (7 Revised)*; National Academies Press: Washington, DC, USA, 2001; p. 25806, ISBN 978-0-309-67777-6.
37. Arora, D.S.; Sharma, R.K. Enhancement in in Vitro Digestibility of Wheat Straw Obtained from Different Geographical Regions during Solid State Fermentation by White Rot Fungi. *BioResources* **2009**, *4*, 909–920. [CrossRef]
38. Van Kuijk, S.J.A.; Sonnenberg, A.S.M.; Baars, J.J.P.; Hendriks, W.H.; Cone, J.W. Fungal Treated Lignocellulosic Biomass as Ruminant Feed Ingredient: A Review. *Biotechnol. Adv.* **2015**, *33*, 191–202. [CrossRef] [PubMed]
39. Cantalapiedra-Hijar, G.; Peyraud, J.L.; Lemosquet, S.; Molina-Alcaide, E.; Boudra, H.; Nozière, P.; Ortigues-Marty, I. Dietary Carbohydrate Composition Modifies the Milk N Efficiency in Late Lactation Cows Fed Low Crude Protein Diets. *Animal* **2014**, *8*, 275–285. [CrossRef]
40. Li, Y.; Wu, Q.; Lv, J.; Jia, X.; Gao, J.; Zhang, Y.; Wang, L. Associations of Protein Molecular Structures with Their Nutrient Supply and Biodegradation Characteristics in Different Byproducts of Seed-Used Pumpkin. *Animals* **2022**, *12*, 956. [CrossRef] [PubMed]
41. Schwab, C.G.; Broderick, G.A. A 100-Year Review: Protein and Amino Acid Nutrition in Dairy Cows. *J. Dairy Sci.* **2017**, *100*, 10094–10112. [CrossRef] [PubMed]
42. Bach, A.; Calsamiglia, S.; Stern, M.D. Nitrogen Metabolism in the Rumen. *J. Dairy Sci.* **2005**, *88*, E9–E21. [CrossRef]
43. Storm, E.; Ørskov, E.R. The Nutritive Value of Rumen Micro-Organisms in Ruminants: 1. Large-Scale Isolation and Chemical Composition of Rumen Micro-Organisms. *Br. J. Nutr.* **1983**, *50*, 463–470. [CrossRef]
44. Broderick, G.A.; Wallace, R.J.; Ørskov, E.R. Control of Rate and Extent of Protein Degradation. In *Physiological Aspects of Digestion and Metabolism in Ruminants*; Elsevier: Amsterdam, The Netherlands, 1991; pp. 541–592. ISBN 978-0-12-702290-1.
45. Jung, H.G.; Deetz, D.A. Cell Wall Lignification and Degradability. In *ASA, CSSA, and SSSA Books*; Jung, H.G., Buxton, D.R., Hatfield, R.D., Ralph, J., Eds.; American Society of Agronomy, Crop Science Society of America, Soil Science Society of America: Madison, WI, USA, 2015; pp. 315–346, ISBN 978-0-89118-238-2.
46. Chalupa, W.; Sniffen, C.J. Protein and Amino Acid Nutrition of Lactating Dairy Cattle. *Vet. Clin. N. Am. Food Anim. Pract.* **1991**, *7*, 353–372. [CrossRef]
47. Elbaz, A.M.; El-sheikh, S.E.; Abdel-Maksoud, A. Growth Performance, Nutrient Digestibility, Antioxidant State, Ileal Histomorphometry, and Cecal Ecology of Broilers Fed on Fermented Canola Meal with and without Exogenous Enzymes. *Trop. Anim. Health Prod.* **2023**, *55*, 46. [CrossRef] [PubMed]
48. Damiran, D.; Yu, P. Molecular Basis of Structural Makeup of Hulless Barley in Relation to Rumen Degradation Kinetics and Intestinal Availability in Dairy Cattle: A Novel Approach. *J. Dairy Sci.* **2011**, *94*, 5151–5159. [CrossRef]
49. Jackson, M.; Mantsch, H.H. The Use and Misuse of FTIR Spectroscopy in the Determination of Protein Structure. *Crit. Rev. Biochem. Mol. Biol.* **1995**, *30*, 95–120. [CrossRef] [PubMed]
50. Yu, P. Protein Secondary Structures (α-Helix and β-Sheet) at a Cellular Level and Protein Fractions in Relation to Rumen Degradation Behaviours of Protein: A New Approach. *Br. J. Nutr.* **2005**, *94*, 655–665. [CrossRef] [PubMed]
51. Nuez-Ortín, W.G.; Yu, P. Effects of Bioethanol Plant and Coproduct Type on the Metabolic Characteristics of the Proteins in Dairy Cattle. *J. Dairy Sci.* **2010**, *93*, 3775–3783. [CrossRef] [PubMed]
52. Samadi; Theodoridou, K.; Yu, P. Detect the Sensitivity and Response of Protein Molecular Structure of Whole Canola Seed (Yellow and Brown) to Different Heat Processing Methods and Relation to Protein Utilization and Availability Using ATR-FT/IR Molecular Spectroscopy with Chemometrics. *Spectrochim. Acta Part A Mol. Biomol. Spectrosc.* **2013**, *105*, 304–313. [CrossRef]
53. Xin, H.; Yu, P. Chemical Profile, Energy Values, and Protein Molecular Structure Characteristics of Biofuel/Bio-Oil Co-Products (Carinata Meal) in Comparison with Canola Meal. *J. Agric. Food Chem.* **2013**, *61*, 3926–3933. [CrossRef]

Disclaimer/Publisher's Note: The statements, opinions and data contained in all publications are solely those of the individual author(s) and contributor(s) and not of MDPI and/or the editor(s). MDPI and/or the editor(s) disclaim responsibility for any injury to people or property resulting from any ideas, methods, instructions or products referred to in the content.

animals

MDPI

Article

The Effects of *Lactobacillus plantarum* and *Lactobacillus buchneri* on the Fermentation Quality, In Vitro Digestibility, and Aerobic Stability of *Silphium perfoliatum* L. Silage

Yitong Jin †, Peng Wang *,†, Fuhou Li , Meng Yu, Jiarui Du, Tianyue Zhao, Qixuan Yi, Hongyu Tang and Bao Yuan *

College of Animal Sciences, Jilin University, Changchun 130062, China; ytjin23@mails.jlu.edu.cn (Y.J.); lifh@jlu.edu.cn (F.L.); yumeng22@mails.jlu.edu.cn (M.Y.); dujr23@mails.jlu.edu.cn (J.D.); tianyue22@mails.jlu.edu.cn (T.Z.); yiqx21@mails.jlu.edu.cn (Q.Y.); tanghy@jlu.edu.cn (H.T.)

* Correspondence: pengwang@jlu.edu.cn (P.W.); yuan_bao@jlu.edu.cn (B.Y.)

† These authors contributed equally to this work.

Simple Summary: Under the shortage of roughage resources in China, *Silphium perfoliatum* L. (SP) can be an ideal roughage choice for ruminants and has good application prospects. Ensiling is an efficient forage treatment technology for retaining the nutrients and extending the shelf life of forage. At present, there have been no reports on the effects of *Lactobacillus plantarum* and *Lactobacillus buchneri*, whether alone or in combination, on *Silphium perfoliatum* L. silage. Therefore, in this experiment, *L. plantarum* and *L. buchneri* were added individually or combined to SP-silage to investigate the effects of different fermentation types of *Lactobacilli* on the fermentation quality, in vitro digestibility, and aerobic stability of SP-silage. It was found that a mixture of *L. plantarum* and *L. buchneri* provided the best silage results. *L. buchneri* can be used as an additive to improve the aerobic stability of SP-silage. The results of our research provide a certain scientific basis and technical support for obtaining high-quality SP-silage in production.

Abstract: In this experiment, *Lactobacillus plantarum* and *Lactobacillus buchneri* were added individually or in combination to *Silphium perfoliatum* L. (SP) silage to investigate the effects of different fermentation types of lactobacilli on the fermentation quality, in vitro digestibility, and aerobic stability of SP-silage, with a view to providing a certain scientific basis and technical support for obtaining high-quality SP-silage in production. The experiment comprised a non-additive group (control), an *L. plantarum* group (LP), an *L. buchneri* group (LB), and an *L. plantarum* and *L. buchneri* mixed treatment group (LPLB). Samples were taken after 60 days of fermentation and analyzed for the fermentation quality, in vitro digestibility, and aerobic stability of the SP-silage. The results showed that the addition of LP, LB, and LPLB significantly reduced the pH and proportion of ammonia nitrogen to total nitrogen and significantly increased the lactic acid, in vitro dry matter digestibility, and in vitro crude protein digestibility in the SP-silage ($p < 0.05$). Compared to the control group, the dry matter and crude protein contents of the LB and LPLB groups were significantly increased, while the neutral detergent fiber and acid detergent fiber contents were significantly reduced ($p < 0.05$). The SP-silage supplemented with LPLB had the highest dry matter and crude protein contents. The gross and digestible energies of the SP-silage in the LB and LPLB groups were significantly higher than those in the control and LP groups ($p < 0.05$). The aerobic stability of the SP-silage was significantly reduced by 24.14% in the LP group and increased by 58.62% and 34.48% in the LB and LPLB groups, respectively, compared to the control group ($p < 0.05$). It was shown that adding a combination of LP and LB resulted in the best fermentation quality, nutritional value, and in vitro digestibility of the SP-silage. LB was effective in improving the aerobic stability of SP-silage.

Keywords: aerobic stability; fermentation quality; in vitro digestibility; *Lactobacillus plantarum*; *Lactobacillus buchneri*; *Silphium perfoliatum* L.

Citation: Jin, Y.; Wang, P.; Li, F.; Yu, M.; Du, J.; Zhao, T.; Yi, Q.; Tang, H.; Yuan, B. The Effects of *Lactobacillus plantarum* and *Lactobacillus buchneri* on the Fermentation Quality, In Vitro Digestibility, and Aerobic Stability of *Silphium perfoliatum* L. Silage. *Animals* **2024**, *14*, 2279. https://doi.org/10.3390/ani14152279

Academic Editor: Alejandro Plascencia

Received: 7 July 2024
Revised: 30 July 2024
Accepted: 3 August 2024
Published: 5 August 2024

Copyright: © 2024 by the authors. Licensee MDPI, Basel, Switzerland. This article is an open access article distributed under the terms and conditions of the Creative Commons Attribution (CC BY) license (https://creativecommons.org/licenses/by/4.0/).

1. Introduction

As one of the key industries indispensable to a country's economy and its people's livelihood, animal husbandry plays a crucial role in maintaining food security, promoting sustainable growth of the rural economy, and facilitating the increase in herdsmen's incomes. However, in recent years, with the rapid development of China's livestock industry, the forage supply problem has become increasingly prominent, and the existing forage resources have made it difficult to support the demand for high-quality development of the livestock industry. Therefore, exploring and broadening new feed resources has become an urgent problem in the current development of animal husbandry. *Silphium perfoliatum* L. (SP) is a perennial dicotyledonous herb belonging to the genus Pine Sedge in the family Asteraceae, which is widely adapted, resilient, of good quality, and easy to cultivate, making it a high-quality and high-yielding forage grass [1]. SP has a high protein content, is rich in various essential amino acids and minerals, and contains biologically active substances such as ascorbic acid, phenolic acids, terpenoids, and flavonoids, which may result in resistance to pH reduction in silage [2]. SP is not only suitable for ruminants such as cows and sheep, but also for monogastric animals such as pigs, chickens and ducks, and can even be used as feed for herbivorous fish. Its wide applicability makes SP important in the global livestock industry. SP is highly resilient to a wide range of soil and climatic conditions, including cold, heat, salinity and flooding tolerance. This resilience makes it easier and more feasible to grow and promote SP globally, providing reliable feed security for the livestock industry.

The growth characteristics of SP show a clear seasonal pattern, with vigorous growth and luxuriant foliage in summer, followed by a gradual slowing down of growth in fall and winter, with foliage yellowing and ceasing to grow and an inability to satisfy continuous supply needs throughout the year [3]. Although SP is cold hardy, its growth can still be limited under extreme low temperature conditions. Ensiling is an efficient forage treatment technology for retaining the nutrients and extending the shelf life of forage. However, fresh SP has a higher moisture content and buffering capacity and a lower water-soluble carbohydrate content, resulting in a poor direct silage effect. Xie et al. [4] showed that alfalfa silage alone had very high pH and proportions of ammonia nitrogen to total nitrogen, low lactic acid concentration, and poor fermentation.

During silage preparation, when no exogenous additives are introduced, the microbial community naturally attached to the raw material plays a dominant role in the fermentation process, which often leads to a loss of dry matter, as well as hydrolysis of the proteins in the silage, thus reducing the nutritional value and utilization efficiency of the feed [5]. Lactic acid bacteria preparations are commonly used as promoters of silage fermentation to effectively control and optimize the silage fermentation process and to reduce unnecessary nutrient loss. Lactic acid bacteria can be divided into two types of fermentation according to their metabolites, namely, homofermentative lactic acid bacteria and heterofermentative lactic acid bacteria. *Lactobacillus plantarum* belongs to homofermentative lactic acid bacteria, which can produce a large amount of lactic acid during silage fermentation, rapidly lowering the pH and improving the fermentation quality [6]. It has been shown that *Leymus chinensis* silage has a higher lactic acid content and a lower pH and butyric acid content after adding *L. plantarum* compared to the control [7]. *L. buchneri* belongs to heterofermentative lactic acid bacteria, which can produce short-chain fatty acids that inhibit the growth and reproduction of harmful microorganisms such as yeasts and molds, thus improving the aerobic stability of silage. Xu et al. [8] showed that *L. buchneri* improves the aerobic stability of corn silage and reduces the risk of oxidative deterioration of nutrients during the aerobic exposure of silage. However, once silage is exposed to air, the lactic acid accumulated within it is readily utilized by yeasts and other harmful microorganisms in the environment, which, in turn, triggers the aerobic spoilage of silage, leading to reduced feed quality and nutrient loss [9]. To address this issue, relying only on homofermentative lactic acid bacteria may increase the risk of aerobic spoilage. A combination of homofermentative and heterofermentative lactic acid bacteria can improve this phenomenon, and the acetic acid produced by heterofermentative lactic acid bacteria can effectively inhibit the growth of harmful microorganisms [10]. Zhang et al. [11] investigated the effect of *L. plantarum* and *L. buchneri* on alfalfa silage and found that the composite addition group had the highest acetic acid and soluble sugar contents and the lowest ammoniacal nitrogen and yeast contents compared to the group where they were added individually. Meanwhile, under aerobic conditions, the combined addition group had the lowest carbon dioxide and mold contents and the best aerobic stability.

At present, there are no reports on the effects of adding *L. plantarum* and *L. buchneri* alone or in combination on SP-silage. There may be differences in the silage quality and aerobic stability of SP-silage inoculated with different types of fermenting lactic acid bacteria. Therefore, in this experiment, *L. plantarum* and *L. buchneri* were added individually or in combination to SP-silage to investigate the effects of different fermentation types of lactic acid bacteria on the fermentation quality, in vitro digestibility, and aerobic stability of SP-silage, which provided a scientific basis for the subsequent optimization of Lactobacillus combinations in order to improve the quality of silage.

2. Materials and Methods

2.1. Experimental Materials and Design

SP was provided by Hulin Fengshan Herbal Medicine Planting Farmers' Specialized Cooperative, Hulin City, and was sown on 16 May 2023, in a high-yield forage base in Tongyu County, Baicheng City, Jilin Province, China (122°2′–123°30′ E longitude, 44°13′–45°16′ N latitude, 170 m above sea level). It was mowed on 20 September, leaving a stubble height of 2–3 cm, and manually mowed and placed in a cool place to wilt overnight so that its moisture modulation was between 65% and 70%. Subsequently, the guillotine was utilized to cut the SP to 1–2 cm, which was then mixed well for silage preparation. LP and LB were purchased from the China Industrial Microbial Strain Preservation and Management Center, and the strain preservation numbers were CICC 6026 and CICC 20293, respectively. The chemical composition, buffering capacity, energy, and microbial population of the silage ingredients are shown in Table 1.

Table 1. Chemical composition, buffering capacity, energy, and microbial counts of *Silphium perfoliatum* L.

Items ‡	*Silphium perfoliatum* L.
Chemical composition and buffering capacity	
Dry matter (% FW)	31.43
Organic matter (% DM)	82.26
Crude protein (% DM)	22.97
Neutral detergent fiber (% DM)	61.50
Acid detergent fiber (% DM)	26.18
Acid detergent lignin (% DM)	9.43
Water-soluble carbohydrate (% DM)	4.68
Buffering capacity (mEq kg^{-1} DM)	242.21
Energy	
GE (MJ kg^{-1} DM)	15.24
DE (MJ kg^{-1} DM)	11.70
ME (MJ kg^{-1} DM)	8.72
NEm (M Jkg^{-1} DM)	6.24
NEl (MJ kg^{-1} DM)	5.33
NEf (MJ kg^{-1} DM)	4.61
Microbial counts	
Lactic acid bacteria (log_{10} cfu g^{-1} FW)	4.31
Yeasts (log_{10} cfu g^{-1} FW)	3.06
Moulds (log_{10} cfu g^{-1} FW)	1.29

‡ FW, fresh weight; DM, dry matter; GE, gross energy; DE, digestible energy; ME, metabolizable energy; NEm, net energy for maintenance; NEl, net energy for lactating cow; NEf, net energy for weight gain; and cfu, colony-forming units.

The following additive groups were set up in this experiment: (1) no additive (Control); (2) *Lactobacillus plantarum* (LP, with a viable count of 5×10^5 cfu g^{-1} FW); (3) *Lactobacillus buchneri* (LB, with a viable count of 5×10^5 cfu g^{-1} FW); (4) mixed preparations of *L. plantarum* and *L. buchneri* (LPLB, with a viable count of 5×10^5 cfu g^{-1} FW). On 21 September 2023, SP-silage preparation was carried out at a high-yield forage base in Tongyu County, Baicheng City, Jilin Province, China (122°2′–123°30′ E longitude, 44°13′–45°16′ N latitude, 170 m above sea level). Lactic acid bacteria were all added at 50 mg kg^{-1} FW. The bacterial solution was loaded into a pre-sterilized hand sprayer (DLX-SD-BJ, Ruichen Plant Protection Co., Ltd., Linyi, China) and evenly sprayed on the short-cut raw materials, while the control group was sprayed with an equal amount of distilled water and stirred thoroughly in a plastic container sterilized with 75% ethanol. The silage materials were then packed into 240 mm × 350 mm polyethylene vacuum-packed bags (Zhongzheng Fangxin Packaging Material Co., Ltd., Chengdu, China) of 500 g each, with six replicates per additive group. The first three replicates were used for fermentation quality, chemical composition, energy, and in vitro digestibility analysis, and the remaining replicates were used for aerobic stability testing. Vacuum sealing was then carried out using a vacuum packaging machine (CS-SP, Changsheng Machinery Co., Ltd., Quanzhou, China). SP-silage was prepared by placing the materials at room temperature, protected from light, for 60 days.

2.2. Fermentation Quality Analysis and Microbial Counting

After the silage was opened, 20 g of the mixed sample was taken, and 180 mL of distilled water was added before sealing and soaking in a refrigerator at 4 °C for 24 h. The extract was obtained by filtration using 4 layers of gauze and used to determine the subsequent fermentation parameters. The pH of the extract was determined using a pH meter (PHBJ-261L, Qinkun Pump Co., Ltd., Shanghai, China). Ammoniacal nitrogen (NH_3-N) was determined using the phenol–sodium hypochlorite colorimetric method [12]. The extracts were subjected to 10 min of centrifugation using a bench-top high-speed refrigerated centrifuge (GL21M, Kaida Scientific Instruments Co., Ltd., Changsha, China) and then filtered using 0.22 um aqueous filter membranes, followed by determination of the lactic acid (LA), acetic acid (AA), propionic acid (PA), and butyric acid (BA) contents using a high-performance liquid chromatograph [13].

Another portion of the filtrate was diluted to 10^{-3}, 10^{-4}, and 10^{-5} with sterile water using the 10-fold dilution method, while the number of lactic acid bacteria, yeasts, and molds in the silage was determined using the plate counting method. Lactic acid bacteria were counted after anaerobic incubation at 37 °C for 48 h using De Man, Rogosa, and Sharpe medium. Yeasts and molds were counted using potato dextrose agar medium at 30 °C for 48 h. The data obtained are expressed as $\log_{10}$ cfu g^{-1} FW.

2.3. Chemical Composition and Energy Analysis

The silage samples were weighed fresh, placed in a constant temperature blast drying oven at 105 °C for 30 min, dried at 65 °C until constant weight, and weighed dry. The dried samples were pulverized through different mesh sieves and mixed well in self-sealing bags to determine the chemical composition. The dry matter (DM) content was determined using the drying method; the crude ash content was determined using the scorching method at 550 °C; the crude protein (CP) content was determined using a fully automated Dumas Nitrogen Determination Instrument; the crude fat content was determined using the Soxhlet extraction method [14]. The acid detergent fiber (ADF), neutral detergent fiber (NDF), acid detergent lignin (ADL), and crude fiber (CF) contents were determined using an ANKOM Model A2000i Fully Automatic Fiber Meter (Ankom Technology, Fairport, NY, USA) [15]. The water-soluble carbohydrate (WSC) content was determined using the anthrone–sulfuric acid colorimetric method [16]. The buffering capacity (BC) was determined via titration [17]. The gross energy (GE) was determined using a fully automated oxygen bomb calorimeter [14], and then the digestible energy (DE), metabolizable energy (ME), net energy for maintenance (NEm), net energy for weight gain (NEf), and net energy for a lactating cow (NEl) were calculated [18].

2.4. Aerobic Stability Analysis

The silage was mixed well and put into a 1 L polyethylene tank and kept at room temperature. The upper layer was sealed with a black plastic sheet and tied with small holes to prevent moisture loss and external contamination and allow air to enter the fermenter freely. At the same time, thermocouples from a multiplexed automatic temperature recorder (RX6000F, Mei-Control Automation Technology Co., Ltd., Hangzhou, China) were inserted in the center of the silage and exposed aerobically for 7 days. The silage and ambient temperatures were recorded automatically at 1 h intervals. The time required for a 2 °C difference between the silage and ambient temperatures was used to assess the aerobic stability of the silage [19].

2.5. In Vitro Fermentation Parameter Analysis

This experiment strictly adhered to the Guidelines for Ethical Review of Laboratory Animal Welfare in China and was approved by the Laboratory Animal Welfare Ethics Committee of Jilin University (license no. SY20209600). The rumen fluid was obtained from four healthy small-tailed chilly sheep (mean weight of 35 kg) of similar body conditions fitted with permanent fistulas located at the Agricultural Experimental Base of Jilin University. Sufficient rumen fluid was collected before morning feeding, placed in plastic bottles preheated at 39 °C and filled with carbon dioxide, mixed well, and then filtered through 4 layers of gauze, with the entire operation being strictly anaerobic. Buffers were prepared using the method of McDougall (1948). At the same time, the rumen fluid was mixed with the buffer at a ratio of 1:4 to prepare a mixed culture solution, and carbon dioxide was passed to ensure anaerobic conditions before placing it in a 39 °C water bath for backup. A total of 52 fiber bags were prepared (12 silage samples × 4 parallel samples + 4 blank controls). A 0.5 g silage sample was weighed in a filter bag (ANKOM F57, aperture 25 um, ANKOM Technology Corporation, Macedon, NY, USA) and sealed in a serum bottle (100 mL). Then, 50 mL of the mixed culture solution was added to the serum bottle and passed through carbon dioxide to maintain an anaerobic environment. This was then immediately put on the rubber stopper and aluminum cap, using special sealing pliers to press tightly, and then it was placed into the constant-temperature culture oscillating shaker (39 °C, 100 r min^{-1}) to incubate for 48 h. All samples were incubated at once. At the end of the timer, the filter bag in the serum bottle was removed and washed with cold distilled water. After washing, it was placed in an electrically heated constant-temperature blast drying oven (65 °C, 48 h) to dry until constant weight. The filter bag containing the samples was removed from the electrically heated constant-temperature blast-drying oven, and the in vitro dry matter digestibility (IVDMD), in vitro organic matter digestibility (IVOMD), in vitro crude protein digestibility (IVCPD), and in vitro neutral detergent fiber digestibility (IVNDFD) were calculated from the weight loss of the samples before and after incubation [20].

2.6. Statistical Analysis

The data were analyzed according to a 2 × 2 factorial treatment design model using the GLM program of SPSS (version 26, International Business Machines Corporation, Armonk, NY, USA):

$$Y_{ijk} = \mu + L_j + B_k + (L \times B)_{jk} + \varepsilon_{ijk}$$

where Y_{ijk} denotes the response variable, μ is the total mean, L_j is a fixed effect of LP 6026 j (j = 1, 2), B_k is a fixed effect of LB 20293 k (k = 1, 2), $(L \times B)_{jk}$ is a fixed interaction effect of LP 6026 j and LB 20293 k, and ε_{ijk} is the error term.

When at least one interaction was significant ($p < 0.05$), significant differences between means were determined by Tukey's method, where $p < 0.05$ was recognized as significant.

3. Results

3.1. Fermentation Quality of Silphium perfoliatum L. Silage

The fermentation quality of SP-silage is shown in Table 2. The pH was significantly lower in all treatment groups compared to the control, with the LP-added SP-silage having the lowest pH of all treatment groups ($p = 0.002$). The LA content of the SP-silage from the three lactic acid bacteria treatment groups was significantly higher than that in the control group ($p < 0.001$). The AA content was highest in the LB group, which was significantly higher than that in the control and LP groups ($p < 0.001$). The BA content of the SP-silage in the LP group was lower than in the other treatment groups, but the difference was not significant ($p = 0.403$). The addition of LP, LB, and LPLB significantly reduced the NH_3-N/TN in the SP-silage compared to the control ($p = 0.003$).

Table 2. Fermentation quality of *Silphium perfoliatum* L. silage prepared with lactic acid bacteria.

Items ‡	Additives †				SEM	p-Value	Significance of Main Effects and Interactions		
	Control	LP	LB	LPLB			LP	LB	LP × LB
pH	4.64 c	4.22 a	4.44 b	4.33 a b	0.071	0.002	0.001	0.399	0.014
LA (%DM)	2.63 a	4.92 c	4.17 b	5.51 c	0.279	<0.001	<0.001	0.001	0.044
AA (%DM)	0.45 a	0.47 a	0.90 b	0.86 b	0.049	<0.001	0.676	<0.001	0.437
PA (%DM)	0.01	0.02	0.03	0.02	0.008	0.064	0.780	0.062	0.078
BA (%DM)	0.49	0.25	0.33	0.32	0.138	0.403	0.235	0.680	0.259
NH_3-N (%TN)	2.96 b	1.60 a	1.76 a	1.47 a	0.284	0.003	0.003	0.011	0.029

a–c Means of the additive treatments within a row with significant differences between the superscript letters ($p < 0.05$). SEM, standard error of the mean. † LB, *Lactobacillus plantarum*; LB, *Lactobacillus buchneri*; LPLB, mixture of *Lactobacillus plantarum* and *Lactobacillus buchneri*. ‡ DM, dry matter; LA, lactic acid; AA, acetic acid; PA, propionic acid; BA, butyric acid; NH_3-N, ammonia nitrogen; and TN, total nitrogen.

LP significantly affected the pH, LA, and NH_3-N/TN ($p < 0.01$). LB had a highly significant effect on LA and AA ($p < 0.01$) and a significant effect on NH_3-N/TN ($p < 0.05$). There was also an LP × LB interaction effect on the pH, LA, and NH_3-N/TN ($p < 0.05$). The pH decreased significantly with the addition of LP, with LB having a smaller effect on the SP-silage (4.22 vs. 4.44; $p = 0.002$). The addition of LB elevated the LA content, but LP was better for the SP-silage (4.92 vs. 4.17; $p < 0.001$). Adding LP and LB individually reduced NH_3-N/TN, but a mixture of LP and LB provided better silage results (2.96 vs. 1.47; $p = 0.003$).

3.2. Chemical Composition of Silphium perfoliatum L. Silage

The chemical compositions of SP-silage are shown in Table 3. The addition of LP, LB, and LPLB significantly increased the DM and CP content of the SP-silage compared to the control group, with the LPLB group having the highest DM and CP content ($p < 0.001$). The OM content of the SP-silage from LB and LPLB was significantly higher than that in the control group ($p = 0.038$), but there was no significant difference between the control and LP groups. The NDF and ADF contents were significantly lower in the LB and LPLB groups compared to the control group, with the SP-silage supplemented with LB having the lowest NDF and ADF contents among all treatment groups ($p = 0.045$; $p < 0.001$). The ADL content of the SP-silage in the LB and LPLB groups was significantly lower than that in the control group ($p = 0.023$), but there was no significant difference between the two groups.

LP had a highly significant effect on the DM and CP content ($p < 0.01$). LB had a highly significant effect on the DM and CP, ADF, and ADL contents ($p < 0.01$), as well as a significant effect on the OM and NDF contents ($p < 0.05$). The was also an LP × LB interaction effect on the ADF content ($p < 0.01$). The addition of LP had no significant effect on the ADF content, while the addition of LB significantly reduced it (24.22 vs. 22.61; $p < 0.001$).

Table 3. Chemical compositions of *Silphium perfoliatum* L. silage prepared with lactic acid bacteria.

Items ‡	Additives †				SEM	p-Value	Significance of Main Effects and Interactions		
	Control	LP	LB	LPLB			LP	LB	LP × LB
DM (%FW)	24.68 ^{a}	26.69 ^{b}	27.38 ^{b}	29.47 ^{c}	0.525	<0.001	0.001	<0.001	0.924
OM (%DM)	77.36 ^{a}	78.06 ^{ab}	78.34 ^{b}	78.42 ^{b}	0.320	0.038	0.123	0.018	0.207
CP (%DM)	16.76 ^{a}	18.21 ^{b}	21.03 ^{c}	21.79 ^{d}	0.310	<0.001	0.001	<0.001	0.162
NDF (%DM)	60.55 ^{b}	57.40 ^{ab}	54.58 ^{a}	55.59 ^{a}	1.799	0.045	0.424	0.016	0.141
ADF (%DM)	24.81 ^{c}	24.22 ^{bc}	22.61 ^{a}	23.70 ^{b}	0.283	<0.001	0.247	<0.001	0.003
ADL (%DM)	7.39 ^{b}	6.67 ^{ab}	6.10 ^{a}	6.08 ^{a}	0.371	0.023	0.200	0.007	0.219

$^{a–d}$ Means of the additive treatments within a row with significant differences between superscript letters ($p < 0.05$). SEM, standard error of the mean. † LP, *Lactobacillus plantarum*; LB, *Lactobacillus buchneri*; LPLB, mixture of *Lactobacillus plantarum* and *Lactobacillus buchneri*. ‡ FW, fresh weight; DM, dry matter; OM, organic matter; CP, crude protein; NDF, neutral detergent fiber; ADF, acid detergent fiber; and ADL, acid detergent lignin.

3.3. Energy of Silphium perfoliatum L. Silage

The energy of the SP-silage is shown in Table 4. The GE and DE of the SP-silage were significantly higher in the LB and LPLB groups than in the control and LP groups ($p = 0.005$; $p < 0.001$), but there was no significant difference between the two groups. The ME was highest in the LB group and significantly higher than in the control and LP groups ($p = 0.017$). The addition of LB and LPLB significantly increased the NEm and NEl of the SP-silage compared to the control ($p = 0.029$; $p = 0.030$). The NEf of the SP-silage was higher in the LB group than in the other treatment groups, but the difference was not significant ($p = 0.063$). LB had a highly significant effect on the GE, DE, ME, NEm, and NEl ($p < 0.01$).

Table 4. Energy of *Silphium perfoliatum* L. silage prepared with lactic acid bacteria.

Items ‡	Additives †				SEM	p-Value	Significance of Main Effects and Interactions		
	Control	LP	LB	LPLB			LP	LB	LP × LB
GE (MJ kg $^{-1}$DM)	15.74 ^{a}	15.77 ^{a}	16.24 ^{b}	16.22 ^{b}	0.068	<0.001	0.919	<0.001	0.570
DE (MJ kg $^{-1}$ DM)	11.85 ^{a}	12.07 ^{a}	12.52 ^{b}	12.64 ^{b}	0.168	0.005	0.184	0.001	0.666
ME (MJ kg $^{-1}$ DM)	9.47 ^{a}	9.61 ^{ab}	9.98 ^{c}	9.90 ^{bc}	0.136	0.017	0.726	0.003	0.294
NEm (MJ kg $^{-1}$ DM)	6.88 ^{a}	7.01 ^{ab}	7.29 ^{b}	7.22 ^{b}	0.118	0.029	0.701	0.006	0.266
NEl (MJ kg $^{-1}$ DM)	5.75 ^{a}	5.86 ^{ab}	6.04 ^{b}	6.10 ^{b}	0.100	0.030	0.272	0.006	0.715
NEf (MJ kg $^{-1}$ DM)	4.49	4.64	4.85	4.78	0.116	0.063	0.653	0.167	0.237

$^{a–c}$ Means of the additive treatments within a row with significant differences between superscript letters ($p < 0.05$). SEM, standard error of the mean. † LP, *Lactobacillus plantarum*; LB, *Lactobacillus buchneri*; LPLB, mixture of *Lactobacillus plantarum* and *Lactobacillus buchneri*. ‡ DM, dry matter; GE, gross energy; DE, digestible energy; ME, metabolizable energy; NEm, net energy for maintenance; NEl, net energy for lactating cow; and NEf, net energy for weight gain.

3.4. In Vitro Digestibility of Silphium perfoliatum L. Silage

The in vitro digestibility of SP-silage is shown in Table 5. The *IV*DMD and *IV*CPD were significantly higher in all treatment groups compared to the control, with the SP-silage supplemented with LPLB having the highest *IV*DMD and *IV*CPD among all treatment groups ($p = 0.001$; $p < 0.001$). The *IV*NDFD was highest in the LB group, which was significantly higher than in the control and LP groups ($p = 0.004$). The *IV*OMD of the SP-silage in the LPLB group was higher than in the other treatment groups, but the difference was not significant ($p = 0.164$).

Table 5. In vitro digestibility of *Silphium perfoliatum* L. silage prepared with lactic acid bacteria.

Items [‡]	Additives [†]				SEM	*p*-Value	Significance of Main Effects and Interactions		
	Control	LP	LB	LPLB			LP	LB	LP × LB
IVDMD (%DM)	63.47 [a]	64.45 [b]	64.33 [b]	65.51 [c]	0.304	0.001	0.001	0.002	0.643
IVOMD (%DM)	67.32	68.19	68.14	68.45	0.465	0.164	0.112	0.136	0.428
IVCPD (%DM)	60.13 [a]	61.69 [b]	64.19 [c]	64.98 [d]	0.240	<0.001	<0.001	<0.001	0.059
IVNDFD (%DM)	52.65 [a]	54.45 [b]	55.44 [b c]	56.14 [c]	0.654	0.004	0.272	0.001	0.027

[a–d] Means of the additive treatments within a row with significant differences between superscript letters ($p < 0.05$). SEM, standard error of the mean. [†] LP, *Lactobacillus plantarum*; LB, *Lactobacillus buchneri*; LPLB, mixture of *Lactobacillus plantarum* and *Lactobacillus buchneri*. [‡] IVDMD, in vitro dry matter digestibility; IVOMD, in vitro organic matter digestibility; IVCPD, in vitro crude protein digestibility; and IVNDFD, in vitro neutral detergent fiber digestibility.

LP had a highly significant effect on the *IVDMD* and *IVCPD* ($p < 0.01$). LB had a highly significant effect on the *IVDMD*, *IVCPD*, and *IVNDFD* ($p < 0.01$). There was also an LP × LB interaction effect on the *IVNDFD* ($p < 0.05$). Adding LB provided better silage results than adding LP. However, the best silage results were obtained with a mixture of LP and LB (52.65 vs. 56.14; $p = 0.004$).

3.5. Aerobic Stability of Silphium perfoliatum L. Silage

The temperature changes of the SP-silage exposed to aerobic conditions as a result of the additives are shown in Figure 1. The temperature of the three lactic acid bacteria additive groups increased slowly from 0 to 48 h and rapidly from 48 to 96 h, with the increase in temperature being more pronounced in the control and LP groups. The LB group temperature took the longest time to reach 2 °C above ambient, while the LP group temperature took the shortest time.

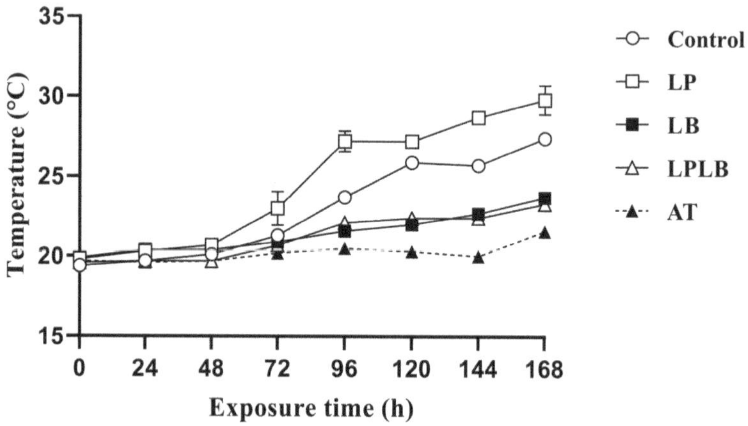

Figure 1. Effect of the application of additives on the temperature of *Silphium perfoliatum* L. silage after exposure to aerobic conditions. Control, no additive; LP, *Lactobacillus plantarum*; LB, *Lactobacillus buchneri*; LPLB, mixture of *Lactobacillus plantarum* and *Lactobacillus buchneri*; and AT, ambient temperature.

The time required for the SP-silage temperature to exceed room temperature by 2 °C after exposure to air is shown in Figure 2: 87 h in the control group, 66 h in the LP group, 138 h in the LB group, and 117 h in the LPLB group. The aerobic stability of the SP-silage was significantly reduced by 24.14% in the LP group and increased by 58.62% and 34.48% in the LB and LPLB groups, respectively, compared to the control group ($p < 0.05$).

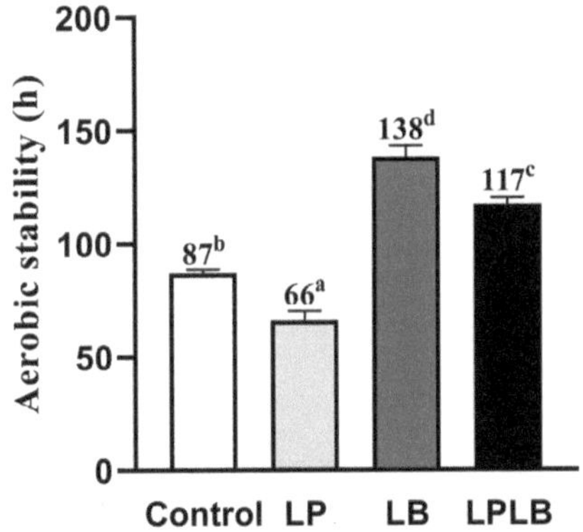

Figure 2. Time required for *Silphium perfoliatum* L. silage temperature to exceed room temperature by 2 °C after exposure to air. Control, no additive; LP, *Lactobacillus plantarum*; LB, *Lactobacillus buchneri*; LPLB, mixture of *Lactobacillus plantarum* and *Lactobacillus buchneri*; [a–d] Means of the additive treatments with significant differences between superscript letters ($p < 0.05$).

4. Discussion

4.1. Effect of Different Fermenting Lactic Acid Bacteria on the Fermentation Quality of Silphium perfoliatum L. Silage

The selection of suitable microbial silage additives can promote the proliferation of lactic acid bacteria, produce LA to make the pH drop rapidly to form an acidic environment suitable for silage fermentation, reduce the loss of nutrients, and improve aerobic stability [21]. In this experiment, different fermentation types of lactic acid bacteria were added to SP-silage to rapidly acidify the fermentation system so as to shorten the fermentation time and improve the fermentation quality [22]. Studies have shown that different fermentation types of lactic acid bacteria positively affect the fermentation quality of silage during the silage process through their degradation of feed substrates and metabolite conversion mechanisms [23].

Silage fermentation is a complex dynamic process involving the mutual transformation and accumulation of many substances. During this process, the pH and LA content within the silage system is a key parameter in measuring the fermentation quality. Generally, the lower the pH and the higher the LA content of silage, the better the fermentation quality [24]. During fermentation, lactic acid bacteria convert sugars to pyruvic acid through glycolytic pathways (e.g., the EMP pathway), which in turn is reduced to LA by lactate dehydrogenase [25]. As the fermentation of lactic acid bacteria continues, LA gradually accumulates in the silage, leading to a significant decrease in the pH of the silage. An acidic environment with low pH has a damaging effect on the cell membranes and enzyme systems of harmful microorganisms, thus inhibiting their growth and reproduction. Also, in addition to the production of LA, lactic acid bacteria may produce other antimicrobial substances (e.g., bacteriocins, hydrogen peroxide, etc.), which have a direct killing or inhibitory effect on harmful microorganisms [25,26]. In this experiment, the addition of LP, LB, and LPLB significantly reduced the pH and significantly increased the LA content of SP-silage compared to the control. This is due to the high activity of exogenously added lactic acid bacteria and the ideal anaerobic environment provided by the compacted silage, which allowed the lactic acid bacteria to proliferate and produce LA. The production of LA further reduced the pH of the silage, creating an acidic environment that inhibited the colonization of other undesirable microorganisms, which, in turn, effectively enhanced the fermentation quality of the silage [27]. Yang et al. [28] showed that the addition of LP significantly increases the LA content, decreases the pH, and improves the fermentation quality in alfalfa silage, which is consistent with the results of this experiment. Romero et al. [29] showed that the addition of LB to silage increases the AA content and LA concentration than the control without any lactic acid bacteria addition. In addition, the LPLB group had the highest LA content in this experiment, and both the LB and LPLB groups had a higher AA content, which is consistent with the findings of Weinters et al.'s study [30]. This was because the homofermentative lactic acid bacteria fermentation end product was LA, whereas the heterofermentative lactic acid bacteria were able to break LA down into AA, 1,2-propanediol, carbon dioxide, and ethanol in an anaerobic environment. BA is a product formed by harmful bacteria during silage fermentation by decomposing and transforming AA and sugar, and a high content of BA represents a strong activity of harmful bacteria, which is unfavorable to silage fermentation. In this experiment, the BA content of all three additive groups was lower than that in the control group, which indicates that the addition of different fermentation types of lactic acid bacteria was effective in improving the fermentation quality of SP-silage.

Spoilage microorganisms decompose proteins and amino acids in silage to produce NH_3-N. In this experiment, all three additive groups significantly reduced the NH_3-N content of SP-silage, indicating that the addition of the three lactic acid bacteria significantly inhibited the protein decomposition of the silage. In SP-silage inoculated with three species of lactic acid bacteria, the pH of the silage was reduced, leading to inhibition of plant protease activity, which, in turn, reduced protein and amino acid catabolism and decreased NH_3-N production [31]. Blajman et al. [32] added a lactic acid bacteria inoculant to alfalfa silage, which reduced the pH and NH_3-N content of the silage compared to the control, in agreement with the results of the present experiment. Filya et al. [33] showed that the combination of LP and LB is effective in lowering the pH of silage and reducing the concentration of NH_3-N, thus reducing nutrient loss during fermentation. The pH of the LPLB group was lower than that in the LB group but higher than that in the LP group, with the highest LA content and the lowest NH_3-N/TN. This may be due to the presence of synergistic effects, as evidenced by the LP $\times$ LB interaction effect on pH, LA, and NH_3-N/TN (Table 2). LP produced LA, which reduced the pH and inhibited the growth of harmful bacteria. LB promoted the conversion of LA to AA, inhibited the activity of aerobic microorganisms, and slowed down the decomposition of amino acids, indicating that the combination of different fermentation types of lactic acid bacteria was more efficient in inhibiting the activity of harmful microorganisms.

4.2. Effect of Different Fermenting Lactic Acid Bacteria on the Energy and Nutritional Value of Silphium perfoliatum L. Silage

Energy is one of the most important indicators for assessing the nutritional value of feeds and is the basis for the study of ME and DE in animals. In this experiment, the LB and LPLB groups significantly increased the GE, DE, ME, and NEl of SP-silage, indicating that the addition of LB and LPLB is an effective method to promote the energy accumulation of silage. This may be because the addition of LB and LPLB inhibits the aerobic spoilage bacterial activity and reduces the consumption of nutrients in SP-silage.

The nutritional content is one of the most important indicators for determining the quality of SP-silage. After silage, the nutrients in SP were reduced compared to pre-silage, which is mainly due to the involvement of a variety of microorganisms in the silage fermentation process. Under the joint action of these microorganisms, nutrients in the substrate were consumed in large quantities, such as CP, which was converted to NH_3-N and amines. In this experiment, the three lactic acid bacteria groups significantly increased the levels of DM and CP compared to the control group, with the LPLB group having the highest levels. This result indicates that both homofermentative and heterofermentative lactic acid bacteria are effective in inhibiting the degradation of DM and CP in SP-silage. In addition, the addition of a mixture of different fermentation types of lactic acid bacteria showed better silage results compared to single strains, which is in line with the findings of Contreras-Govea et al. [34]. This is because the addition of lactic acid bacteria accelerated the fermentation process, leading to a rapid drop in pH, which effectively inhibited the activity of aerobic microorganisms and proteolytic enzymes, thus reducing nutrient loss [35].

In addition, the LB and LPLB groups were more effective than the LP group in enhancing the DM and CP contents, which may be related to the difference in the fermentation end-products of LB as a heterofermentative lactic acid bacteria. The reason may be that LB produces ethanol and AA during the decomposition of LA. Ethanol has a fungicidal effect, while AA inhibits the reproduction of some fungi. Compared to LP, LB was much better at retaining DM and CP. Therefore, the addition of LB and LPLB is effective in reducing the loss of DM and CP in SP-silage, which is in agreement with the findings of Reich et al. [36].

NDF and ADF are key nutritional parameters for forage quality. A low NDF content indicates good feed palatability, while a low ADF content implies high feed utilization. Studies have shown that the addition of different fermentation types of lactic acid bacteria to barley, spent mushroom substrate, and perennial ryegrass results in a decrease in the levels of NDF and ADF, which is in agreement with the results of the present experiment [37–39]. The was also an LP × LB interaction effect on the ADF content (p = 0.003). In SP-silage inoculated with LP alone, there was no significant change in the content of ADF compared to the control. LB and LPLB were particularly effective in reducing the NDF and ADF contents. This may be due to the production of ferulic acid esterase during LB reproduction. Ferulic acid esterase breaks down the ferulic ester bond between lignin and hemicellulose, destroying the fiber structure, which, in turn, reduces the fiber content and enhances the fiber digestibility of the feed to some extent [39].

4.3. Effect of Different Fermenting Lactic Acid Bacteria on the In Vitro Digestibility of Silphium perfoliatum L. Silage

In vitro simulation of rumen fermentation provides a more accurate estimate of the potential digestibility of forage in ruminants [40]. By determining the amount of *IV*DMD in the rumen, it is possible to estimate the energy distribution potential in the feed [41]. The results of this experiment showed that the addition of different fermentation types of lactic acid bacteria to SP-silage significantly increased the *IV*DMD and *IV*CPD. Tian et al. [42] found that the introduction of lactic acid bacteria in *Leymus chinensis* silage results in significantly higher *IV*DMD and *IV*CPD compared to the control. This finding is supported by Wan et al. [43], where the addition of lactic acid bacteria to wilted sudangrass silage similarly elevated the *IV*DMD and *IV*CPD, consistent with the results of this experiment. This is due to the production of LA by lactic acid bacteria to form an acidic environment, which inhibits the activity of harmful microorganisms and reduces nutrient consumption, resulting in an increase in the *IV*DMD and *IV*CPD. Research has shown that feeding animals silage supplemented with lactic acid bacteria improves animal performance to some extent. Nkosi et al. [44] showed that lambs fed whole-crop maize silage supplemented with *L. buchneri NCIMB 40788* improve their crude protein intake, final body weight, and average daily gain compared to the control. Inoculation of alfalfa silage with *L. buchneri 40788* results in increased milk production in lactating dairy cows compared to the control [45]. Addah et al. [46] found that barley silage fed to growing cattle with the addition of a composite fungicide (*L. buchneri LN4017* + *L. plantarum LP7109* + *L. casei LC3200*) is superior to the control in terms of feed conversion ratio and average daily weight gain, suggesting that the compound bacterial agent has a positive effect on improving the performance of animal production. Lactic acid bacteria in the fermentation process can degrade the large molecule proteins in silage into free amino acids and small-molecule polypeptides, providing nutrients for the organism and promoting digestion and absorption in the gastrointestinal tract. Lactic acid bacteria can also secrete and produce a variety of digestive enzymes in the host body to decompose and transform antinutritional factors, thus improving the production performance of animals [47]. LP and LB, as probiotics, can colonize and multiply in the animal's intestinal tract to form a dominant flora. They inhibit the growth of harmful microorganisms such as E. coli and Salmonella, promote intestinal health, reduce the incidence of gastrointestinal diseases, and increase economic benefits by reducing deaths and treatment costs due to disease. At the same time, the combined addition of LP and LB to SP-silage improved the nutrient utilization of silage and reduced the animal's feed requirement, thus reducing feeding costs. Therefore, based on the results of this experimental study, we can expect that if SP-silage with added LPLB was introduced into ruminant diets, it could effectively make up for the deficiencies of a single inoculant and contribute to the improvement of ruminant production performance and economic benefits for farmers. In addition, the combined addition of LP and LB to SP-silage needs to take into account the fact that lactic acid bacteria may be affected by temperature, humidity, pH and other factors during the silage process, leading to a decrease in survival. Faced with such a problem, we can optimize the silage conditions, such as controlling the appropriate temperature and humidity, and adding the right amount of sugar substances as a fermentation substrate to improve the survival rate of lactic acid bacteria.

The fibrous matter content and composition of feed are important factors affecting the internal environment of the rumen in ruminants. Herein, the NDF content of feeds affected the *IV*NDFD, which led to different *IV*NDFDs for the different feeds. In addition, there was an LP × LB interaction effect on the *IV*NDFD, suggesting that the combined synergistic effect of LP and LB significantly improved the in vitro digestibility of the silage. Although the addition of either LP or LB alone increased the *IV*NDFD of the silage, the LPLB group had the highest *IV*NDFD. Weinberg et al. [48] found that the addition of specific microbial formulations to silage may lead to a decrease in LA levels due to competition with rumen microorganisms, further contributing to an increase in rumen pH and providing a more favorable environment for the growth of fibrolytic bacteria in the rumen. Therefore, inoculation of silage with microbial preparations can improve fiber digestion in the rumen. Studies have shown that the addition of LB to corn silage improves the *IV*NDFD, consistent with the results of this experiment [48]. Nsereko et al. [39] added lactic acid bacteria with ferulic acid esterase expression to silage and found them to be effective in improving the fiber digestibility of silage. Ferulic acid esterase is an enzyme capable of hydrolyzing ferulic acid ester bonds, which are widely found in non-starch polysaccharides (e.g., hemicellulose) in plant cell walls. The possible mechanism of action of this enzyme lies in the weakening of the structural stability of the cell wall by acting on the ferulic acid ester bond in hemicellulose [39]. At the same time, ferulic acid esterase strips hemicellulose from lignin, thereby increasing the fiber digestibility of the feed. Kang et al. [49] showed that *L. buchneri PTA6138* has ferulic acid esterase activity. When *L. buchneri PTA6138* was used in combination with *L. casei PTA6315*, it was effective in improving the *IV*NDFD of corn silage. In this assay, we did not verify whether LB had ferulic acid esterase activity. However, both LB and LPLB were observed to increase the *IV*NDFD of SP-silage. Based on this finding, we believe that a more in-depth study of this phenomenon is warranted.

4.4. *Effect of Different Fermenting Lactic Acid Bacteria on the Aerobic Stability of Silphium perfoliatum L. Silage*

Once the silage bag was opened, the original anaerobic state was rapidly transformed into an aerobic state, which, in turn, activated the activity of aerobic microorganisms. In general, spoilage of silage is mainly due to the activity of aerobic microorganisms such as yeasts and molds in an aerobic environment. Yeasts and molds utilize the LA, remaining sugars, and CP in the silage after it is opened, triggering a continuous release of heat from the silage, which ultimately leads to spoilage of the silage, a phenomenon confirmed in the study of Hu et al. [50].

In this experiment, the aerobic stability of the LP group was significantly lower than that in the control group, while the aerobic stability of the LB and LPLB groups was significantly higher, which was prolonged by 51 and 30 h, respectively, compared to the control group. During fermentation, homofermentative lactic acid bacteria dominated the process, producing LA as the main fermentation product. Short-chain fatty acids inhibited the growth of yeast and mold. Wang et al. [51] showed that LA produced by homofermentative lactic acid bacteria can be used as a substrate for yeast growth under aerobic exposure conditions while producing fewer short-chain fatty acids. In addition,

LA itself is not an effective antimould agent. Consequently, SP-silage inoculated with homofermentative lactic acid bacteria is more susceptible to spoilage during silage storage, which has been confirmed in previous studies [52]. Weinberg et al. [35] showed that LP reduces the aerobic stability of wheat silage and is highly susceptible to spoilage, which is in agreement with the results of this experiment. Heterofermentative lactic acid bacteria, after becoming the dominant flora in silage, improve the aerobic stability of silage by increasing the concentration of AA and 1,2-propanediol and producing bacteriocins [53]. AA has a strong fungus-inhibiting ability and can prevent feed spoilage. In this experiment, both the LB and LPLB groups had a higher AA content, which was significantly higher than the control and LP groups. This is one of the main reasons why the LB and LPLB groups were able to improve the aerobic stability of SP-silage [54]. According to Wang et al. [51], inoculation with either LB or LPLB significantly improves the aerobic stability of whipgrass silage. Filya et al. [33] showed that wheat, sorghum, and corn silages supplemented with LB alone have a lower WSC, LA content, and pH compared to LP alone, yet they are more aerobically stable. In addition, they found that the addition of LPLB not only helps to reduce the loss of DM but also enhances the aerobic stability of silage, a finding that coincides with the results of this experiment.

5. Conclusions

The addition of LP and LB alone or in combination improved the silage quality and in vitro digestibility of SP-silage, with the best silage results obtained using a LPLB inoculant. The LB inoculant alone can improve the aerobic stability of SP-silage. Further in vivo trials are needed to validate and evaluate the actual effects of LPLB-inoculated SP-silage on ruminant growth performance before recommending our findings to farmers.

Author Contributions: Conceptualization, Y.J., B.Y. and P.W.; methodology, Y.J., M.Y., J.D., Q.Y. and H.T.; software, Y.J., T.Z., M.Y., F.L., J.D. and Q.Y.; validation, B.Y., P.W., J.D., Q.Y., M.Y. and T.Z.; formal analysis, Y.J., P.W. and J.D.; investigation, M.Y. and F.L.; resources, B.Y. and H.T.; data curation, Y.J., P.W. and M.Y; writing—original draft preparation, Y.J. and P.W.; writing—review and editing, Y.J., P.W. and T.Z; visualization, B.Y. and P.W.; supervision, B.Y. and P.W.; project administration, B.Y. and P.W.; funding acquisition, P.W. and F.L. All authors have read and agreed to the published version of the manuscript.

Funding: This study was supported by funds from the China Agriculture Research System (CARS-37) and the Jilin Province Agricultural Key Core Technology Demonstration and Promotion (Industrial Technology System) Project (JARS-2024-0703).

Institutional Review Board Statement: This study strictly followed the Chinese Laboratory Animal Welfare Ethical Review Guidelines and was approved by the Laboratory Animal Welfare Ethics Committee of Jilin University (permit number: SY20209600).

Informed Consent Statement: Not applicable, as this research did not involve any humans. The small-tailed cold sheep used in the trial were from Changchun Haoyue Halal Meat Co. in Jilin Province, not from private farms.

Data Availability Statement: The original contributions presented in the study are included in the article, further inquiries can be directed to the corresponding author.

Conflicts of Interest: The authors declare no conflicts of interest.

References

1. Van Tassel, D.L.; Albrecht, K.A.; Bever, J.D.; Boe, A.A.; Brandvain, Y.; Crews, T.E.; Gansberger, M.; Gerstberger, P.; González-Paleo, L.; Hulke, B.S.; et al. Accelerating *Silphium* domestication: An opportunity to develop new crop ideotypes and breeding strategies informed by multiple disciplines. *Crop Sci.* **2017**, *57*, 1274–1284. [CrossRef]
2. Kowalska, G.; Pankiewicz, U.; Kowalski, R. Evaluation of chemical composition of some *Silphium* L. species as alternative raw materials. *Agriculture* **2020**, *10*, 132. [CrossRef]
3. El-Sayed, N.H.; Wojcińska, M.; Drost-Karbowska, K.; Matławska, I.; Williams, J.; Mabry, T.J. Kaempferol triosides from *Silphium perfoliatum*. *Phytochemistry* **2002**, *60*, 835–838. [CrossRef]

4. Xie, Y.; Bao, J.; Li, W.; Sun, Z.; Gao, R.; Wu, Z.; Yu, Z. Effects of applying lactic acid bacteria and molasses on the fermentation quality, protein fractions and in vitro digestibility of baled alfalfa silage. *Agronomy* **2021**, *11*, 91. [CrossRef]
5. Ren, H.; Feng, Y.; Pei, J.; Li, J.; Wang, Z.; Fu, S.; Zheng, Y.; Li, Z.; Peng, Z. Effects of *Lactobacillus plantarum* additive and temperature on the ensiling quality and microbial community dynamics of cauliflower leaf silages. *Bioresour. Technol.* **2020**, *307*, 123238. [CrossRef]
6. Filya, I. The effect of *Lactobacillus buchneri* and *Lactobacillus plantarum* on the fermentation, aerobic stability, and ruminal degradability of low dry matter corn and sorghum silages. *J. Dairy Sci.* **2003**, *86*, 3575–3581. [CrossRef] [PubMed]
7. Zhang, Q.; Yu, Z.; Yang, H.; Na, R.S. The effects of stage of growth and additives with or without cellulase on fermentation and in vitro degradation characteristics of *Leymus chinensis* silage. *Grass Forage Sci.* **2016**, *71*, 595–606. [CrossRef]
8. Xu, D.; Ding, W.; Ke, W.; Li, F.; Zhang, P.; Guo, X. Modulation of metabolome and bacterial community in whole crop corn silage by inoculating homofermentative *Lactobacillus plantarum* and heterofermentative *Lactobacillus buchneri*. *Front. Microbiol.* **2019**, *9*, 3299. [CrossRef]
9. Woolford, M.K. The detrimental effects of air on silage. *J. Appl. Bacteriol.* **1990**, *68*, 101–116. [CrossRef]
10. Xu, S.; Yang, J.; Qi, M.; Smiley, B.; Rutherford, W.; Wang, Y.; McAllister, T.A. Impact of saccharomyces cerevisiae and *Lactobacillus buchneri* on microbial communities during ensiling and aerobic spoilage of corn silage. *J. Anim. Sci.* **2019**, *97*, 1273–1285. [CrossRef]
11. Zhang, T.; Li, L.; Wang, X.; Zeng, Z.; Hu, Y.; Cui, Z. Effects of *Lactobacillus buchneri* and *Lactobacillus plantarum* on fermentation, aerobic stability, bacteria diversity and ruminal degradability of alfalfa silage. *World J. Microbiol. Biotechnol.* **2009**, *25*, 965–971. [CrossRef]
12. Kung, L.; Shaver, R.D.; Grant, R.J.; Schmidt, R.J. Silage review: Interpretation of chemical, microbial, and organoleptic components of silages. *J. Dairy Sci.* **2018**, *101*, 4020–4033. [CrossRef] [PubMed]
13. Cao, Y.; Cai, Y.; Hirakubo, T.; Fukui, H.; Matsuyama, H. Fermentation characteristics and microorganism composition of total mixed ration silage with local food by-products in different seasons. *Anim. Sci. J.* **2011**, *82*, 259–266. [CrossRef] [PubMed]
14. AOAC. *Official Methods of Analysis*, 18th ed.; AOAC Int.: Gaithersburg, MD, USA, 2005.
15. Van Soest, P.J.; Robertson, J.B.; Lewis, B.A. Methods for dietary fiber, neutral detergent fiber, and nonstarch polysaccharides in relation to animal nutrition. *J. Dairy Sci.* **1991**, *74*, 3583–3597. [CrossRef] [PubMed]
16. Deriaz, R.E. Routine analysis of carbohydrates and lignin in herbage. *J. Sci. Food Agric.* **1961**, *12*, 152–160. [CrossRef]
17. Playne, M.J.; McDonald, P. The buffering constituents of herbage and of silage. *J. Sci. Food Agric.* **1966**, *17*, 264–268. [CrossRef]
18. Huang, Y.Y.; Si, R.; Chen, X.B. *Feeding of Cattle, Sheep and Goats*, 3rd ed.; China Agricultural University Press: Beijing, China, 2013; pp. 140–154.
19. Kung, L.J.; Sheperd, A.C.; Smagala, A.M.; Endres, K.M.; Bessett, C.A.; Ranjit, N.K.; Glancey, J. The effect of preservatives based on propionic acid on the fermentation and aerobic stability of corn silage and a total mixed ration. *J. Dairy Sci.* **1998**, *81*, 1322–1330. [CrossRef] [PubMed]
20. He, L.; Zhou, W.; Wang, Y.; Wang, C.; Chen, X.; Zhang, Q. Effect of applying lactic acid bacteria and cellulase on the fermentation quality, nutritive value, tannins profile and in vitro digestibility of *Neolamarckia cadamba* leaves silage. *J. Anim. Physiol. Anim. Nutr.* **2018**, *102*, 1429–1436. [CrossRef]
21. Zhang, J.; Guo, G.; Chen, L.; Li, J.; Yuan, X.; Yu, C.; Shimojo, M.; Shao, T. Effect of applying lactic acid bacteria and propionic acid on fermentation quality and aerobic stability of oats-common vetch mixed silage on the T ibetan plateau. *Anim. Sci. J.* **2015**, *86*, 595–602. [CrossRef]
22. Bai, J.; Ding, Z.; Ke, W.; Xu, D.; Wang, M.; Huang, W.; Zhang, Y.; Liu, F.; Guo, X. Different lactic acid bacteria and their combinations regulated the fermentation process of ensiled alfalfa: Ensiling characteristics, dynamics of bacterial community and their functional shifts. *Microb. Biotechnol.* **2021**, *14*, 1171–1182. [CrossRef]
23. Xu, D.; Wang, N.; Rinne, M.; Ke, W.; Weinberg, Z.G.; Da, M.; Bai, J.; Zhang, Y.; Li, F.; Guo, X. The bacterial community and metabolome dynamics and their interactions modulate fermentation process of whole crop corn silage prepared with or without inoculants. *Microb. Biotechnol.* **2021**, *14*, 561–576. [CrossRef] [PubMed]
24. Tharangani, R.M.H.; Yakun, C.; Zhao, L.S.; Shen, Y.F.; Ma, L.; Bu, D.P. Proposal and validation of integrated alfalfa silage quality index (ASQI) method for the quality assessment of alfalfa silage for lactating dairy cows. *Anim. Feed Sci. Technol.* **2022**, *289*, 115339. [CrossRef]
25. Soundharrajan, I.; Park, H.S.; Rengasamy, S.; Sivanesan, R.; Choi, K.C. Application and future prospective of lactic acid bacteria as natural additives for silage production—A review. *Appl. Sci.* **2021**, *11*, 8127. [CrossRef]
26. Okoye, C.O.; Wang, Y.; Gao, L.; Wu, Y.; Li, X.; Sun, J.; Jiang, J. The performance of lactic acid bacteria in silage production: A review of modern biotechnology for silage improvement. *Microbiol. Res.* **2023**, *266*, 127212. [CrossRef] [PubMed]
27. Teixeira Franco, R.; Buffière, P.; Bayard, R. Ensiling for biogas production: Critical parameters. A review. *Biomass Bioenergy* **2016**, *94*, 94–104. [CrossRef]
28. Yang, F.; Wang, Y.; Zhao, S.; Wang, Y. *Lactobacillus plantarum* inoculants delay spoilage of high moisture alfalfa silages by regulating bacterial community composition. *Front. Microbiol.* **2020**, *11*, 1989. [CrossRef]
29. Romero, J.J.; Zhao, Y.; Balseca-Paredes, M.A.; Tiezzi, F.; Gutierrez-Rodriguez, E.; Castillo, M.S. Laboratory silo type and inoculation effects on nutritional composition, fermentation, and bacterial and fungal communities of oat silage. *J. Dairy Sci.* **2017**, *100*, 1812–1828. [CrossRef] [PubMed]

30. Winters, A.L.; Cockburn, J.E.; Dhanoa, M.S.; Merry, R.J. Effects of lactic acid bacteria in inoculants on changes in amino acid composition during ensilage of sterile and non-sterile ryegrass. *J. Appl. Microbiol.* **2000**, *89*, 442–452. [CrossRef] [PubMed]
31. Cai, Y.; Benno, Y.; Ogawa, M.; Kumai, S. Effect of applying lactic acid bacteria isolated from forage crops on fermentation characteristics and aerobic deterioration of silage. *J. Dairy Sci.* **1999**, *82*, 520–526. [CrossRef]
32. Blajman, J.E.; Vinderola, G.; Páez, R.B.; Signorini, M.L. The role of homofermentative and heterofermentative lactic acid bacteria for alfalfa silage: A meta-analysis. *J. Agric. Sci.* **2020**, *158*, 107–118. [CrossRef]
33. Filya, I.; Sucu, E.; Karabulut, A. Effect of *Lactobacillus buchneri* on the fermentation, aerobic stability and ruminal degradability of maize silage. *J. Appl. Microbiol.* **2006**, *101*, 1216–1223. [CrossRef] [PubMed]
34. Contreras-Govea, F.E.; Muck, R.E.; Broderick, G.A.; Weimer, P.J. *Lactobacillus plantarum* effects on silage fermentation and in vitro microbial yield. *Anim. Feed Sci. Technol.* **2013**, *179*, 61–68. [CrossRef]
35. Weinberg, Z.; Ashbell, G.; Hen, Y.; Azrieli, A.; Szakacs, G.; Filya, I. Ensiling whole-crop wheat and corn in large containers with *Lactobacillus plantarum* and *Lactobacillus buchneri*. *J. Ind. Microbiol. Biotechnol.* **2002**, *28*, 7–11. [CrossRef] [PubMed]
36. Reich, L.J.; Kung, L. Effects of combining *Lactobacillus buchneri* 40788 with various lactic acid bacteria on the fermentation and aerobic stability of corn silage. *Anim. Feed Sci. Technol.* **2010**, *159*, 105–109. [CrossRef]
37. Kung, L.; Carmean, B.R.; Tung, R.S. Microbial inoculation or cellulase enzyme treatment of barley and vetch silage harvested at three maturities. *J. Dairy Sci.* **1990**, *73*, 1304–1311. [CrossRef]
38. Yi, Q.X.; Wang, P.; Tang, H.Y.; Zhao, T.Y.; Sheng, Z.Y.; Luo, H.L. Fermentation quality, in vitro digestibility, and aerobic stability of ensiling spent mushroom substrate with microbial additives. *Animals* **2023**, *13*, 920. [CrossRef]
39. Nsereko, V.L.; Smiley, B.K.; Rutherford, W.M.; Spielbauer, A.; Forrester, K.J.; Hettinger, G.H.; Harman, E.K.; Harman, B.R. Influence of inoculating forage with lactic acid bacterial strains that produce ferulate esterase on ensilage and ruminal degradation of fiber. *Anim. Feed Sci. Technol.* **2008**, *145*, 122–135. [CrossRef]
40. Madsen, J.; Hvelplund, T.; Weisbjerg, M.R. Appropriate methods for the evaluation of tropical feeds for ruminants. *Anim. Feed Sci. Technol.* **1997**, *69*, 53–66. [CrossRef]
41. Mertens, D.R. Creating a system for meeting the fiber requirements of dairy cows. *J. Dairy Sci.* **1997**, *80*, 1463–1481. [CrossRef] [PubMed]
42. Tian, J.; Yu, Y.; Yu, Z.; Shao, T.; Na, R.; Zhao, M. Effects of lactic acid bacteria inoculants and cellulase on fermentation quality and in vitro digestibility of *Leymus chinensis* silage. *Grassl. Sci.* **2014**, *60*, 199–205. [CrossRef]
43. Wan, J.C.; Xie, K.Y.; Wang, Y.X.; Liu, L.; Yu, Z.; Wang, B. Effects of wilting and additives on the ensiling quality and rumen fermentation characteristics of sudangrass silage. *Anim. Biosci.* **2021**, *34*, 56–65. [CrossRef] [PubMed]
44. Nkosi, B.D.; Meeske, R.; Palic, D.; Langa, T.; Leeuw, K.; Groenewald, I.B. Effects of ensiling whole crop maize with bacterial inoculants on the fermentation, aerobic stability, and growth performance of lambs. *Anim. Feed Sci. Technol.* **2009**, *154*, 193–203. [CrossRef]
45. Kung, L.; Taylor, C.C.; Lynch, M.P.; Neylon, J.M. The effect of treating alfalfa with *Lactobacillus buchneri* 40788 on silage fermentation, aerobic stability, and nutritive value for lactating dairy cows. *J. Dairy Sci.* **2003**, *86*, 336–343. [CrossRef] [PubMed]
46. Addah, W.; Baah, J.; Okine, E.K.; McAllister, T.A. A third-generation esterase inoculant alters fermentation pattern and improves aerobic stability of barley silage and the efficiency of body weight gain of growing feedlot cattle. *J. Anim. Sci.* **2012**, *90*, 1541–1552. [CrossRef] [PubMed]
47. Deng, Z.; Hou, K.; Zhao, J.; Wang, H. The probiotic properties of lactic acid bacteria and their applications in animal husbandry. *Curr. Microbiol.* **2022**, *79*, 22. [CrossRef] [PubMed]
48. Weinberg, Z.G.; Shatz, O.; Chen, Y.; Yosef, E.; Nikbahat, M.; Ben-Ghedalia, D.; Miron, J. Effect of lactic acid bacteria inoculants on in vitro digestibility of wheat and corn silages. *J. Dairy Sci.* **2007**, *90*, 4754–4762. [CrossRef] [PubMed]
49. Kang, T.W.; Adesogan, A.T.; Kim, S.C.; Lee, S.S. Effects of an esterase-producing inoculant on fermentation, aerobic stability, and neutral detergent fiber digestibility of corn silage. *J. Dairy Sci.* **2009**, *92*, 732–738. [CrossRef]
50. Hu, W.; Schmidt, R.J.; McDonell, E.E.; Klingerman, C.M.; Kung, L. The effect of *Lactobacillus buchneri* 40788 or *Lactobacillus plantarum* MTD-1 on the fermentation and aerobic stability of corn silages ensiled at two dry matter contents. *J. Dairy Sci.* **2009**, *92*, 3907–3914. [CrossRef] [PubMed]
51. Wang, M.; Yang, C.; Jia, L.; Yu, K. Effect of *Lactobacillus buchneri* and *Lactobacillus plantarum* on the fermentation characteristics and aerobic stability of whipgrass silage in laboratory silos. *Grassl. Sci.* **2014**, *60*, 233–239. [CrossRef]
52. Adesogan, A.T.; Salawu, M.B. The effect of different additives on the fermentation quality, aerobic stability and in vitro digestibility of pea/wheat bi-crop silages containing contrasting pea to wheat ratios. *Grass Forage Sci.* **2002**, *57*, 25–32. [CrossRef]
53. Zhang, F.; Miao, F.; Wang, X.; Lu, W.; Ma, C. Effects of homo- and hetero-fermentative lactic acid bacteria on the quality and aerobic stability of corn silage. *Can. J. Anim. Sci.* **2021**, *101*, 761–770. [CrossRef]
54. Kung, L.; Ranjit, N.K. The effect of *Lactobacillus buchneri* and other additives on the fermentation and aerobic stability of barley silage. *J. Dairy Sci.* **2001**, *84*, 1149–1155. [CrossRef] [PubMed]

Disclaimer/Publisher's Note: The statements, opinions and data contained in all publications are solely those of the individual author(s) and contributor(s) and not of MDPI and/or the editor(s). MDPI and/or the editor(s) disclaim responsibility for any injury to people or property resulting from any ideas, methods, instructions or products referred to in the content.

animals

MDPI

Article

Impact of 5–20% Hydroponic Wheat Sprouts Inclusion on Growth and Metabolic Parameters of Growing Ewes

Li Min [1,2,†], Yong Tuo [1,3,†], Dagang Li [2], Changjiang Zang [3], Guzalnur Amat [1], Zhijun Zhang [1] and Tongjun Guo [1,*]

[1] Key Laboratory of Xinjiang Feed Biotechnology, Feed Research Institute, Xinjiang Academy of Animal Science, Urumqi 830011, China; minli@gdaas.cn (L.M.); t777yvx@126.com (Y.T.); 18140749758@163.com (G.A.); zzj850916@sina.com (Z.Z.)
[2] Ministry of Agriculture Key Laboratory of Animal Nutrition and Feed Science in South China, Institute of Animal Science, Guangdong Academy of Agricultural Sciences, Guangzhou 510640, China; lidagang@gdaas.cn
[3] College of Animal Science, Xinjiang Agricultural University, Urumqi 830052, China; zcj780@126.com
* Correspondence: guotaoxj@126.com
† These authors contributed equally to this work.

Simple Summary: Simple Summary: Hydroponic wheat sprouts could be used as a potential sustainable feedstuff resource for livestock production. Based on its nutrient composition, it could exert good palatability and digestibility, thereby improving the feed intake and productive performance of livestock. We have evaluated the responses of four-month-old growing Hu ewes, while their diets are replaced with different proportions (5–20%) of hydroponic wheat sprouts. In this experiment, substitution 15% of the basal diet with hydroponic wheat sprouts would obtain the optimum production performance and net profit for growing ewes. Inversely, the higher substitution ratio (20%) will result in the decline of production performance and economic effectiveness, and show negative effect on rumen fermentation.

Abstract: The aim of this study was to assess the impact of varying proportions (5–20%) of hydroponic wheat sprouts in the diet of growing four-month-old Hu ewes on their productive performance, metabolic profiles, rumen fermentation, and alterations in microflora. Compared with the control group (CON), the optimum final weight of ewes has been presented in the group of substitution 15% (S15) of the basal diet with hydroponic wheat sprouts. Furthermore, 1–30 d the average daily gain (ADG), 31–60 d ADG, and average feed intake were both significantly improved in S15 compared with CON ($p < 0.05$). Feeding hydroponic wheat sprouts can significantly increase high-density lipoprotein and interleukin-2 ($p < 0.05$) accompanied by the numerical increase of the content of interferon-γ, suggesting its positive effect on ewes' health and immune systems. In this process, it is noteworthy that feeding hydroponic wheat sprouts results in an increase in relative abundance of *Olsenella*, *Limosilactobacillus*, *Shuttleworthia*, and *Prevotella_7*, and a decrease in relative abundance of *Succinimonas*, *Pseudobutyrivibrio*, and *Anaerovibrio* in the rumen of growing ewes. It implies that the response of rumen microflora adapted to the change of dietary ingredients, as well as the relationship between rumen microflora changes and the improvement of productive performance and immune system in growing ewes. Considering the usage cost and application effect, S15 of the basal diet with hydroponic wheat sprouts could be the appropriate application solution for growing ewes.

Keywords: unconventional feeds; alternative forages; economic effectiveness; rumen manipulation; growing ewes

Citation: Min, L.; Tuo, Y.; Li, D.; Zang, C.; Amat, G.; Zhang, Z.; Guo, T. Impact of 5–20% Hydroponic Wheat Sprouts Inclusion on Growth and Metabolic Parameters of Growing Ewes. *Animals* **2024**, *14*, 1630. https://doi.org/10.3390/ani14111630

Academic Editors: Dingfu Xiao and Tiantian Meng

Received: 7 May 2024
Revised: 23 May 2024
Accepted: 27 May 2024
Published: 30 May 2024

Copyright: © 2024 by the authors. Licensee MDPI, Basel, Switzerland. This article is an open access article distributed under the terms and conditions of the Creative Commons Attribution (CC BY) license (https://creativecommons.org/licenses/by/4.0/).

1. Introduction

The growing population coupled with the decreasing arable land worldwide have intensified the challenge of sustainable food supply chains [1]. Currently, hydroponic fodder has been used as a potential sustainable feedstuff resource for livestock production, which could be beneficial to reducing the food-feed competition and even improving productive performance [2]. Furthermore, owing to the rising prices of protein and energy feedstuff, feeding hydroponic fodder has also been used as an alternative solution extensively worldwide to reduce feed costs [3]. Therein, sprouted wheat is wheat grain that has been soaked in water, placed in trays, and allowed to germinate and ultimately convert into green feed within 6 to 8 days [4]. It requires soilless culture, less water and labor, and enables production at any time of the year. More importantly, compared with wheat grain, hydroponic wheat sprouts can provide more biomass, and also provide high-quality roughage for ruminants, which would be beneficial to rumen function. Ma et al. [5] indicated that hydroponic wheat sprouts are rich in protein, multi-amino acids, vitamins, and other nutritional components, which could exert good palatability and digestibility, thereby improving the feed intake and productive performance of livestock. However, limited research is available on the application of feeding hydroponic wheat sprouts on growing ewes.

To date, the application of hydroponic fodder is mainly focused on hydroponic barley sprouts in sheep production. Feeding hydroponic barley sprouts to sheep could result in better digestibility and rumen fermentation, as well as high animal performance [2]. Supplements of sprouted barley at different levels can improve the digestibility and rumen fermentation characteristics of Awassi lambs [6]. Feeding hydroponic barley sprouts could also be beneficial to lactating ewes, and replacing 5–15% of the basal diet could alleviate the oxidative stress of lactating ewes and improve the milk quality [7].

As regards to hydroponic wheat sprouts, Guerrero-Cervantes, et al. [8] found that hydroponic wheat sprouts are a suitable solution for alternative corn and cottonseed meals in the diets of Katahdin ewes (up to 30%) during gestation and lactation without having a negative effect, but instead having a beneficial effect on their reproductive performance, indicated by the growth and development of their lambs. Currently, little information is available about hydroponic wheat sprout food on productive performance, metabolic profiles, rumen fermentation, and the microflora changes of growing ewes. Therefore, the objective of this study was to evaluate the responses of four-month-old growing Hu ewes, while their diets are replaced with different proportions (5–20%) of hydroponic wheat sprouts.

2. Materials and Methods

2.1. Animals and Experimental Design

All animal procedures performed in this study were approved and guided by the Animal Care and Use Committee of the Xinjiang Academy of Animal Sciences (FRI-2023002, Urumqi, China). Fifty healthy four-month-old growing Hu ewes with a homogeneous genetic background and similar body weight (30.83 ± 0.52 kg) were selected for the experiment. All of the ewes were fed the same diet to maintain the consistency of the rumen function before the experiment. The experiment was conducted under the same feeding and management conditions. They were randomly divided into five groups in a randomized complete block design, each consisting of 10 ewes, and fed with different total mixed ration diets. The nutrient requirements of growing ewes were formulated according to feeding standards for meat-producing sheep and goats in China (NY/T 816-2021 [9]). The control group (CON) was fed with the basal diet including: corn straw, alfalfa, corn, cottonseed meal, wheat bran, calcium carbonate, sodium bicarbonate, salt, and premix. As we measured in the present study, the dry matter content of hydroponic wheat sprouts is 16.94%, the crude protein content is 15.36%, the crude fat content is 4.60%, the neutral detergent fiber is 30.12%, the acid detergent fiber is 18.40%, the calcium content is 0.49%, and the phosphorus content is 0.79%. The diets of the treatment groups were used in an equal replacement method by substitution 5% (S5), 10% (S10), 15% (S15) and 20% (S20) of the basal diet (dry matter basis) with hydroponic wheat sprouts, respectively (Table 1). Briefly, hydroponic wheat sprouts can partially replace the expensive protein feed and roughage. The application of hydroponic wheat sprouts may reduce feed cost and the food-feed competition (owing to having more biomass than wheat grain), and also may improve productive performance.

Table 1. Dietary ingredients and nutrient composition (dry matter basis) for four-month-old growing Hu ewes in this experiment.

Item	Groups				
	CON	S5	S10	S15	S20
	Ingredients (%)				
Hydroponic wheat sprouts	0	5	10	15	20
Corn straw	32.0	31.5	29.0	26.0	24.5
Alfalfa	10	7	6	5	3
Corn	30.0	29.0	28.8	28.8	28.0
Cottonseed meal	19.5	18.9	17.59	16.4	15.6
Wheat bran	4	4	4	4	4
Calcium carbonate	0	0.1	0.2	0.3	0.4
Sodium bicarbonate	0.7	0.7	0.7	0.7	0.7
Salt	0.8	0.8	0.8	0.8	0.8
Premix [1]	3	3	3	3	3
	Nutrient composition (%)				
Dry matter content	87.09	83.59	80.39	77.06	73.68
Crude protein	16.01	15.93	15.92	16.08	16.09
Crude fat	2.69	2.91	3.18	3.41	3.55
Neutral detergent fiber	27.90	27.77	26.88	23.00	23.47
Acid detergent fiber	16.30	14.72	14.68	11.58	11.00
Ash	11.59	12.18	11.78	11.77	11.38
Calcium	0.84	0.85	0.88	0.91	0.93
Phosphorus	0.53	0.55	0.57	0.59	0.61
Metabolic energy (MJ/kg)	11.19	11.19	11.19	11.19	11.19

[1] Premix contained the following per kg of diet: VA 1,000,000 IU, VD3 65,000 IU, VE 5000 mg, Fe 2000 mg, Cu 1750 mg, Zn 5500 mg, Mn 2550 mg, Se 75 mg, I 70 mg, Co 40 mg.

2.2. Production Data and Sample Collection

The experiment lasted two months. Ewes in each group were housed in individual units with ad libitum access to corresponding diets. The amount of feedstuff offered and refused was weighed daily, thus calculating the feed intake. At the beginning (0 d), mid-term (30 d), and end (60 d) of the experiment, ewes were weighed after fasting for 12 h. At the end of the experiment, blood samples were collected, and then serum samples were obtained via jugular venipuncture and centrifuged at 3000× *g* for 15 min (4 °C). Meanwhile, five ewes were randomly selected from each group, and rumen fluid samples were collected via esophageal tubing and filtered through four layers of sterilized cheesecloth at the end of the experiment after fasting for 12 h. The pH value was measured immediately by pH meter, and the remaining parts were stored at −80 °C for the determination of rumen fermentation and microflora changes.

2.3. Serum Biochemistry Analysis

The concentrations of total protein, albumin, glucose, blood urea nitrogen, triglyceride, total cholesterol, high-density lipoprotein, low-density lipoprotein, superoxide dismutase, glutathion peroxidase, malonaldehyde, lactic dehydrogenase, creatine kinase, glutamic pyruvic transaminase, glutamic oxalacetic transaminase, interleukin-1β, interleukin-2, interferon-γ, and CD4+ in serum were determined and analyzed by Thermo Scientific Microplate Reader using commercially available enzyme-linked immunosorbent assay kits (Nanjing Jiancheng Biological Engineering Institute, Nanjing, China).

2.4. Rumen Fermentation Characteristics Determination

After thawing, ruminal supernatant was obtained by centrifugation (1000× *g* at 4 °C for 10 min) for the analysis of volatile fatty acids. The concentrations of volatile fatty acids in ruminal samples were analyzed as described by Mohammed, et al. [10] using a gas chromatograph (GC-2010 Shimadzu, Kyoto, Japan). The concentration of ammoniacal nitrogen was detected based on alkaline sodium hypochlorite-phenol spectrophotometry [11].

2.5. Rumen Microflora Sequencing and Analysis

Microbial DNA was extracted from rumen fluid samples using cetyltrimethylammonium bromide method accompanied by bead beating [12]. The bacterial 16S rRNA V3-V4 region was amplified using referenced primers 338F (ACTCCTACGGGAGGCAGCA) and 806R (GGACTACHVGGGTWTCTAAT) [13]. Purified amplicons were pooled in equimolar amounts and paired-end sequenced on an Illumina PE300 platform (Illumina, San Diego, CA, USA) according to the standard protocols. The raw sequencing reads were deposited into the NCBI Sequence Read Archive (SRA) data: PRJNA1076690.

Bioinformatic analysis of the rumen microflora was conducted by the BMK cloud platform (http://www.biocloud.net/ accessed on 30 January 2024). According to the quality of a single nucleotide, raw data were primarily filtered by Trimmomatic (version 0.33). Identification and removal of primer sequences was processed by Cutadapt (version 1.9.1). PE reads obtained from previous steps were assembled by USEARCH (version 10) and followed by chimera removal using UCHIME (version 8.1). Sequences with similarity >97% were clustered into the same operational taxonomic unit (OTU) by USEARCH (v10), and the OTUs counts less than 2 in all samples were filtered.

The qualified sequences with more than 97% similarity thresholds were allocated to one OTU using USEARCH (version 10.0). Taxonomy annotation of the OTUs was performed based on the Naive Bayes classifier in QIIME2 using the SILVA database (release 138.1) with a confidence threshold of 70%. Alpha and beta diversity were calculated and displayed by the QIIME2. One-way analysis of variance was used to compare bacterial abundance and diversity. Linear discriminant analysis (LDA) coupled with effect size (LEfSe) was applied to evaluate the differentially abundant taxa (LDA score > 2, $p < 0.05$).

2.6. Statistical Analysis

Statistical analysis was performed using SPSS 23.0 statistical program. All experiment data were tested and confirmed for normal distribution using the Shapiro-Wilk test. Subsequently, the data of productive performance, serum biochemistry, and rumen fermentation characteristics were analyzed statistically using the one-way ANOVA procedure accompanied by Tukey's test. The results of the test were displayed as the mean and standard error of the mean (SEM), and $p < 0.05$ as significant level of difference.

3. Results

3.1. Effect of Feeding Hydroponic Wheat Sprouts on Productive Performance and Economic Effectiveness of Growing Ewes

Although there was no significant difference in final weight of ewes among the diet of five groups, the optimum final weight of ewes has been presented in S15, with the increased of average 3.14 kg compared with CON (Table 2). Furthermore, 1–30 d ADG, 31–60 d ADG, and average feed intake were both significantly improved ($p < 0.05$) in S15 compared with CON. Finally, the feed to gain ratio is 10.07, 10.15, 9.02, 8.87, and 9.76 from the groups of CON to S5, S10, S15, and S20 in turn.

Table 2. Effect of feeding hydroponic wheat sprouts on productive performance of growing ewes.

Item	Groups					SEM	*p* Value
	CON	S5	S10	S15	S20		
Initial weight (kg)	30.53	30.54	30.96	30.93	31.18	0.52	0.99
Final weight (kg)	36.96	37.89	38.50	40.10	39.39	0.56	0.44
1–30 d ADG (g/d)	115.95 ^{b}	121.67 ^{b}	134.52 ^{b}	165.95 ^{a}	137.38 ^{b}	5.00	<0.01
31–60 d ADG (g/d)	93.06 ^{b}	105.56 ^{ab}	120.28 ^{ab}	139.05 ^{a}	130.55 ^{a}	5.16	0.02
ADG (g/d)	106.25	111.31	128.57	145.42	136.31	6.74	0.32
Average feed intake (kg/d)	1.07 ^{c}	1.13 ^{bc}	1.16 ^{b}	1.29 ^{a}	1.33 ^{a}	0.01	<0.01
Feed to gain ratio	10.07	10.15	9.02	8.87	9.76		

ADG: average daily gain. Within rows, means with different letter superscripts ($^{a–c}$) are significantly different ($p < 0.05$).

The other important purpose of the application of feeding hydroponic wheat sprouts is to reduce feeding costs. As shown in Table 3, the minimum of feed unit price and the maximum net profit have been obtained in S15 according to the current market price in Kashgar, Xinjiang, China. Furthermore, if the price of hydroponic wheat sprouts is increased to more than 0.8 CNY/kg in future, it is inappropriate to use hydroponic wheat sprouts as a feedstuff resource to partially replace the basal diet for livestock production (Table 3).

Table 3. Economic effectiveness analysis of the application of feeding hydroponic wheat sprouts in growing ewes.

Item	Groups				
	CON	S5	S10	S15	S20
Total weight gain (kg/piece)	6.43	7.35	7.54	9.17	8.21
Feed unit price (CNY/kg) [1]	2.32	1.89	1.61	1.42	1.27
Total feed intake (kg/piece)	71.01	93.28	121.41	150.08	177.23
Feed cost (CNY/piece)	164.78	176.15	195.64	212.51	224.27
Weight gain profit (CNY/piece) [2]	210.38	220.39	254.57	287.93	269.89
Net profit (CNY/piece)	45.59	44.24	58.93	75.42	45.63
The price change of hydroponic wheat sprouts [3]					
0.5 CNY/kg	45.59	44.24	58.93	75.42	45.63
0.6 CNY/kg	45.59	41.99	54.06	67.67	34.96
0.7 CNY/kg	45.59	39.74	49.18	59.92	24.29
0.8 CNY/kg	45.59	37.48	44.30	52.17	13.62
0.9 CNY/kg	45.59	35.23	39.42	44.42	2.95

CNY: China Yuan. [1] The price of feed raw materials is calculated according to the current market price in Kashgar, Xinjiang (May 2022), which is used in the present study as follows: hydroponic wheat sprouts 0.5 CNY/kg, corn straw 0.85 CNY/kg, alfalfa 1.8 CNY/kg, corn 2.65 CNY/kg, cottonseed meal 3.75 CNY/kg, wheat bran 2.8 CNY/kg, calcium carbonate 2.5 CNY/kg, sodium bicarbonate 2.8 CNY/kg, salt 0.85 CNY/kg, premix 5.8 CNY/kg. [2] Unit price of live weight of Hu sheep is calculated according to the current market price in Kashgar, Xinjiang (August 2022): 33 CNY/kg. [3] If the price of hydroponic wheat sprouts is increased from 0.5 CNY/kg to 0.9 CNY/kg, the net profit of the application of feeding hydroponic wheat sprouts in growing ewes is presented among each group.

3.2. Effect of Feeding Hydroponic Wheat Sprouts on Serum Metabolic Profiles of Growing Ewes

Most blood biochemical indexes do not significantly change after feeding different proportions of hydroponic wheat sprouts to growing ewes (Table 4). It is noteworthy that feeding hydroponic wheat sprouts causes a varying degree of increase ($p < 0.05$) in glutamic oxalacetic transaminase, urea nitrogen, high-density lipoprotein, and interleukin-2 in growing ewes.

Table 4. Effect of feeding hydroponic wheat sprouts on serum metabolic profiles of growing ewes.

Item	Groups					SEM	*p* Value
	CON	S5	S10	S15	S20		
Total protein (g/L)	53.21	61.16	64.60	61.39	59.81	1.42	0.13
Albumin (g/L)	13.66	16.16	16.42	13.85	15.56	0.40	0.07
Superoxide mutase (U/mL)	156.79	166.43	162.27	154.78	149.56	5.06	0.88
Malonaldehyde (nmol/mL)	3.50	3.36	3.03	2.74	2.97	0.11	0.12
Glutathione peroxidase (U/mL)	22.44	24.87	23.43	24.49	25.92	0.97	0.83
Lactate dehydrogenase (U/L)	3333.3	3586.2	3697.9	3348.7	3738.5	75.2	0.26
Creatine kinase (U/L)	0.101	0.091	0.110	0.124	0.101	0.006	0.51
Alanine aminotransferase (U/L)	2.22	2.64	2.56	2.81	2.40	0.09	0.27
Glutamic oxalacetic transaminase (U/L)	8.04 ^{b}	10.63 ^{a}	11.35 ^{a}	10.40 ^{ab}	11.03 ^{a}	0.32	<0.01
Urea nitrogen (mmol/L)	5.80 ^{b}	6.60 ^{a}	7.30 ^{a}	6.12 ^{b}	6.41 ^{ab}	0.14	0.01
Glucose (mmol/L)	2.14	2.25	2.71	2.24	2.6	0.10	0.28
Triglyceride (mmol/L)	0.26	0.31	0.34	0.25	0.27	0.02	0.52
Total cholesterol (mmol/L)	1.08	1.23	1.35	1.18	1.21	0.05	0.52
High-density lipoprotein (mmol/L)	0.35 ^{c}	0.41 ^{bc}	0.54 ^{a}	0.51 ^{ab}	0.54 ^{a}	0.02	<0.01
Low-density lipoprotein (mmol/L)	0.39	0.43	0.48	0.4	0.43	0.02	0.59
Interleukin-1β (pg/mL)	25.47	28.47	29.24	26.56	29.2	0.57	0.12
Interleukin-2 (pg/mL)	28.91 ^{b}	34.33 ^{ab}	35.83 ^{a}	34.96 ^{a}	34.65 ^{a}	0.7	0.01
Interferon-γ (pg/mL)	65.02	67.00	73.60	72.21	74.61	1.82	0.38
CD4+ (ng/mL)	203.8	236	263.9	236.1	229.6	7.20	0.14

Within rows, means with different letter superscripts ($^{a–c}$) are significantly different ($p < 0.05$).

3.3. *Effect of Feeding Hydroponic Wheat Sprouts on Rumen Fermentation and Microflora Changes of Growing Ewes*

S20 of the basal diet with hydroponic wheat sprouts exerted the lowest value of rumen fermentation parameters in growing ewes among the diets of five groups, including ammoniacal nitrogen, total volatile fatty acids, acetate, and valerate (Table 5). In addition to S20, there was no significant difference in the results of ammoniacal nitrogen and volatile fatty acids among other groups.

Table 5. Effect of feeding hydroponic wheat sprouts on rumen fermentation parameters of growing ewes.

Item	Groups					SEM	*p* Value
	CON	S5	S10	S15	S20		
pH	6.17	6.14	6.24	6.22	6.11	0.02	0.16
Ammoniacal nitrogen (mol/100 mL)	23.30	24.43	23.10	26.10	19.33	0.83	0.12
Acetate (mmol/L)	57.62	69.52	56.39	60.53	54.95	2.01	0.07
Propionate (mmol/L)	17.47	18.38	14.52	18.33	15.77	0.65	0.24
Butyrate (mmol/L)	12.10	14.15	12.05	12.73	12.22	0.45	0.56
Valerate (mmol/L)	2.28 ^{ab}	2.01 ^{ab}	2.09 ^{ab}	2.48 ^{a}	1.92 ^{b}	0.06	0.03
Total volatile fatty acids (mmol/L)	89.47	104.06	85.06	94.07	84.86	2.77	0.10

Within rows, means with different letter superscripts (^{a,b}) are significantly different ($p < 0.05$).

With respect to microflora changes, the alpha diversity analysis, beta diversity analysis, and rumen microorganism composition changes were presented in Figures 1–4. S15 and S20 of the basal diet with hydroponic wheat sprouts resulted in a decrease in the richness indexes of ACE (Figure 1a) and Chao 1 index (Figure 1b), and also diversity index of Shannon index (Figure 1c). No significant difference was observed in the Simpson index (Figure 1d).

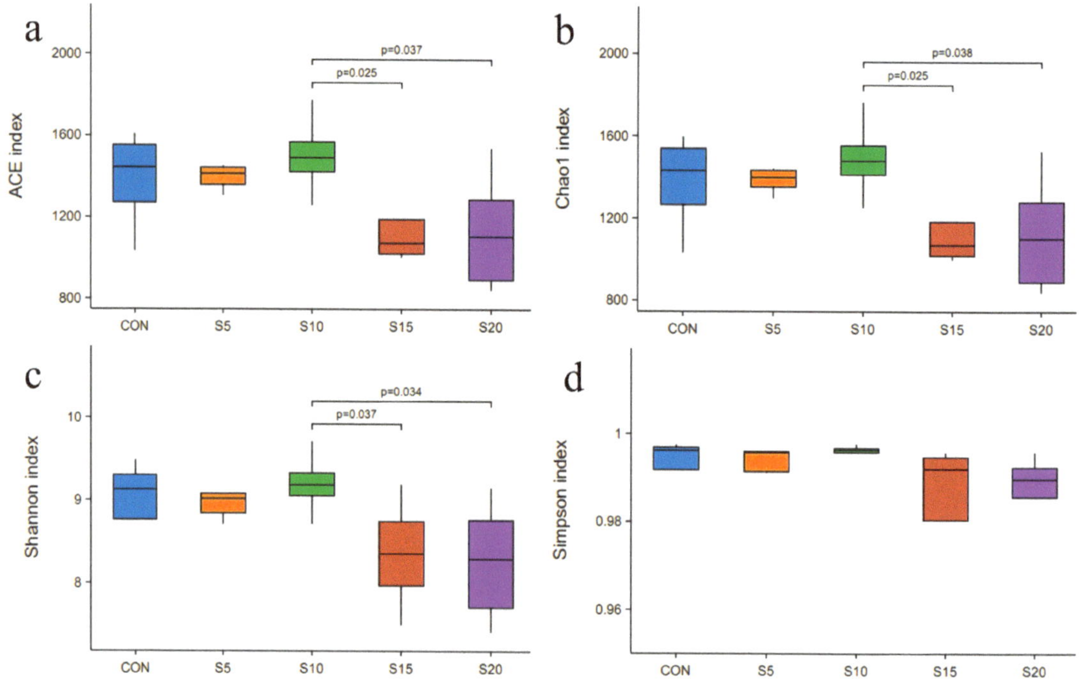

Figure 1. Effect of feeding hydroponic wheat sprouts on the alpha diversity index of rumen bacteria in growing ewes. (**a**): ACE index, (**b**): Chao 1 index, (**c**): Shannon index, (**d**): Simpson index.

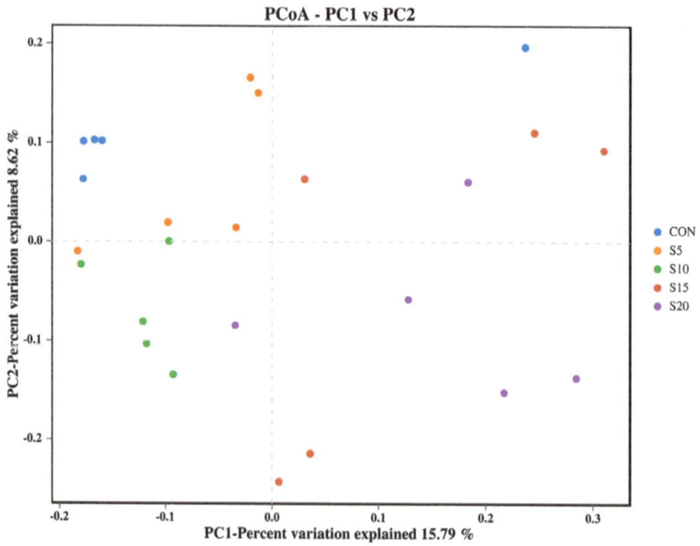

Figure 2. Principal coordinate analysis of the effect of feeding hydroponic wheat sprouts on the rumen bacterial community based on the unweighted unifrac distance.

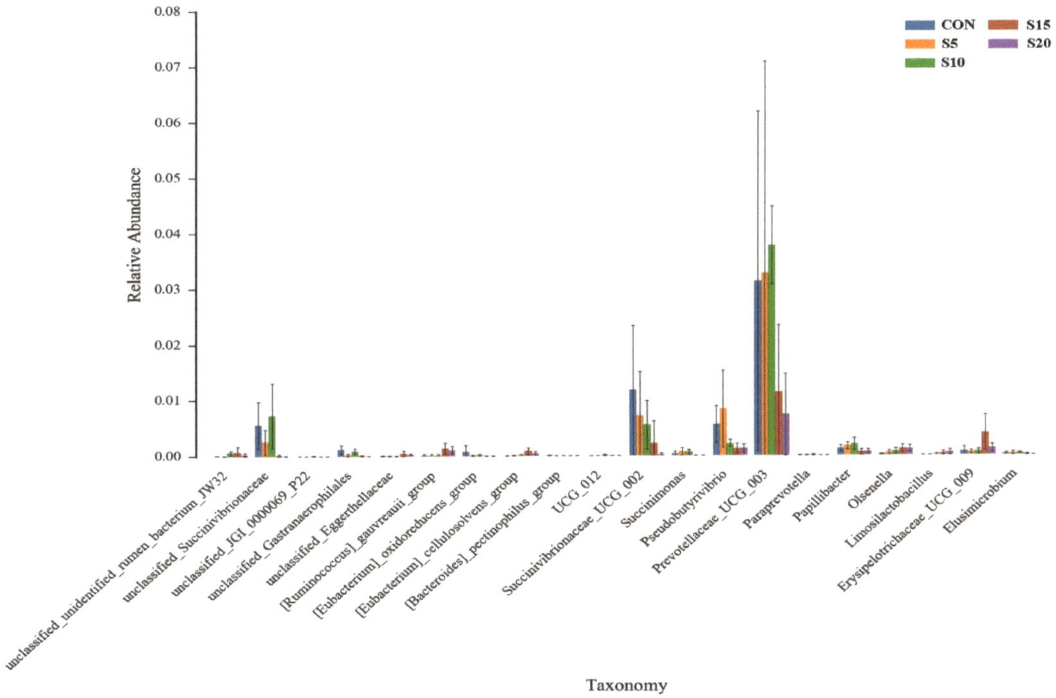

Figure 3. Effect of feeding hydroponic wheat sprouts on the microflora changes of growing ewes in the rumen bacterial community using the Wilcoxon rank-sum test.

Principal coordinate analysis (PCoA) based on the unweighted unifrac distance was performed to compare the bacterial community (Figure 2). There is a relatively clear separation among the diets of five groups.

Based on the Wilcoxon rank-sum test, we found that with the increasing of substitution with hydroponic wheat sprouts, there were obviously increases in the relative abundance of [*Ruminococcus*]_gauvreauii_group, *Olsenella*, *Limosilactobacillus*, and Erysipelotrichaceae_UCG_009 (Figure 3). Meanwhile, there were obviously decreases in the relative abundance of unclassified_Succinivibrionaceae, Succinivibrionaceae_UCG_002, *Succinimonas*, *Pseudobutyrivibrio*, Prevotellaceae_UCG_003, and *Papillibacter* (Figure 3).

Furthermore, LEfSe analysis revealed that compared with CON, S15 of the basal diets with hydroponic wheat sprouts have given rise to a significant increase in the relative abundance of *Saccharofermentans*, Erysipelotrichaceae_UCG_009, Succinivibrionaceae_UCG_001, *Shuttleworthia*, unclassified_Muribaculaceae, *Ruminococcus*__gauvreauii_group, *Olsenella*, *Eubacterium*__cellulosolvens_group, Lachnospiraceae_UCG_002, and *Limosilactobacillus*, Prevotella_7 (Figure 4). Simultaneously, there was a reduction in the relative abundance of *Eubacterium*__oxidoreducens_group, Lachnospiraceae_UCG_004, *Fretibacterium*, unclassified_Gastranaerophilales, Lachnospiraceae_UCG_009, *Monoglobus*, *Pseudobutyrivibrio*, *Anaerovibrio*, Succinivibrionaceae_UCG_002, and Prevotellaceae_UCG_001 (Figure 4).

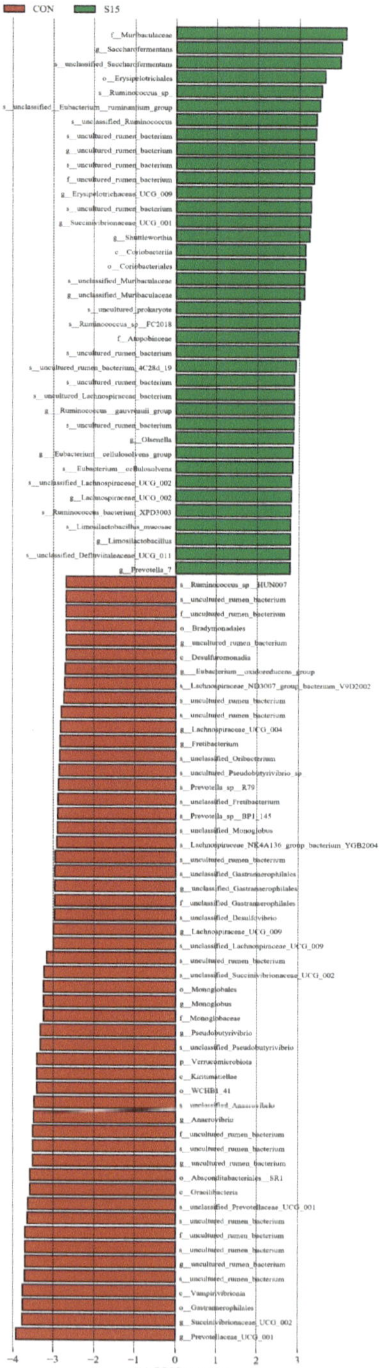

Figure 4. Linear discriminant analysis effect size (LEfSe) analysis of the microflora changes of growing ewes in the rumen bacterial community by substitution 15% (S15) of the basal diet with hydroponic wheat sprouts.

4. Discussion

Hydroponic wheat sprouts were found to contain a high protein content and high-quality fiber, suggesting that it could be used as a high-quality roughage resource for growing ewes. In this experiment, S15 of the basal diet with hydroponic wheat sprouts would obtain the optimum production performance and net profit. Nevertheless, the higher substitution ratio (S20) will result in the decline of production performance, net profit, and even negative effect on rumen fermentation. Guerrero-Cervantes, et al. [8] indicated that hydroponic wheat sprouts are a suitable substitute for the basal diet in ewes' diets during gestation and lactation without having a detrimental effect. However, owing to the excessive substitution ratio (30%) in this process, the replacement strategy did not improve the feed intake of the ewes during gestation and daily gain of the lambs [8]. Hence, based on the response of using different proportions of hydroponic wheat sprouts in the present study, we recommended that S15 of the basal diet with hydroponic wheat sprouts might be the appropriate application solution for growing ewes. The diet with 15% hydroponic wheat sprouts had lower fiber values, which would influence the growth performance, metabolism, and ruminal parameters of the ewes.

However, there was no significant impact on the metabolism of the growing ewes after feeding them hydroponic wheat sprouts, as reflected by most blood biochemical indexes. Intriguingly, the content of glutamic oxalacetic transaminase and urea nitrogen had a tendency to increase in all hydroponic wheat sprouts substitution groups, which might suggest the promotion of liver metabolism [7] and nitrogen metabolism [14] in growing ewes. Theoretically, the increase in serum glutamic oxalacetic transaminase might not be beneficial to health. However, glutamic oxaloacetic transaminase is involved in the catalysis in the process of ammonia transfer on wheat, soybean, corn, and other major crops [15]. It plays an important role in catalyzing the conversion of glutamate to aspartic in plants. The activity of glutamic oxaloacetic transaminase and its related metabolites in S5-S20 might be higher than CON, which might also result in the increase of serum glutamic oxalacetic transaminase in growing ewes in this study. Similar to this phenomenon, feeding hydroponic barley seedlings to lactating ewes will significantly increase the blood glutamic oxalacetic transaminase value in lambs [7].

It is well known that the role of high-density lipoprotein consists of mobilizing circulating cholesterol to the liver to be finally metabolized [16]. In this sense, the elevation of high-density lipoprotein as a result of feeding hydroponic wheat sprouts here would avoid the accumulation of cholesterol in peripheral tissues impacting positively on ewes' health and welfare. Chang et al. [17] elucidated that wheat sprouts could significantly attenuate the accumulation of lipid droplets in liver cells and non-alcoholic fatty liver disease-induced mice through the AMPK pathway activity, accompanied by the decrease of total cholesterol in the blood. Another study indicated that the extraction of wheat sprouts could suppress acetaminophen-induced hepatotoxicity in mice, manifested by inhibiting oxidative liver damage and reducing the expression of inflammatory cytokines [18]. The observed effects on metabolic profiles in growing ewes might reveal the hepatoprotective effect of feeding hydroponic wheat sprouts.

Interleukin-2 is a key cytokine with broad immunomodulatory activity. It could prevent the uncontrolled expansion of immune responses and limit overall inflammation [19]. Additionally, interleukin-2 induces interferon-γ production, which is a cytokine that also plays a vital role in inducing and modulating immune systems [20]. As shown in Table 4, feeding hydroponic wheat sprouts can numerically increase the content of interferon-γ accompanied by the significantly increase of interleukin-2. In the present study, the increase in interleukin-2 and interferon-γ suggests that feeding hydroponic wheat sprouts might improve immune systems in growing ewes.

The microflora changes of growing ewes in this study also supports the idea that feeding them hydroponic wheat sprouts might improve their immune systems. It was found that an increase in the abundance of *Olsenella* was considered to improve immunity in calves [21]. Moreover, *Olsenella* was also considered to strengthen the immunoregulation capacity of a goat model [22] and mouse model [23] host. Similarly, *Limosilactobacillus* has been proved to have antibacterial and intestinal immunity-enhancing effects [24]. An increase in the abundance of *Limosilactobacillus* would facilitate the expression of colon immune factors in rats, which indicated that the gut flora changes of *Limosilactobacillus* can also improve the body's immune system [25]. Taken together, the increase in the abundance of *Olsenella* and *Limosilactobacillus* (Figures 3 and 4) after feeding hydroponic wheat sprouts might show an improvement in the immune systems of growing ewes as demonstrated by the increase of interferon-γ and interleukin-2 (Table 4).

Additionally, the change of diet structure by replacement with hydroponic wheat sprouts is likely to lead to the change of rumen microflora in growing ewes. As shown in Table 1, with an increasing substitution of hydroponic wheat sprouts (S15), there is a reduction of usage amount of corn straw and a decrease in the nutrient composition of neutral detergent fiber and acid detergent fiber. *Pseudobutyrivibrio* is an important member of the rumen microbiome contributing to the degradation and utilization of lignocellulosic plant material [26]. Li et al. [27] demonstrated that the abundance of *Pseudobutyrivibrio* has increased linearly with increasing neutral detergent fiber and acid detergent fiber in the diet of sheep. In addition to *Pseudobutyrivibrio*, Ma et al. [28] indicated that *Succinimonas* in the rumen also play a critical role in forage degradation. Conversely, owing to the reduction in the usage of corn straw (containing more xylan, hemicellulose, and fiber) in this study, there were decreases in the abundance of *Pseudobutyrivibrio* and *Succinimonas*, suggesting the rumen microflora adapted to the change in dietary ingredients. Analogously, it is worth noting that the increase in the abundance of *Shuttleworthia* and *Prevotella_7* occurred in the hydroponic wheat sprouts-fed rumen microbiome, which is similar to what happened in a previous study. *Shuttleworthia* and *Prevotella_7* genera were significantly enhanced in the wheat-fed rumen microbiome [29].

The improvement of the productive performance in growing ewes might also be related to rumen microflora changes. There was a greater relative abundance of *Shuttleworthia* in the rumen of low-residual feed intake (high-efficiency) heifers [30] and dairy cows [31]. Moreover, Kim [32] has summarized that *Anaerovibrio* in the rumen was less abundant in the high-efficiency group of ruminants. It was also found that the presence of Succinivibrionaceae_UCG_002 was inversely proportional to feed efficiency [33]. Therefore, the increase in the abundance of *Shuttleworthia* and decrease in the abundance of *Anaerovibrio* and Succinivibrionaceae_UCG_002 in the rumen could support the results of the improvement of final weight, ADG, feed intake, and feed-to-gain ratio in the present study. However, the underlying mechanism during this process still needs further research.

5. Conclusions

In this experiment, S15 of the basal diet with hydroponic wheat sprouts would obtain an optimum production performance and net profit for growing ewes. However, the higher substitution ratio (20%) will result in the decline of production performance and economic effectiveness, and have a negative effect on rumen fermentation. Feeding hydroponic wheat sprouts might improve immune system and feed efficiency in growing ewes, indicated by the blood biochemical indexes and rumen microflora changes. Due to the function of these rumen microflora, the increase in the abundance of *Olsenella* and *Limosilactobacillus* after feeding hydroponic wheat sprouts might evidence the improvement of immune systems in growing ewes. Furthermore, the increase in the abundance of *Shuttleworthia* and decrease in the abundance of *Anaerovibrio* and Succinivibrionaceae_UCG_002 in the rumen could support the results of the improvement of feed efficiency. With the reducing usage of corn straw, there were decreases in the abundance of the rumen microbiome (*Pseudobutyrivibrio* and *Succinimonas*) contributing to the degradation and utilization of lignocellulosic plant material. Overall, we recommended that S15 of the basal diet with hydroponic wheat sprouts could be the appropriate application solution for growing ewes. The application of hydroponic wheat sprouts (S15) is likely to improve productive performance, reduce feeding costs, and alleviate the food competition between humans and animals. It could be demonstrated and promoted in Xinjiang and throughout the country for sheep production.

Author Contributions: Conceptualization, T.G.; methodology, L.M.; software, L.M.; validation, Y.T.; formal analysis, L.M.; investigation, Y.T.; resources, Y.T.; data curation, Y.T.; writing—original draft preparation, L.M.; writing—review and editing, D.L., C.Z., G.A. and Z.Z.; visualization, T.G.; supervision, T.G.; project administration, T.G.; funding acquisition, T.G. All authors have read and agreed to the published version of the manuscript.

Funding: This study was supported by the Key Laboratory of Xingjiang Feed Biotechnology, and Guangdong Modern Agro-industry Technology Research System.

Institutional Review Board Statement: All animal procedures performed in this study were approved and guided by the Animal Care and Use Committee of the Xinjiang Academy of Animal Sciences (FRI-2023002, Urumqi, China).

Informed Consent Statement: Not applicable.

Data Availability Statement: The raw sequencing reads were deposited into the NCBI Sequence Read Archive (SRA) data: PRJNA1076690.

Conflicts of Interest: The authors declare no conflicts of interest.

References

1. Gumisiriza, M.S.; Ndakidemi, P.A.; Mbega, E.R. A simplified non-greenhouse hydroponic system for small-scale soilless urban vegetable farming. *MethodsX* **2022**, *9*, 101882. [CrossRef]
2. Farghaly, M.M.; Abdullah, M.A.M.; Youssef, I.M.I.; Abdel-Rahim, I.R.; Abouelezz, K. Effect of feeding hydroponic barley sprouts to sheep on feed intake, nutrient digestibility, nitrogen retention, rumen fermentation and ruminal enzymes activity. *Livest. Sci.* **2019**, *228*, 31–37. [CrossRef]
3. Hafla, A.N.; Soder, K.J.; Brito, A.F.; Rubano, M.D.; Dell, C.J. Effect of sprouted barley grain supplementation of an herbage-based or haylage-based diet on ruminal fermentation and methane output in continuous culture. *J. Dairy Sci.* **2014**, *97*, 7856–7869. [CrossRef]
4. Harerimana, T.; Hatungimana, E.; Hirwa, C.d.A.; Niyireba, R.T. Effect of hydroponic wheat fodder supplementation on dry matter intake and growth performances of piglets. *Sci. Afr.* **2023**, *19*, e01507. [CrossRef]
5. Ma, Y.; Wang, J.; Sun, Y.; Dong, Y.; Cai, H.; Raja, I.H.; Guo, T.; Zhang, S. Effects of Compound Salt Concentration on Growth, Physiological and Nutritional Value of Hydroponic Forage Wheat. *Agriculture* **2023**, *13*, 1833. [CrossRef]
6. Al-Baadani, H.H.; Alowaimer, A.N.; Al-Badwi, M.A.; Abdelrahman, M.M.; Soufan, W.H.; Alhidary, I.A. Evaluation of the Nutritive Value and Digestibility of Sprouted Barley as Feed for Growing Lambs: In Vivo and In Vitro Studies. *Animals* **2022**, *12*, 1206. [CrossRef] [PubMed]
7. Ma, Y.; Guo, T.; Zhang, Z.; Amat, G.; Jing, Y.; Tuo, Y.; Hou, L. Effect of feeding hydroponic barley seedlings to lactating ewes on blood biochemical indexes and growth performance of lambs. *Front. Vet. Sci.* **2024**, *10*, 1280544. [CrossRef]
8. Guerrero-Cervantes, M.; Cerrillo-Soto, M.A.; Plascencia, A.; Salem, A.Z.M.; Estrada-Angulo, A.; Rios-Rincón, F.G.; Luginbuhl, J.M.; Bernal-Barragán, H.; Abdalla, A.L. Productive and reproductive performance and metabolic profiles of ewes supplemented with hydroponically grown green wheat (*Triticum aestivum* L.). *Anim. Feed. Sci. Technol.* **2016**, *221*, 206–214. [CrossRef]
9. *NY/T 816-2021*; Nutrient Requirements of Meat-Type Sheep and Goat. Ministry of Agriculture and Rural Affairs of the People's Republic of China: Beijing, China, 2021.
10. Mohammed, N.; Ajisaka, N.; Lila, Z.A.; Hara, K.; Mikuni, K.; Hara, K.; Kanda, S.; Itabashi, H. Effect of Japanese horseradish oil on methane production and ruminal fermentation in vitro and in steers1. *J. Anim. Sci.* **2004**, *82*, 1839–1846. [CrossRef]
11. Broderick, G.A.; Kang, J.H. Automated Simultaneous Determination of Ammonia and Total Amino Acids in Ruminal Fluid and In Vitro Media1. *J. Dairy Sci.* **1980**, *63*, 64–75. [CrossRef]
12. Minas, K.; McEwan, N.R.; Newbold, C.J.; Scott, K.P. Optimization of a high-throughput CTAB-based protocol for the extraction of qPCR-grade DNA from rumen fluid, plant and bacterial pure cultures. *FEMS Microbiol. Lett.* **2011**, *325*, 162–169. [CrossRef] [PubMed]
13. Shen, J.; Li, Z.; Yu, Z.; Zhu, W. Effects of dietary replacement of soybean meal with dried distillers grains with solubles on the microbiota occupying different ecological niches in the rumen of growing Hu lambs. *J. Anim. Sci. Biotechnol.* **2020**, *11*, 93. [CrossRef] [PubMed]
14. Saro, C.; Mateo, J.; Caro, I.; Carballo, D.E.; Fernández, M.; Valdés, C.; Bodas, R.; Giráldez, F.J. Effect of Dietary Crude Protein on Animal Performance, Blood Biochemistry Profile, Ruminal Fermentation Parameters and Carcass and Meat Quality of Heavy Fattening Assaf Lambs. *Animals* **2020**, *10*, 2177. [CrossRef] [PubMed]
15. Zhu, A.; Liu, H.; Wang, Y.; Sun, H.; Han, G. Grazing intensity changed the activities of nitrogen assimilation related enzymes in desert Steppe Plants. *BMC Plant Biol.* **2021**, *21*, 436. [CrossRef] [PubMed]

16. Contreras-Solís, I.; Porcu, C.; Sotgiu, F.D.; Chessa, F.; Pasciu, V.; Dattena, M.; Caredda, M.; Abecia, J.A.; Molle, G.; Berlinguer, F. Effect of Strategic Supplementation of Dietary By-Pass Linseed Oil on Fertility and Milk Quality in Sarda Ewes. *Animals* **2023**, *13*, 280. [CrossRef]
17. Chang, B.; Bae, J.; Yun, S.; Kim, Y.; Park, S.; Kim, S. Wheat sprouts (*Triticum aestivum* Linn.) cultured by a smart farm system ameliorate NAFLD through the AMPK-mediated SREBP signaling pathway. *Food Sci. Biotechnol.* **2023**, *32*, 1585–1594. [CrossRef]
18. Lim, J.-Y.; Yun, D.-H.; Lee, J.-H.; Kwon, Y.-B.; Lee, Y.-M.; Lee, D.-H.; Kim, D.-K. Extract of Triticum aestivum Sprouts Suppresses Acetaminophen-Induced Hepatotoxicity in Mice by Inhibiting Oxidative Stress. *Molecules* **2021**, *26*, 6336. [CrossRef]
19. Mao, C.; Xu, Y.; Shi, L.; Guo, S.; Jin, X.; Yan, S.; Shi, B. Effects of Photoperiod Change on Melatonin Secretion, Immune Function and Antioxidant Status of Cashmere Goats. *Animals* **2019**, *9*, 766. [CrossRef] [PubMed]
20. Ashrafi, H.; Sadeghi, A.A.; Chamani, M. Effect of Organic Selenium Supplementation on the Antioxidant Status, Immune Response, and the Relative Expression of IL-2 and IFN-γ Genes in Ewes During the Hot Season. *Biol. Trace Elem. Res.* **2023**, *202*, 2052–2061. [CrossRef]
21. Wang, X.; Guo, C.; Xu, X.; Zhang, L.; Li, S.; Dai, D.; Du, W. Effect of Alkaline Mineral Complex Buffer Supplementation on Rumen Fermentation, Rumen Microbiota and Rumen Epithelial Transcriptome of Newborn Calves. *Fermentation* **2023**, *9*, 973. [CrossRef]
22. Zhong, Z.; Zhang, Y.; Li, X.; Li, L.; Zhang, R.; Zhang, S. Differential Responses of Digesta- and Mucosa-Associated Jejunal Microbiota of Hu Sheep to Pelleted and Non-Pelleted High-Grain Diets. *Animals* **2022**, *12*, 1695. [CrossRef] [PubMed]
23. Chen, Q.; Yin, Q.; Xie, Q.; Jiang, C.; Zhou, L.; Liu, J.; Li, B.; Jiang, S. 2′-Fucosyllactose Promotes the Production of Short-Chain Fatty Acids and Improves Immune Function in Human-Microbiota-Associated Mice by Regulating Gut Microbiota. *J. Agric. Food Chem.* **2022**, *70*, 13615–13625. [CrossRef]
24. Liu, B.; Yang, L.; Wu, Y.; Zhao, X. Protective effect of Limosilactobacillus fermentum HFY06 on dextran sulfate sodium-induced colitis in mice. *Front. Microbiol.* **2022**, *13*, 935792. [CrossRef]
25. Qu, Q.; Zhao, C.; Yang, C.; Zhou, Q.; Liu, X.; Yang, P.; Yang, F.; Shi, X. Limosilactobacillus fermentum-fermented ginseng improved antibiotic-induced diarrhoea and the gut microbiota profiles of rats. *J. Appl. Microbiol.* **2022**, *133*, 3476–3489. [CrossRef] [PubMed]
26. Palevich, N.; Kelly, W.J.; Leahy, S.C.; Denman, S.; Altermann, E.; Rakonjac, J.; Attwood, G.T. Comparative Genomics of Rumen *Butyrivibrio* spp. Uncovers a Continuum of Polysaccharide-Degrading Capabilities. *Appl. Environ. Microbiol.* **2019**, *86*, e01993-01919. [CrossRef] [PubMed]
27. Li, R.; Teng, Z.; Lang, C.; Zhou, H.; Zhong, W.; Ban, Z.; Yan, X.; Yang, H.; Farouk, M.H.; Lou, Y. Effect of different forage-to-concentrate ratios on ruminal bacterial structure and real-time methane production in sheep. *PLoS ONE* **2019**, *14*, e0214777. [CrossRef] [PubMed]
28. Ma, Y.; Chen, X.; Khan, M.Z.; Xiao, J.; Cao, Z. A Combination of Novel Microecological Agents and Molasses Role in Digestibility and Fermentation of Rice Straw by Facilitating the Ruminal Microbial Colonization. *Front. Microbiol.* **2022**, *13*, 948049. [CrossRef] [PubMed]
29. Savin, K.W.; Moate, P.J.; Williams, S.R.O.; Bath, C.; Hemsworth, J.; Wang, J.; Ram, D.; Zawadzki, J.; Rochfort, S.; Cocks, B.G. Dietary wheat and reduced methane yield are linked to rumen microbiome changes in dairy cows. *PLoS ONE* **2022**, *17*, e0268157. [CrossRef] [PubMed]
30. Clemmons, B.A.; Mulon, P.-Y.; Anderson, D.E.; Ault-Seay, T.B.; Henniger, M.T.; Schneider, L.G.; Staton, M.; Voy, B.H.; Donohoe, D.R.; Campagna, S.R.; et al. Ruminal Bacterial Communities and Metabolome Variation in Beef Heifers Divergent in Feed Efficiency. *Ruminants* **2022**, *2*, 282–296. [CrossRef]
31. Jewell, K.A.; McCormick, C.A.; Odt, C.L.; Weimer, P.J.; Suen, G. Ruminal Bacterial Community Composition in Dairy Cows Is Dynamic over the Course of Two Lactations and Correlates with Feed Efficiency. *Appl. Environ. Microbiol.* **2015**, *81*, 4697–4710. [CrossRef]
32. Kim, M. Invited Review—Assessment of the gastrointestinal microbiota using 16S ribosomal RNA gene amplicon sequencing in ruminant nutrition. *Anim. Biosci.* **2023**, *36*, 364–373. [CrossRef] [PubMed]
33. Zhang, Y.K.; Zhang, X.X.; Li, F.D.; Li, C.; Li, G.Z.; Zhang, D.Y.; Song, Q.Z.; Li, X.L.; Zhao, Y.; Wang, W.M. Characterization of the rumen microbiota and its relationship with residual feed intake in sheep. *Animal* **2021**, *15*, 100161. [CrossRef] [PubMed]

Disclaimer/Publisher's Note: The statements, opinions and data contained in all publications are solely those of the individual author(s) and contributor(s) and not of MDPI and/or the editor(s). MDPI and/or the editor(s) disclaim responsibility for any injury to people or property resulting from any ideas, methods, instructions or products referred to in the content.

animals

MDPI

Article

Effect of Dietary Benzoic Acid Supplementation on Growth Performance, Rumen Fermentation, and Rumen Microbiota in Weaned Holstein Dairy Calves

Haonan Dai [1], Qi Huang [1], Shujing Li [2], Dewei Du [1], Wenli Yu [2], Jia Guo [1], Zengyuan Zhao [2], Xin Yu [1], Fengtao Ma [1] and Peng Sun [1,*]

1 State Key Laboratory of Animal Nutrition and Feeding, Institute of Animal Science, Chinese Academy of Agricultural Sciences, Beijing 100193, China; daihaonan0325@163.com (H.D.); workhq@163.com (Q.H.); dudw120459@163.com (D.D.); guojia_work@163.com (J.G.); 82101212373@caas.cn (X.Y.); fengtaoma@163.com (F.M.)

2 Shijiazhuang Tianquan Elite Dairy Ltd., Shijiazhuang 050200, China; embryochina@163.com (S.L.); anboyuwenli@126.com (W.Y.); zhaozengyuanjsdx@163.com (Z.Z.)

* Correspondence: sunpeng02@caas.cn

Simple Summary: Weaning stress affects the health of calves, inhibiting growth and disrupting gastrointestinal function. Benzoic acid (BA) is known for promoting growth and intestinal health. This study aims to investigate the effects of BA on the growth performance, rumen fermentation, and rumen microbiota of weaned Holstein dairy calves. Thirty-two Holstein dairy calves (60 days of age) were randomly divided into four groups (n = 8) supplemented with 0%, 0.25%, 0.50%, and 0.75% BA to the basal diet (dry matter basis) for 42 days, respectively. The results showed that supplementation with 0.50% BA promoted growth performance by increasing feed intake and average daily gain and reducing feed-to-gain ratio, improved rumen fermentation parameters, and increased the relative abundance of probiotics in the rumen. This suggests that 0.50% BA supplementation might be an effective method for the efficient breeding of weaned calves.

Citation: Dai, H.; Huang, Q.; Li, S.; Du, D.; Yu, W.; Guo, J.; Zhao, Z.; Yu, X.; Ma, F.; Sun, P. Effect of Dietary Benzoic Acid Supplementation on Growth Performance, Rumen Fermentation, and Rumen Microbiota in Weaned Holstein Dairy Calves. *Animals* **2024**, *14*, 2823. https://doi.org/10.3390/ani14192823

Academic Editor: Adriana Bonanno

Received: 29 July 2024
Revised: 15 September 2024
Accepted: 27 September 2024
Published: 30 September 2024

Copyright: © 2024 by the authors. Licensee MDPI, Basel, Switzerland. This article is an open access article distributed under the terms and conditions of the Creative Commons Attribution (CC BY) license (https://creativecommons.org/licenses/by/4.0/).

Abstract: Supplementation with benzoic acid (BA) in animal feed can reduce feeds' acid-binding capacity, inhibit pathogenic bacterial growth, enhance nutrient digestion, and increase intestinal enzyme activities. This study aimed to investigate the effects of different doses of BA on the growth performance, rumen fermentation, and rumen microbiota of weaned Holstein dairy calves. Thirty-two Holstein calves at 60 days of age were randomly assigned into four groups (n = 8): a control group (fed with a basal diet without BA supplementation; CON group) and groups that were supplemented with 0.25% (LBA group), 0.50% (MBA group), and 0.75% (HBA group) BA to the basal diet (dry matter basis), respectively. The experiment lasted for 42 days, starting at 60 days of age and ending at 102 days of age, with weaning occurring at 67 days of age. Supplementation with BA linearly increased the average daily gain of the weaned dairy calves, which was significantly higher in the LBA, MBA, and HBA groups than that in the CON group. The average daily feed intake was quadratically increased with increasing BA supplementation, peaking in the MBA group. Supplementation with BA linearly decreased the feed-to-gain (F/G) ratio, but did not affect rumen fermentation parameters, except for the molar proportion of butyrate and iso-butyrate, which were linearly increased with the dose of BA supplementation. Compared with the CON group, the molar proportions of iso-butyrate in the LBA, MBA, and HBA groups and that of butyrate in the HBA group were significantly higher than those in the CON group. Supplementation with BA had no significant effect on the alpha and beta diversity of the rumen microbiota, but significantly increased the relative abundances of beneficial bacteria, such as *Bifidobacterium*, and reduced those of the harmful bacteria, such as *unclassified_o__Gastranaerophilales* and *Oscillospiraceae_UCG-002*, in the rumen. Functional prediction analysis using the MetaCyc database revealed significant variations in the pathways associated with glycolysis across groups, including the GLYCOLYSIS-TCA-GLYOX-BYPASS, GLYCOL-GLYOXDEG-PWY, and P105-PWY pathways. In conclusion, BA supplementation improved the composition and function of rumen microbiota, elevated the production of butyrate and iso-butyrate, and increased the growth performance of weaned Holstein dairy calves.

Keywords: benzoic acid; growth performance; butyrate; iso-butyrate; *Bifidobacterium*; dairy calf

1. Introduction

Early weaning without negative effects has been considered to decrease the cost of raising dairy heifer calves [1]. However, the reason that early weaning has not been widely used in most dairy farms is due to the subsequently exacerbated weaning stresses, which affect the health of the calves, inhibiting growth and disrupting gastrointestinal function [1,2]. Therefore, within the context of modern intensive dairy farming, mitigating weaning stress in calves emerges as an urgent imperative [3]. In addition to the determination of the optimum age and appropriate weaning method [1,2], many studies have proved that supplementation with feed additives is an effective strategy that helps alleviate weaning stress [3,4].

Organic acids, characterized by their carboxylic acid group (R-COOH), serve as feed additives and are known for enhancing the growth performance and intestinal health of livestock and poultry. They are extensively incorporated into diets for these animals and are being explored as potential alternatives to antibiotic growth promoters [5,6]. In addition, benzoic acid (BA) is an organic acid compound and the simplest aromatic acid structurally [7,8]. As a monocarboxylic acid, BA exhibits broad-spectrum antibacterial properties and notable effectiveness against various bacteria. Its applications span across diverse industries, including industrial production, cosmetics, pharmaceuticals, food, and feed industries, and it is made into organic acidifiers, fragrances, plasticizers, and preservatives [9–14].

Previous research has demonstrated the beneficial effects of BA supplementation in various livestock species. In nursery and grower-finisher pigs, dietary supplementation with 0.3% and 0.5% BA significantly enhances their growth performance [15]. Nursery pigs exhibit a linear response to BA up to 0.5%, while the optimal average daily gain (ADG) in grower-finisher pigs is achieved with a calculated supplementation level of 0.36% [15]. Furthermore, combining BA and *Bacillus coagulans* improves the growth performance and nutrient digestibility in piglets while maintaining a balance of intestinal microflora [16]. In turkeys, adding 0.1% BA to their diet enhances caecal function by increasing lactic acid bacteria populations, reducing coliform bacteria populations, and lowering the buffering capacity and pH values in caecal contents [17]. For laying hens, supplementation with 1000 mg/kg BA significantly improves egg quality without affecting production performance [18]. The BA also promotes intestinal health by improving intestinal morphology and enriching microbial composition [18]. In the context of beef production, adding 0.5% BA in the diet for finishing steers does not affect the shelf life of beef post slaughter, nor does it impact beef color or lipid oxidation [19]. Similarly, adding 0.5% BA to the diet of fattening cattle does not alter carcass characteristics and performance, but significantly increases eating rate and dry matter intake (DMI) [20].

Considering its efficacy of promoting growth and intestinal health in monogastric animals, we hypothesized that supplementation with BA might potentially promote the growth and improve the rumen fermentation as well as the composition and function of the rumen microbiota of dairy calves during and after weaning. Therefore, this study aimed to investigate the effects of varying doses of BA supplementation on the growth performance, rumen fermentation, and rumen microbiota of weaned Holstein dairy calves. The findings of this study may provide insights into optimizing the rearing practices of weaned calves efficiently.

2. Materials and Methods

The animal protocol for this research was approved by the Experimental Animal Ethics Committee of the Institute of Animal Science, Chinese Academy of Agricultural Sciences (IAS, CAAS, Beijing, China) (approval number IAS2023-138). All the procedures involved were carried out in accordance with the "Guidelines for the Management and Use of Laboratory Animals" of the IAS, CAAS.

2.1. *Animals, Diets, and Experimental Design*

A total of 32 Holstein heifer calves (60 days of age) with similar body weight (91.8 ± 0.6 kg) were selected for this study and randomly assigned to 4 groups (n = 8) using a random number generator. The control group (CON group) received a basal diet, while the low BA group (LBA group), medium BA group (MBA group), and high BA group (HBA group) were supplemented with 0.25%, 0.50%, and 0.75% BA in the basal diet (as dry matter basis), respectively.

This study was conducted at Xingtai Junchang Pastoral Industry Co., Ltd. (Xingtai, Hebei Province, China). Benzoic acid (100% purity) was purchased from Guangzhou Huayu Biotechnology Co., Ltd. (Guangzhou, China). Newborn calves were fed 4 L colostrum within one hour of birth. Thereafter, 2 L of colostrum was fed three times a day for two consecutive days. At 3 days of age, the calves were transferred from the nursery pens to individual calf hutches (3.7 × 1.4 × 1.5 m, enclosed by iron railings and lined with hay). The heifers were fed with 8 L heated raw milk and starter from 4 days of age. Oat grass had been supplemented to the heifers at 21 days of age. The dairy calves started weaning at 60 days of age, involving a 7-day transitional period during which milk feeding gradually decreased by 1 L per day until its cessation at 67 days of age. Solid feed which meets the growth and nutritional needs of weaned calves from 68 days of age was then provided continuously until the conclusion of the study at 102 days of age. The experimental period lasted for a total of 42 days. During the transition period, the dairy calves were fed milk three times a day (05:00, 12:00, and 19:00), with the addition of starter once (08:30). After weaning, growth feed was added three times a day (08:30, 14:30, and 21:30), and calves had ad libitum access to feed and water. The composition and nutrient levels of the basal diet are shown in Table 1, and meet or exceed the nutritional requirements recommended by the National Research Council (NRC, 2021) [21].

2.2. *Sample Collection and Analysis*

2.2.1. Body Weight and Feed Intake

The initial body weight (IBW) and final body weight (FBW) of the dairy calves were recorded at 60 and 102 days of age, respectively. The ADG was obtained by subtracting the IBW from the FBW and dividing the difference by the experimental period. The amount of leftover feed was recorded every day, which was used to calculate the average daily feed intake (ADFI) and feed-to-gain (F/G) ratio.

2.2.2. Determination of Nutrient Levels in Diet

During the experimental period, feed was collected for analysis every two weeks using the quartering method. The measurements of dry matter (DM, AOAC, 2005; method 930.15) [22], crude protein (CP, AOAC, 2000, method 976.05) [23], and ether extract (EE, AOAC, 2003, method 4.5.05) [24] were carried out according to the standard protocol of the Association of Official Analytical Chemists. The measurements of neutral detergent fiber (NDF) content and acid detergent fiber (ADF) contents were carried out according to the method of Van Soest et al. [25]. The contents of Ca and P were measured by atomic absorption spectroscopy (AOAC International,1990; method 985.35) [26] and spectrophotometry (AOAC International, 1990; method 986.24) [26], respectively.

Table 1. Diet composition and nutrient levels.

Items	Proportion (%)	
	Starter	Growth Feed
Diet composition (as-fed basis)		
Corn	32.00	39.73
Soybean Meal	17.00	9.73
Cottonseed Meal	2.73	3.27
DDGS	13.18	13.63
Bran	10.55	16.09
Puffed Soybean Flour	2.73	
Wheat Flour	4.54	
$NaHCO_3$	0.45	0.73
[1] Premix	7.73	7.73
Oat Grass	9.09	9.09
Total	100	100
Nutrient Levels (as dry matter basis)		
Dry Matter	89.03	88.64
Crude Protein	19.13	17.63
Ether Extract	2.88	2.48
Ash	9.47	6.40
Neutral Detergent Fiber	27.92	34.57
Acid Detergent Fiber	13.36	13.27
Ca	1.10	0.97
P	0.73	0.71

[1] Supplied per kilogram of diet: Vitamin A 800 kIU, Vitamin D_3 240 kIU, Vitamin E 7 kIU, Vitamin C 200 mg, Vitamin K_3 15 mg, Zn 11,000 mg, Mn 7000 mg, Cu 1500 mg, I 40 mg, Se 25 mg, Co 15 mg.

2.2.3. Rumen Fluid Sampling

At 101 days of age, rumen fluid samples were collected from each dairy calf using an oral cavity collector precisely 2 h before the morning feeding. The initial 30 mL of rumen fluid was discarded to ensure the sample's integrity. Subsequently, 80 mL of rumen fluid was extracted and filtered through four layers of gauze, then quick frozen in liquid nitrogen and stored in liquid nitrogen for the determination of volatile fatty acids (VFAs) and rumen microbiota.

2.2.4. Determination of VFAs

After thoroughly shaking the rumen fluid sample, 2 mL of the sample was taken and centrifuged at 10,000× *g* and 4 °C for 10 min. The supernatant was collected and added 0.15 mL of 25% metaphosphoric acid for fixation. The concentrations of VFAs were determined using a gas chromatograph (7890B-7000D, Agilent Technologies, Santa Clara, CA, USA) [27].

2.2.5. DNA Extraction, PCR Amplification, and 16S rRNA Gene Sequencing

The total microbial DNA was extracted from rumen fluid samples using the YH Soil DNA Extraction Kit (Meiji Biopharmaceutical Technology Co., Ltd., Shanghai, China), following the manufacturer's instructions. DNA concentration and purity were assessed using a NanoDrop 2000 (Thermo Fisher Scientific, Waltham, MA, USA) and quality was verified via 1% agarose gel electrophoresis.

The V3-V4 region of the bacterial 16S rRNA genes was amplified using the ABI GeneAmp 9700 PCR thermocycler (ABI, Los Angeles, CA, USA) with the forward primer 338F (5′-ACTCCTACGGGAGGCAGCAG-3′) and reverse primer 806R (3′-GGACTACHVGGGT WTCTAAT-5′) [28]. The PCR reaction was performed in triplicate with a reaction mixture consisting 4 μL 5 × FastPfu Buffer, 2 μL 2.5 mM dNTPs, 0.8 μL forward primer (5 μM), 0.8 μL reverse primer (5 μM), 0.4 μL FastPfu Polymerase, 0.2 μL BSA, 10 ng template DNA, and added ddH_2O to 20 μL [29]. The PCR cycling conditions comprised initial denaturation at 95 °C for 3 min, followed by 30 cycles at 95°C for 30 s, 60 °C for 30 s, 72 °C for 45 s, and a final extension at 72 °C for 10 min.

Following PCR, the amplicons were extracted using 2% agarose gel and further purified with the AxyPrep DNA Gel Extraction Kit (Axygen Biosciences, Union City, CA, USA). The DNA concentrations were quantified using the QuantiFluor-ST (Promega, Madison, WI, USA). Purified amplicons were pooled in equimolar concentrations and subjected to paired-end sequencing (PE250) on the NextSeq2000 platform (Illumina, San Diego, CA, USA) according to the standard protocols. The 16S rRNA sequencing data for all the samples were deposited into the NCBI Sequence Read Archive (SRA) under accession number PRJNA1128666 (https://www.ncbi.nlm.nih.gov/sra/PRJNA1128666, accessed on 26 June 2024).

Raw sequence tags were quality-filtered using QIIME (version 1.9.1, Denver, CO, USA) to obtain high-quality clean tags, and FLASH (version 1.2.11, Baltimore, MD, USA) was used for splicing. Sequences with low quality (scores < 20), reads containing ambiguous bases or unmatched to primer sequences, and reads < 200 bp were filtered and removed, and the respective barcodes were removed. Operational taxonomic units (OTUs) were generated at ≥97% similarity using UPARSE (version 7.0.1090, San Diego, CA, USA) [30]. Then the representative sequences were taxonomically classified against the SILVA Database (Release 138, Bremen, HB, Germany) using the RDP Classifier (version 2.11, East Lansing, MI, USA). The alpha diversity analyses (Ace, Chao, Coverage, Shannon, Simpson, Sobs) were performed using the Mothur (version.1.30.2, Michigan, MI, USA). The principal coordinates analysis (PCoA) was performed using a Bray–Curtis distance matrix [31]. The functional predictive analysis was carried out using PICRUSt2 (version 2.2.0, Cambridge, MA, USA).

2.2.6. Statistical Analysis

The growth performance and rumen fermentation parameters among groups were analyzed using one-way ANOVA followed by Tukey's test in SPSS software (IBM SPSS statistics 26, Chicago, IL, USA). Differences in the alpha diversity indices and relative abundances of rumen microbiota were carried out using the Kruskal–Wallis H test method and stats package in R software (version 3.3.1, Vienna, Austria). Spearman's correlation analysis was used to analyze the differential pathways prediction based on the MetaCyc database, and relationships between differential rumen microbiota, differential pathways, and VFAs. Results were presented as the mean ± standard error of the mean (SEM), and statistical significance was defined as $p < 0.05$.

3. Results

3.1. Growth Performance

As shown in Table 2, the ADFI showed a quadratic increase with BA supplementation, peaking in the MBA group ($p = 0.01$). Compared to the CON group, ADFI was significantly higher in the LBA and MBA groups ($p = 0.01$). Similarly, BA supplementation linearly increased the ADG in weaned dairy calves, with significantly higher values observed in the LBA, MBA, and HBA groups compared to the CON group ($p = 0.03$). Correspondingly, the FBW of the dairy calves also increased quadratically with BA supplementation ($p = 0.02$). In addition, the F/G ratio linearly decreased with the supplementation with BA ($p = 0.03$), but there was no significant difference on treatment ($p = 0.17$).

3.2. Rumen Fermentation Parameters

Supplementation with BA did not affect rumen fermentation parameters except for the molar proportions of butyrate and iso-butyrate, which showed a linearly increase with BA dose ($p < 0.01$). Specifically, compared to the CON group, the molar proportion of iso-butyrate was significantly higher in the LBA, MBA, and HBA groups ($p < 0.01$), and that of butyrate was higher in the HBA group ($p = 0.04$). The acetate-to-propionate (A/P) ratio linearly increased with increasing BA supplementation levels ($p = 0.02$), but there was no significant difference on treatment ($p = 0.07$), as shown in Table 3.

Table 2. Effect of different levels of BA on the growth performance in weaned Holstein dairy calves.

Items	Treatment [1]				SEM [2]	p-Value		
	CON	LBA	MBA	HBA		Treatment	Linear	Quadratic
IBW [3] (d 60, kg)	91.0	93.1	91.4	91.9	0.6	0.68	0.83	0.58
FBW [4] (d 102, kg)	137.1	146.6	144.6	140.9	1.4	0.08	0.56	0.02
ADG [5] (kg/d)	1.07 ^{b}	1.22 ^{a}	1.27 ^{a}	1.24 ^{a}	0.03	0.03	0.02	0.06
ADFI [6] (kg/d DM)	3.05 ^{b}	3.30 ^{a}	3.34 ^{a}	3.18 ^{ab}	0.03	0.01	0.27	0.02
F/G ratio [7]	2.85	2.70	2.63	2.56	0.05	0.17	0.03	0.64

[1] CON, the control group, which was fed with the basal diet; LBA (low BA group), MBA (medium BA group), and HBA (high BA group) are the experimental groups, which were supplemented with 0.25%, 0.50%, and 0.75% BA in the basal diet (as dry matter basis), respectively. [2] SEM, standard error of the mean; [3] IBW, initial body weight; [4] FBW, final body weight; [5] ADG, average daily gain; [6] ADFI, average daily feed intake; [7] F/G ratio, feed-to-gain ratio. $^{a\,b}$ Within a row, means with different superscripts differ, $p < 0.05$.

Table 3. Effect of different levels of BA on rumen fermentation parameters in weaned Holstein dairy calves.

Items	Treatment [1]				SEM [2]	p-Value		
	CON	LBA	MBA	HBA		Treatment	Linear	Quadratic
VFAs [3] (mmol/L)								
Acetate	49.20	49.54	52.12	52.90	1.77	0.86	0.41	0.95
Propionate	32.26	28.28	28.82	30.67	1.85	0.94	0.97	0.54
Iso-butyrate	0.46 ^{b}	0.71 ^{a}	0.83 ^{a}	0.89 ^{a}	0.05	<0.01	<0.01	0.26
Butyrate	6.90 ^{b}	7.89 ^{ab}	8.56 ^{ab}	10.52 ^{a}	0.47	0.04	<0.01	0.58
Iso-valerate	0.87	1.09	1.22	1.19	0.12	0.76	0.34	0.62
Valerate	1.83	1.98	2.17	1.80	0.15	0.82	0.95	0.40
Acetate/Propionate	1.43	1.72	1.83	1.81	0.06	0.07	0.02	0.16
TVFAs [4] (mmol/L)	91.33	89.48	93.82	98.26	3.85	0.86	0.50	0.70

[1] CON, the control group, which was fed with the basal diet; LBA (low BA group), MBA (medium BA group), and HBA (high BA group) are the experimental groups, which were supplemented with 0.25%, 0.50%, and 0.75% BA in the basal diet (as dry matter basis), respectively. [2] SEM, standard error of the mean; [3] VFAs, volatile fatty acids; [4] TVFAs, total volatile fatty acids. $^{a\,b}$ Within a row, means with different superscripts differ, $p < 0.05$.

3.3. Composition of Rumen Microbiota

A total of 2,030,939 high-quality sequences were acquired, with the average of 63,483 ± 6186 sequences per sample which belonged to 439 ± 94 OTU. Figure 1A demonstrated that all samples achieved coverage indices above 99.0%, indicating comprehensive species diversity and community structure. The richness index (Chao, Ace, Sobs, Shannon, Simpson, Sobs) did not significantly differ among groups (Figure 1, $p > 0.05$).

As shown in Figure 2, there was no distinct separation between or within groups ($p > 0.05$).

Taxonomic analysis identified 18 phyla in the rumen fluid, predominantly *Bacteroidetes*, *Firmicutes*, and *Proteobacteria*, which collectively accounted for 96.8% (CON), 97.1% (LBA), 98.3% (MBA), and 97.9% (HBA) of the total phyla (Figure 3A). *Prevotella*, *Prevotella_7* and *Erysipelotrichaceae UCG-002* were the dominant genera in all groups (Figure 3B), with varying relative abundance: *Prevotella* (33.3% CON, 44.9% LBA, 29.9% MBA, and 28.6% HBA), *Prevotella_7* (5.9% CON, 6.8% LBA, 12.2% MBA, and 19.5% HBA), and *Erysipelotrichaceae UCG-002* (11.7% CON, 4.9% LBA, 5.6% MBA, and 13.8% HBA).

3.4. Comparison of Rumen Microbiota at the Genus Level

At the genus level, 14 genera showed a significant difference in their relative abundance among the CON, LBA, MBA, and HBA groups (Figure 4, $p < 0.05$), spanning the *Bacteroideta*, *Firmicutes*, *Cyanobacteria*, *Actinobacteriota*, and *Proteobacteria* phyla. The BA supplementation reduced the relative abundances of *Shuttleworthia* ($p = 0.01$), *unclassified_o__Gastranaerophilales* ($p = 0.02$), *Oscillospiraceae_UCG-002* ($p = 0.04$), and *Prevotellaceae_UCG-003* ($p = 0.04$) while increasing that of *Bifidobacterium* ($p = 0.01$).

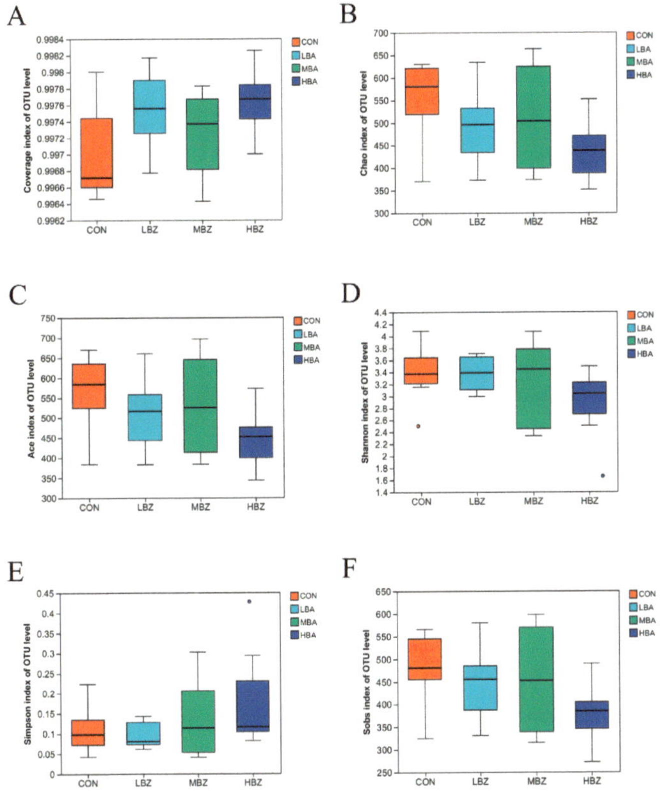

Figure 1. Alpha diversity analysis of rumen microbiota in weaned Holstein dairy calves. (**A**) Coverage index. (**B**) Chao index. (**C**) Ace index. (**D**) Shannon index. (**E**) Simpson index. (**F**) Sobs index. CON, the control group, which was fed with basal diet; LBA (low BA group), MBA (medium BA group), and HBA (high BA group) are the experimental groups, which were supplemented with 0.25%, 0.50%, and 0.75% BA in the basal diet (as dry matter basis), respectively.

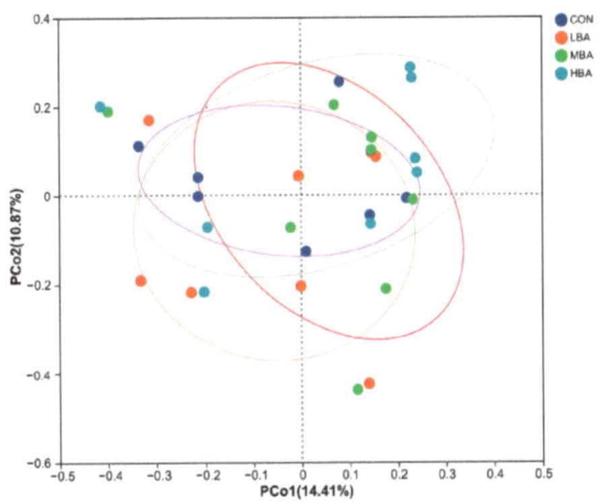

Figure 2. Principal coordinate analysis (PCoA) of rumen microbiota in weaned Holstein dairy calves based on Bray–Curtis dissimilarity. CON, the control group, which was fed with basal diet; LBA (low BA group), MBA (medium BA group), and HBA (high BA group) are the experimental groups, which were supplemented with 0.25%, 0.50%, and 0.75% BA in the basal diet (as dry matter basis), respectively.

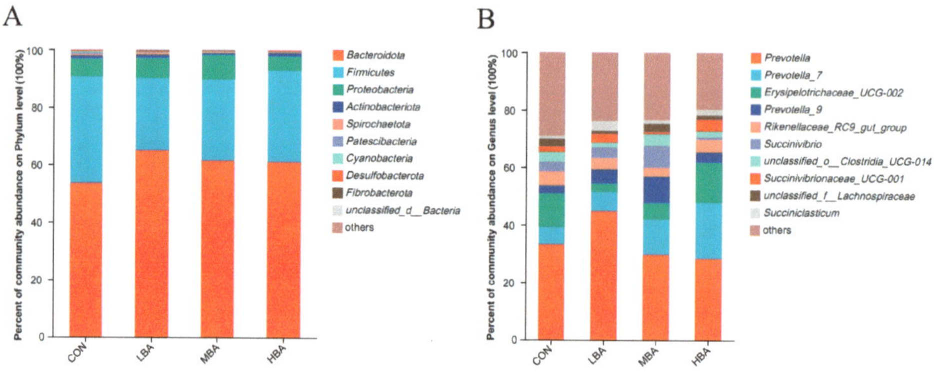

Figure 3. Relative abundances of rumen microbiota at phylum and genus levels in weaned Holstein dairy calves. (**A**) Phylum level. (**B**) Genus level. CON, the control group, which was fed with basal diet; LBA (low BA group), MBA (medium BA group), and HBA (high BA group) are the experimental groups, which were supplemented with 0.25%, 0.50%, and 0.75% BA on the basal diet (as dry matter basis), respectively.

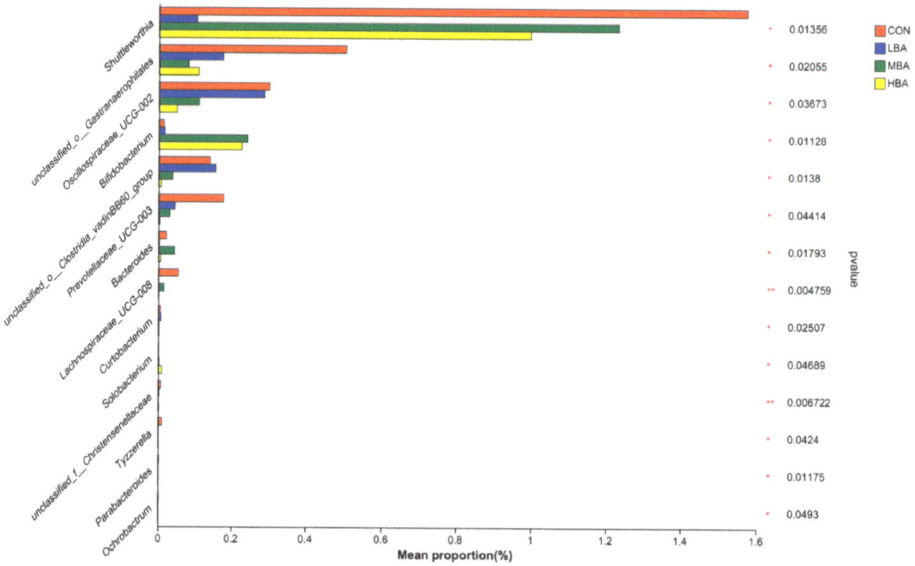

Figure 4. Comparison of rumen bacteria at the genus level in weaned Holstein dairy calves. CON, the control group, which was fed with basal diet; LBA (low BA group), MBA (medium BA group), and HBA (high BA group) are the experimental groups, which were supplemented with 0.25%, 0.50%, and 0.75% BA on the basal diet (as dry matter basis), respectively. * $0.01 < p < 0.05$; ** $0.001 < p < 0.01$.

3.5. Functional Differences in Rumen Microbiota

Functional profiles constructed using the MetaCyc database revealed nine significantly different pathways related to the biosynthesis of key metabolites in glycolysis and amino acid metabolism among the groups (Figure 5, $p < 0.05$). The BA supplementation down-regulated eight metabolic pathways: P105-PWY, GLYCOLYSIS-TCA-GLYOX-BYPASS, TCA-GLYOX-BYPASS, GLYCOL-GLYOXDEG-PWY, PWY-6728, PWY0-42, PWY-6562, and PWY1G-0.

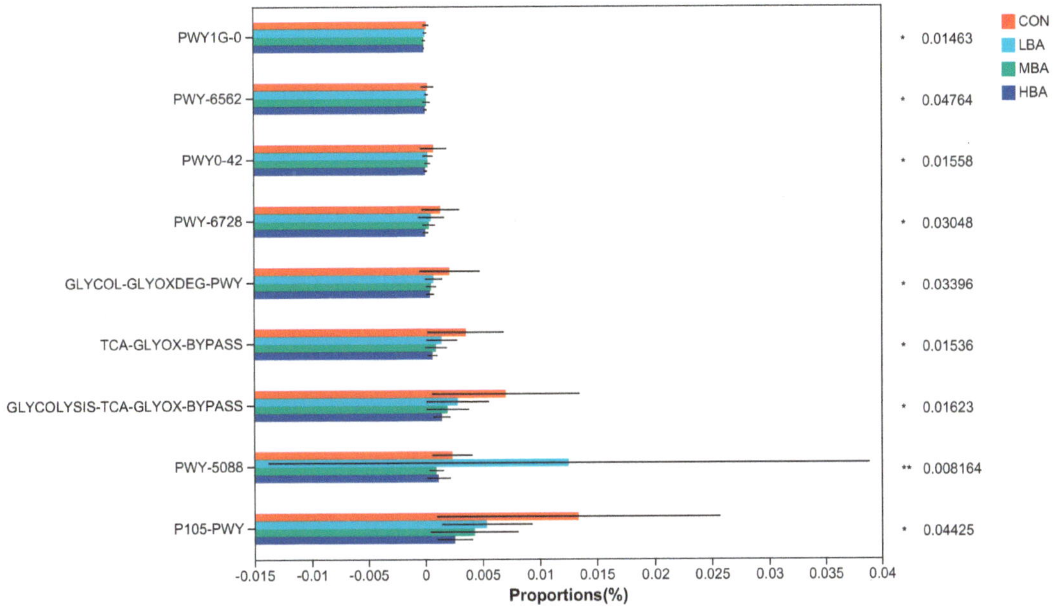

Figure 5. Functional changes in rumen microbiota of weaned Holstein dairy calves. CON, the control group, which was fed with basal diet; LBA (low BA group), MBA (medium BA group), and HBA (high BA group) are the experimental groups, which were supplemented with 0.25%, 0.50%, and 0.75% BA in the basal diet (as dry matter basis), respectively. * $0.01 < p < 0.05$; ** $0.001 < p < 0.01$.

3.6. Correlation Analysis of Differential Pathways, Differential Rumen Microbiota, and VFAs

Spearman's correlation analysis was conducted to explore the relationships among differential pathways, differential rumen microbiota, and VFAs. As shown in Figure 6A, *Unclassified_o__Gastranaerophilales* showed positive correlations with GLYCOLYSIS–TCA-GLYOX-BYPASS (R = 0.52, $p < 0.01$), P105-PWY (R = 0.47, $p < 0.01$), PWY0-42 (R = 0.49, $p < 0.01$), TCA-GLYOX-BYPASS (R = 0.52, $p < 0.01$), and GLYCOL-GLYOXDEG-PWY (R = 0.44, $p < 0.05$). *Bifidobacterium* was negatively correlated with PWY-5088 (R = −0.39, $p < 0.05$) and PWY-6562 (R = −0.37, $p < 0.05$).

Figure 6B highlighted that iso-butyrate exhibited negative correlations with PWY-5088 (R = −0.42, $p < 0.05$), PWY-6562 (R = −0.37, $p < 0.05$), GLYCOL-GLYOXDEG-PWY (R = −0.50, $p < 0.01$), P105-PWY (R = −0.50, $p < 0.01$), GLYCOLYSIS–TCA-GLYOX-BYPASS (R = −0.58, $p < 0.001$), PWY0-42 (R = −0.57, $p < 0.001$), and TCA-GLYOX-BYPASS (R = −0.58, $p \leq 0.001$). Butyrate showed negative correlations with GLYCOL-GLYOXDEG-PWY (R = −0.41, $p < 0.05$), P105-PWY (R = −0.44, $p < 0.05$), GLYCOLYSIS–TCA-GLYOX-BYPASS (R = −0.52, $p < 0.01$), PWY0-42 (R = −0.49, $p < 0.01$), PWY-6728 (R = −0.54, $p < 0.01$), and TCA-GLYOX-BYPASS (R = −0.52, $p < 0.01$).

In Figure 6C, it can be seen that iso-butyrate demonstrated a positive correlation with *Bifidobacterium* (R = 0.42, $p < 0.05$) and a negative correlation with *unclassified_o__Gastranaerophilales* (R = −0.58, $p \leq 0.001$). Butyrate was negatively correlated with *Unclassified_o__Gastranaerophilales* (R = −0.43, $p < 0.05$), *unclassified_o__Clostridia_vadinBB60_group* (R = −0.43, $p < 0.05$), and *Prevotellaceae_UCG-003* (R = −0.48, $p < 0.01$).

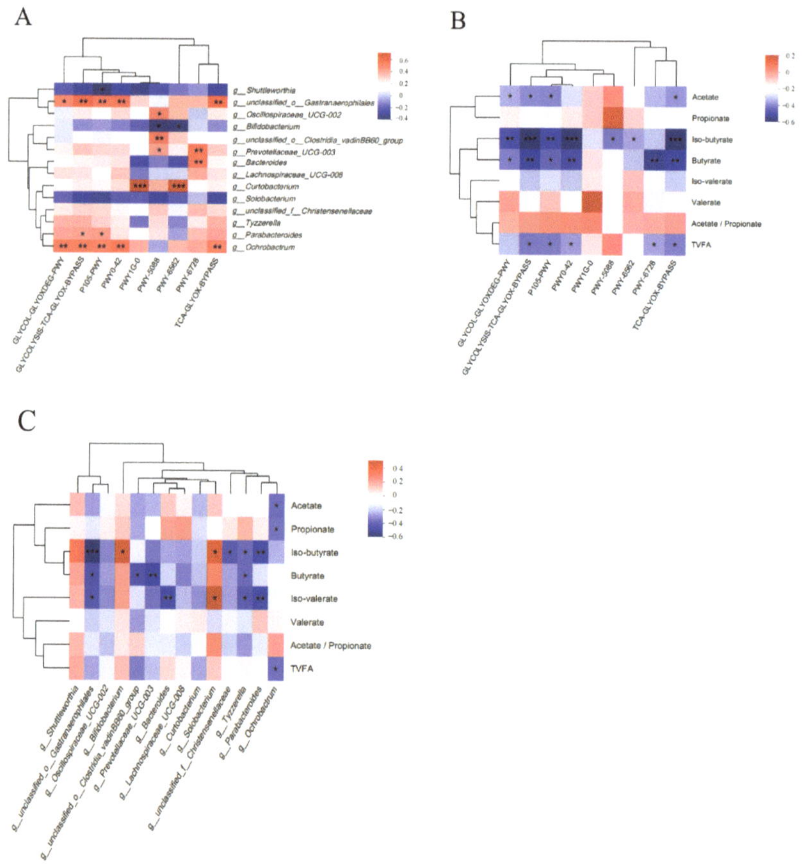

Figure 6. Correlation analysis. (**A**) Correlation analysis of differential pathways and differential rumen microbiota; (**B**) correlation analysis of differential pathways and VFAs; (**C**) correlation analysis of differential rumen microbiota and VFAs. The red represents positive correlation, blue represents negative correlation, and the depth of the color represents the degree of correlation. * $0.01 < p < 0.05$; ** $0.001 < p < 0.01$, *** $p \leq 0.001$.

4. Discussion

Weaning stress may affect the health and wellbeing of dairy calves, reducing their growth performance and gastrointestinal development [4]. The ADG and F/G ratios are related to feed types, animal varieties, and management levels, while ADFI is influenced by the energy content and NDF content of diet, feed palatability, and feeding method [32–36]. Although numerous studies have indicated that supplementation with BA could improve the growth performance of livestock and poultry, few studies have reported the effect of BA on the growth performance of weaned Holstein dairy calves. Diao et al. [37] found that supplementation with 5000 mg/kg BA increased the ADG and ADFI in young pigs. Another study suggested that 10,000 mg/kg BA supplementation significantly improved the growth performance in weaned pigs and significantly decreased their F/G ratio [38]. Józefiak et al. [39] found that 0.1% BA supplantation resulted in an increase in the BWG of chickens in the first 2 weeks of fattening [39]. Although BA has been extensively studied in monogastric animals, research on its effects in ruminants, particularly weaned calves, remains limited. Only Williams et al. [20] found that adding 0.5% BA to the diet of fattening cattle significantly increased eating rate and DMI. The present study showed that the 0.5% BA supplementation significantly improved the ADG and ADFI, and linearly decreased the F/G ratio of weaned Holstein dairy calves, results which were consistent with the conclusions of previous research [20,37–39].

Dietary carbohydrates and other plant fibers, after being anaerobically fermented by a variety of bacteria and enzymes in the rumen, are broken down and then transform into VFAs, methane, CO_2, and hydrogen. It is known that ruminal VFAs can meet 50–70% of the energy requirements of dairy cows [40,41]. Both dietary types and feed additives affect the rumen environment, including the number and activity of rumen microbiota, thereby changing the production of VFAs [42,43]. The VFAs produced by the rumen microbiota mainly include acetate, propionate, and butyrate [44]. Numerous studies have shown that butyrate promoted the development of the rumen, as it was converted into β-hydroxybutyrate and acetoacetate in the rumen epithelium and used as a direct energy donor for rumen epithelial cells and papilla development [45,46]. In the rumen of young ruminants, administration of any of the three main VFAs can promote the growth and functional maturation of rumen epithelium, with butyrate having the most remarkable effect on rumen papillae proliferation [47–50]. Previous studies have shown that butyrate was the most effective inducer of cellular functional changes in VFAs, which played an important role in regulating DNA histone modifications and gene networks, controlling cell differentiation and proliferation, inducing cell cycle arrest and apoptosis, and altering histone acetylation and methylation [51–54]. Available evidence has shown that iso-butyrate is essential for the development of the rumen in ruminants, which not only stimulates the growth of rumen organisms but promotes cellulose digestion [55]. As a branched-chain fatty acid (BCFA), iso-butyrate is typically present in the rumen as a metabolic product of protein bacterial catabolism [56]. In the rumen, fiber degradation in the diet produces acetate, and its linear increase in concentration may also be related to the improvement of digestion of structural carbohydrates, while fiber-degrading bacteria and starch-degrading bacteria increase the deposition of propionate. Different types of rumen fermentation have different A/P ratios. The higher the proportion of acetate, the more inclined the fermentation type is towards the acetate type, while the higher the proportion of propionate, the more inclined the fermentation type is towards the propionate type [57]. In the present study, supplementation of BA linearly increased the content of butyrate, iso-butyrate, and A/P ratio, indicating that BA might have the potential to promote rumen development in weaned Holstein dairy calves and to improve the digestion of structural carbohydrates.

The balance of the ecological environment of rumen is critical for ruminant nutrition, and the rumen's ecological environment may be influenced by factors such as feeding methods, diet, and environment [57]. *Bifidobacterium*, an anaerobic probiotic that colonizes the intestines of humans and animals, participates in various physiological processes such as immunity, digestion, and absorption, playing an important role in maintaining the balance of the gut microbiota, inhibiting pathogen growth, and promoting gastrointestinal health [58]. Members of *Bifidobacterium* can evade rumen degradation through various dietary carbohydrates, a large amount of which typically helps promote intestinal and host health [59,60]. *Gastranaerophilales* (which belongs to the phylum *Cyanobacteria*), an early microbial biomarker reflecting the changes of flora in the gastrointestinal tract, is related to inflammatory response. As a fermentation bacterium, it may produce formate, lactate, ethanol, and CO_2 under anaerobic conditions [61–63]. Wang et al. [64] found that *Gastranaerophilales* and leukotriene B_4 (LTB_4) were significantly increased in milk from clinical mastitis (CM) cows, and *Gastranaerophilales* was positively correlated with LTB_4. Bach et al. [65] found that the relative abundance of *Gastranaerophilales* in the rumen was negatively correlated with milk yield. Perea et al. [66] found that *Gastranaerophilales* was associated with poor growth performance in lambs. In addition, previous research indicated that *Oscillospiraceae_UCG-002* was associated with abnormal health conditions [67,68]. Previous research also showed that when stocking density increased, *Oscillospiraceae_UCG-002* significantly enriched in the gut of hens, which might pose a threat to host health [69,70]. However, there are few studies which have reported the effect of BA on rumen microbiota of weaned Holstein dairy calves. In the present research, with the supplementation of BA, the relative abundance of *Bifidobacterium* significantly increased, but those of *Gastranaerophilales* and *Oscillospiraceae_UCG-002* significantly decreased. This might stimulate the rumen development of calves, improve rumen fermentation parameters, and promote their health, but the mechanism still needs further exploration.

The glyoxylate cycle and TCA cycle are sub-pathways of GLYCOLYSIS-TCA-GLYOX-BYPASS and GLYCOL-GLYOXDEG-PWY, and the latter is a variant of P105-PWY [71]. The up-regulation of the glyoxylate cycle is consistent with the fact that butyrate degradation leads to the production of two molecules of acetyl-CoA [72]. Indeed, the strong up-regulation also exists in enzymes (such as catalase) involved in antioxidant activity and cofactor synthesis [73]. Lacroux et al. [73] found that in the peroxisomes, butyrate underwent assimilation reaction, producing acetyl-CoA through the β-oxidation pathway and further producing tri/dicarboxylic acid in the glyoxylate cycle. Within the TCA cycle, butyrate has been shown to be the more effective oxidation substrate compared to ketone 3-hydroxybutyrate (3-OHB) [74]. Sanchez et al. [75] found that butyrate was decomposed into acetyl-CoA by the β-oxidation pathway, and then entered the TCA cycle and served as an energy substrate to produce citrate. Mathewson et al. [76] proved that most carbon from butyrate was bound into the TCA cycle, leading to the rapid metabolism of butyrate. At present, the specific functions of *Gastranaerophilales* are not clear, but studies have shown that it promotes host digestion and serves as a source of vitamins B and K [77]. Soo et al. [62] found that *Gastranaerophilales* only supported metabolism through fermentation because it seemed to lack genes for anaerobic and aerobic respiration. In the present research, *Gastranaerophilales* was positively correlated with the GLYCOLYSIS-TCA-GLYOX-BYPASS, GLYCOL-GLYOXDEG-PWY, and P105-PWY pathways, and negatively correlated with butyrate. Additionally, the GLYCOLYSIS-TCA-GLYOX-BYPASS, GLYCOL-GLYOXDEG-PWY, and P105-PWY pathways were negatively correlated with butyrate. With the supplementation of BA, the relative abundance of *Gastranaerophilales* decreased, the GLYCOLYSIS-TCA-GLYOX-BYPASS, GLYCOL-GLYOXDEG-PWY, and P105-PWY pathways were down-regulated, and the yield of butyrate increased. The relationship between the above three pathways and butyrate was consistent with previous research [71,72,74], while further research is needed to confirm the relationship between *Gastranaerophilales* and the pathways.

5. Conclusions

The present research indicated that supplementation with BA helps to promote the grow performance and improved rumen fermentation and the composition and function of rumen microbiota of weaned Holstein dairy calves. The addition of different doses of BA linearly increased the ADG and quadratically enhanced the ADFI and FBW, but linearly decreased the F/G ratio. Simultaneously, increasing doses of BA supplementation promoted rumen fermentation by linearly increasing the production of butyrate and isobutyrate and the A/P ratio. In the rumen, the relative abundance of *Bifidobacterium* was increased, but those of the harmful bacteria, such as *unclassified_o__Gastranaerophilales*, and *Oscillospiraceae_UCG-002* were decreased. Finally, supplementation with different doses of BA down-regulated the GLYCOLYSIS-TCA-GLYOX-BYPASS, GLYCOL-GLYOXDEG-PWY, and P105-PWY pathways, which are associated with glycolysis. Under the conditions of this research, supplementation of 0.50% BA was optimal, which might be used as an effective method in the efficient breeding of weaned calves.

Author Contributions: Conceptualization, P.S. and H.D.; methodology, H.D., F.M., S.L., W.Y., D.D. and Z.Z.; software, Q.H., X.Y., J.G. and H.D.; validation, P.S. and H.D.; formal analysis, S.L., W.Y., Z.Z. and H.D.; investigation, D.D., J.G. and H.D.; resources, P.S., S.L., W.Y. and Z.Z.; data curation, F.M., D.D., X.Y., J.G. and H.D.; writing—original draft preparation, H.D.; writing—review and editing, P.S., Q.H. and H.D. All authors have read and agreed to the published version of the manuscript.

Funding: This research was funded by the National Key Research and Development Program of China (2022YFD1301101; 2022YFD1300505), the earmarked fund for China Agriculture Research

www.ingramcontent.com/pod-product-compliance
Lightning Source LLC
Chambersburg PA
CBHW041357210926
51850CB00013B/3102
* 9 7 8 3 7 2 5 8 2 6 8 9 6 *

MDPI AG
Grosspeteranlage 5
4052 Basel
Switzerland
Tel.: +41 61 683 77 34

Animals Editorial Office
E-mail: animals@mdpi.com
www.mdpi.com/journal/animals

Disclaimer/Publisher's Note: The title and front matter of this reprint are at the discretion of the Guest Editors. The publisher is not responsible for their content or any associated concerns. The statements, opinions and data contained in all individual articles are solely those of the individual Editors and contributors and not of MDPI. MDPI disclaims responsibility for any injury to people or property resulting from any ideas, methods, instructions or products referred to in the content.

18. Gong, H.; Yang, Z.; Celi, P.; Yan, L.; Ding, X.; Bai, S.; Zeng, Q.; Xu, S.; Su, Z.; Zhuo, Y.; et al. Effect of benzoic acid on production performance, egg quality, intestinal morphology, and cecal microbial community of laying hens. *Poult. Sci.* **2021**, *100*, 196–205. [CrossRef]
19. Wang, L.M.; Huang, S.; Chalupa-Krebzdak, S.; Vásquez Mejía, S.M.; Mandell, I.B.; Bohrer, B.M. Effects of essential oils and(or) benzoic acid in beef finishing cattle diets on the fatty acid profile and shelflife stability of ribeye steaks and ground beef. *Meat Sci.* **2020**, *168*, 108195. [CrossRef]
20. Williams, M.S.; Mandell, I.B.; Bohrer, B.M.; Wood, K.M. The effects of feeding benzoic acid and/or live active yeast (Saccharomyces cerevisiae) on beef cattle performance, feeding behavior, and carcass characteristics. *Transl. Anim. Sci.* **2021**, *5*, txab143. [CrossRef]
21. National Research Council. *Nutrient Requirements of Dairy Cattle*; National Academy Press: Washington, DC, USA, 2021.
22. AOAC International. *Official Methods of Analysis*, 18th ed.; AOAC International: Washington, DC, USA, 2005.
23. AOAC International. *Official Methods of Analysis*, 17th ed.; AOAC International: Arlington, MA, USA, 2000.
24. AOAC International. *Official Methods of Analysis*, 17th ed.; AOAC International: Gaithersburg, MD, USA, 2003.
25. Van Soest, P.J.; Robertson, J.B.; Lewis, B.A. Methods for dietary fiber, neutral detergent fiber, and nonstarch polysaccharides in relation to animal nutrition. *J. Dairy Sci.* **1991**, *74*, 3583–3597. [CrossRef]
26. AOAC International. *Official Methods of Analysis*, 15th ed.; AOAC International: Arlington, MA, USA, 1990.
27. Li, Z.; Wang, M.; Usman, S.; Khan, A.; Zhang, Y.; Li, F.; Bai, J.; Chen, M.; Zhang, Y.; Guo, X. Effects of nisin on bacterial community and fermentation profiles, in vitro rumen fermentation, microbiota, and methane emission of alfalfa silage. *J. Sci. Food Agric.* **2023**, *103*, 6706–6718. [CrossRef] [PubMed]
28. Parada, A.E.; Needham, D.M.; Fuhrman, J.A. Every base matter: Assessing small subunit rRNA primers for marine microbiomes with mock communities, time series and global field samples. *Environ. Microbiol.* **2016**, *18*, 1403–1414. [CrossRef] [PubMed]
29. Wang, Y.; Nan, X.; Zhao, Y.; Jiang, L.; Wang, H.; Hua, D.; Zhang, F.; Wang, Y.; Liu, J.; Yao, J.; et al. Dietary supplementation with inulin improves lactation performance and serum lipids by regulating the rumen microbiome and metabolome in dairy cows. *Anim. Nutr.* **2021**, *7*, 1189–1204. [CrossRef] [PubMed]
30. Liu, K.; Xu, Q.; Wang, L.; Wang, J.; Guo, W.; Zhou, M. The impact of diet on the composition and relative abundance of rumen microbes in goat. *Asian-Australas. J. Anim. Sci.* **2017**, *30*, 531–537. [CrossRef]
31. Yu, X.; Wo, Y.; Ma, F.; Shan, Q.; Wei, J.; Sun, P. Zinc methionine improves the lipid metabolism that is associated with the alteration of intestine mucosal proteomes and microbiota of newborn Holstein dairy calves. *Anim. Res. One Health* **2024**, *2*, 71–85. [CrossRef]
32. Duthie, C.A.; Haskell, M.; Hyslop, J.J.; Waterhouse, A.; Wallace, R.J.; Roehe, R.; Rooke, J.A. The impact of divergent breed types and diets on methane emissions, rumen characteristics and performance of finishing beef cattle. *Animal* **2017**, *11*, 1762–1771. [CrossRef]
33. Ma, Y.; Chen, X.; Khan, M.Z.; Xiao, J.; Cao, Z. A Combination of Novel Microecological Agents and Molasses Role in Digestibility and Fermentation of Rice Straw by Facilitating the Ruminal Microbial Colonization. *Front. Microbiol.* **2022**, *13*, 948049. [CrossRef]
34. Ali, O.; Szabo-Fodor, J.; Febel, H.; Mézes, M.; Balogh, K.; Glávits, R.; Kovács, M.; Zantomasi, A.; Szabó, A. Porcine Hepatic Response to Fumonisin B(1) in a Short Exposure Period: Fatty Acid Profile and Clinical Investigations. *Toxins* **2019**, *11*, 655. [CrossRef]
35. Roques, S.; Deborde, C.; Guimas, L.; Marchand, Y.; Richard, N.; Jacob, D.; Skiba-Cassy, S.; Moing, A.; Fauconneau, B. Integrative Metabolomics for Assessing the Effect of Insect (*Hermetia illucens*) Protein Extract on Rainbow Trout Metabolism. *Metabolites* **2020**, *10*, 83. [CrossRef]
36. Seibt, K.D.; Ghaffari, M.H.; Scheu, T.; Koch, C.; Sauerwein, H. Characteristics of the Oxidative Status in Dairy Calves Fed at Different Milk Replacer Levels and Weaned at 14 Weeks of Age. *Antioxidants* **2021**, *10*, 260. [CrossRef] [PubMed]
37. Diao, H.; Gao, Z.; Yu, B.; Zheng, P.; He, J.; Yu, J.; Huang, Z.; Chen, D.; Mao, X. Effects of benzoic acid (VevoVitall(R)) on the performance and jejunal digestive physiology in young pigs. *J. Anim. Sci. Biotechnol.* **2016**, *7*, 32. [CrossRef] [PubMed]
38. Kluge, H.; Broz, J.; Eder, K. Effect of benzoic acid on growth performance, nutrient digestibility, nitrogen balance, gastrointestinal microflora and parameters of microbial metabolism in piglets. *J. Anim. Physiol. Anim. Nutr.* **2006**, *90*, 316–324. [CrossRef]
39. Jozefiak, D.; Kaczmarek, S.; Rutkowski, A. The effects of benzoic acid supplementation on the performance of broiler chickens. *J. Anim. Physiol. Anim. Nutr.* **2010**, *94*, 29–34. [CrossRef] [PubMed]
40. Chen, D.; Chen, X.; Tu, Y.; Wang, B.; Lou, C.; Ma, T.; Diao, Q. Effects of mulberry leaf flavonoid and resveratrol on methane emission and nutrient digestion in sheep. *Anim. Nutr.* **2015**, *1*, 362–367. [CrossRef]
41. Bergman, E.N. Energy contributions of volatile fatty acids from the gastrointestinal tract in various species. *Physiol. Rev.* **1990**, *70*, 567–590. [CrossRef]
42. Li, L.P.; Qu, L.; Li, T. Supplemental dietary Selenohomolanthionine affects growth and rumen bacterial population of Shaanbei white cashmere wether goats. *Front. Microbiol.* **2022**, *13*, 942848. [CrossRef]
43. Li, Y.; Wang, Y.; Lv, J.; Dou, X.; Zhang, Y. Effects of Dietary Supplementation with *Clostridium butyricum* on the Amelioration of Growth Performance, Rumen Fermentation, and Rumen Microbiota of Holstein Dairys. *Front. Nutr.* **2021**, *8*, 763700. [CrossRef] [PubMed]
44. Bell, A.W.; Bauman, D.E. Adaptations of glucose metabolism during pregnancy and lactation. *J. Mammary Gland. Biol. Neoplasia* **1997**, *2*, 265–278. [CrossRef]

System (CARS-37), and the Agricultural Science and Technology Innovation Program (cxgc-ias-07, Beijing, China).

Institutional Review Board Statement: The animal study protocol was approved by the Ethics Committee of the Chinese Academy of Agricultural Sciences (approval number IAS2023-138).

Informed Consent Statement: Written informed consent has been obtained from the owner of the animals involved in this study.

Data Availability Statement: The 16S rRNA sequencing data for all the samples were deposited into the NCBI Sequence Read Archive (SRA) under accession number PRJNA1128666 (https://www.ncbi.nlm.nih.gov/sra/PRJNA1128666, accessed on 26 June 2024).

Conflicts of Interest: The co-authors Shujing Li, Wenli Yu and Zengyuan Zhao are employees of Shijiazhuang Tianquan Elite Dairy Ltd. The remaining authors have no relevant financial or non-financial competing interests.

References

1. Wickramasinghe, J.; Kaya, C.A.; Beitz, D.; Appuhamy, R. Early Stepdown Weaning of Dairy Calves with Glutamine and Branched-Chain Amino Acid Supplementations. *Animals* **2022**, *12*, 1474. [CrossRef] [PubMed]
2. Wickramasinghe, H.; Kaya, C.A.; Baumgard, L.H.; Appuhamy, J. Early step-down weaning of dairy calves from a high milk volume with glutamine supplementation. *J. Dairy Sci.* **2022**, *105*, 1186–1198. [CrossRef] [PubMed]
3. Ma, J.; Shah, A.M.; Shao, Y.; Wang, Z.; Zou, H.; Kang, K. Dietary supplementation of yeast cell wall improves the gastrointestinal development of weaned calves. *Animal Nutrition.* **2020**, *6*, 507–512. [CrossRef]
4. Wang, S.; Wang, F.; Kong, F.; Cao, Z.; Wang, W.; Yang, H.; Wang, Y.; Bi, Y.; Li, S. Effect of Supplementing Different Levels of L-Glutamine on Holstein Calves during Weaning. *Antioxidants* **2022**, *11*, 542. [CrossRef]
5. Dai, D.; Qi, G.; Wang, J.; Zhang, H.; Qiu, K.; Han, Y.; Wu, Y.; Wu, S. Dietary organic acids ameliorate high stocking density stress-induced intestinal inflammation through the restoration of intestinal microbiota in broilers. *J. Anim. Sci. Biotechnol.* **2022**, *13*, 124. [CrossRef]
6. Abu, E.N.; Ragaa, N.M. Eubiotic effect of a dietary acidifier (*potassium diformate*) on the health status of cultured Oreochromis niloticus. *J. Adv. Res.* **2015**, *6*, 621–629.
7. Willing, B.P.; Pepin, D.M.; Marcolla, C.S.; Forgie, A.J.; Diether, N.E.; Bourrie, B.C.T. Bacterial resistance to antibiotic alternatives: A wolf in sheep's clothing. *Anim. Front.* **2018**, *8*, 39–47. [CrossRef]
8. Burgess, K.; Li, H.; Abo-Zeid, Y.; Fatimah; Williams, G.R. The Effect of Molecular Properties on Active Ingredient Release from Electrospun Eudragit Fibers. *Pharmaceutics* **2018**, *10*, 103. [CrossRef]
9. Ivanovic, M.; Mirkovic, N.; Mirkovic, M.; Miocinovic, J.; Radulovic, A.; Solevic Knudsen, T.; Radulovic, Z. Autochthonous Enterococcus durans PFMI565 and Lactococcus lactis subsp. lactis BGBU1-4 in Bio-Control of Listeria monocytogenes in Ultrafiltered Cheese. *Foods* **2021**, *10*, 1448. [CrossRef]
10. Zhang, Y.; Wei, J.; Qiu, Y.; Niu, C.; Song, Z.; Yuan, Y.; Yue, T. Structure-Dependent Inhibition of Stenotrophomonas maltophilia by Polyphenol and Its Impact on Cell Membrane. *Front. Microbiol.* **2019**, *10*, 2646. [CrossRef]
11. Stratford, M.; Vallieres, C.; Geoghegan, I.A.; Archer, D.B.; Avery, S.V. The Preservative Sorbic Acid Targets Respiration, Explaining the Resistance of Fermentative Spoilage Yeast Species. *mSphere* **2020**, *5*, 10–1128. [CrossRef] [PubMed]
12. Moore, J.P.; Li, H.; Engmann, M.L.; Bischof, K.M.; Kunka, K.S.; Harris, M.E.; Tancredi, A.C.; Ditmars, F.S.; Basting, P.J.; George, N.S.; et al. Inverted Regulation of Multidrug Efflux Pumps, Acid Resistance, and Porins in Benzoate-Evolved *Escherichia coli* K-12. *Appl. Environ. Microbiol.* **2019**, *85*, e00966-19. [CrossRef] [PubMed]
13. Song, J.H.; Kim, H.; Jeong, M.; Kong, M.J.; Choi, H.K.; Jun, W.; Kim, Y.; Choi, K.C. In Vivo Evaluation of Dendropanax morbifera Leaf Extract for Anti-Obesity and Cholesterol-Lowering Activity in Mice. *Nutrients* **2021**, *13*, 1424. [CrossRef] [PubMed]
14. Yue, L.; Li, Y.; Zhong, M.; Chai, X.; Zhao, P.; Huang, R.; Kang, Y.; Yang, X. Benzoic Acid, Chlorine Dioxide, and 1-Methylcyclopropene Induce Flavonoid Metabolic Shifts in Postharvest Flowering Chinese Cabbage Revealed by High-Dimensional Analytical Data. *Int. J. Mol. Sci.* **2022**, *23*, 6011. [CrossRef]
15. Zhai, H.; Ren, W.; Wang, S.; Wu, J.; Guggenbuhl, P.; Kluenter, A.M. Growth performance of nursery and grower-finisher pigs fed diets supplemented with benzoic acid. *Anim. Nutr.* **2017**, *3*, 232–235. [CrossRef] [PubMed]
16. Pu, J.; Chen, D.; Tian, G.; He, J.; Zheng, P.; Mao, X.; Yu, J.; Huang, Z.; Luo, J.; Luo, Y.; et al. Effects of benzoic acid, Bacillus coagulans and oregano oil combined supplementation on growth performance, immune status and intestinal barrier integrity of weaned piglets. *Anim. Nutr.* **2020**, *6*, 152–159. [CrossRef]
17. Giannenas, I.; Papaneophytou, C.P.; Tsalie, E.; Tsalie, E.; Pappas, I.; Triantafillou, E.; Tontis, D.; Kontopidis, G.A. Dietary Supplementation of Benzoic Acid and Essential Oil Compounds Affects Buffering Capacity of the Feeds, Performance of Turkey Poults and Their Antioxidant Status, pH in the Digestive Tract, Intestinal Microbiota and Morphology. *Asian-Australas. J. Anim. Sci.* **2014**, *27*, 225–236. [CrossRef]

72. Pietrocola, F.; Galluzzi, L.; Bravo-San, P.J.; Madeo, F.; Kroemer, G. Acetyl coenzyme A: A central metabolite and second messenger. *Cell Metab.* **2015**, *21*, 805–821. [CrossRef]
73. Lacroux, J.; Atteia, A.; Brugiere, S.; Couté, Y.; Vallon, O.; Steyer, J.P.; van Lis, R. Proteomics unveil a central role for peroxisomes in butyrate assimilation of the heterotrophic Chlorophyte alga *Polytomella* sp. *Front. Microbiol.* **2022**, *13*, 1029828. [CrossRef]
74. Carley, A.N.; Maurya, S.K.; Fasano, M.; Wang, Y.; Selzman, C.H.; Drakos, S.G.; Lewandowski, E.D. Short-Chain Fatty Acids Outpace Ketone Oxidation in the Failing Heart. *Circulation* **2021**, *143*, 1797–1808. [CrossRef]
75. Sanchez, H.N.; Moroney, J.B.; Gan, H.; Shen, T.; Im, J.L.; Li, T.; Taylor, J.R.; Zan, H.; Casali, P. B cell-intrinsic epigenetic modulation of antibody responses by dietary fiber-derived short-chain fatty acids. *Nat. Commun.* **2020**, *11*, 60. [CrossRef]
76. Mathewson, N.D.; Jenq, R.; Mathew, A.V.; Koenigsknecht, M.; Hanash, A.; Toubai, T.; Oravecz-Wilson, K.; Wu, S.R.; Sun, Y.; Rossi, C.; et al. Gut microbiome-derived metabolites modulate intestinal epithelial cell damage and mitigate graft-versus-host disease. *Nat. Immunol.* **2016**, *17*, 505–513. [CrossRef] [PubMed]
77. Di Rienzi, S.C.; Sharon, I.; Wrighton, K.C.; Koren, O.; Hug, L.A.; Thomas, B.C.; Goodrich, J.K.; Bell, J.T.; Spector, T.D.; Banfield, J.F.; et al. The human gut and groundwater harbor non-photosynthetic bacteria belonging to a new candidate phylum sibling to Cyanobacteria. *eLife* **2013**, *2*, e1102. [CrossRef] [PubMed]

Disclaimer/Publisher's Note: The statements, opinions and data contained in all publications are solely those of the individual author(s) and contributor(s) and not of MDPI and/or the editor(s). MDPI and/or the editor(s) disclaim responsibility for any injury to people or property resulting from any ideas, methods, instructions or products referred to in the content.

45. Kamke, J.; Kittelmann, S.; Soni, P.; Li, Y.; Tavendale, M.; Ganesh, S.; Janssen, P.H.; Shi, W.; Froula, J.; Rubin, E.M.; et al. Rumen metagenome and metatranscriptome analyses of low methane yield sheep reveals a Sharpea-enriched microbiome characterised by lactic acid formation and utilisation. *Microbiome* **2016**, *4*, 56. [CrossRef]
46. Laarman, A.H.; Ruiz-Sanchez, A.L.; Sugino, T.; Guan, L.L.; Oba, M. Effects of feeding a calf starter on molecular adaptations in the ruminal epithelium and liver of Holstein dairy calves. *J. Dairy Sci.* **2012**, *95*, 2585–2594. [CrossRef] [PubMed]
47. Lane, M.A.; Jesse, B.W. Effect of volatile fatty acid infusion on development of the rumen epithelium in neonatal sheep. *J. Dairy Sci.* **1997**, *80*, 740–746. [CrossRef]
48. Sakata, T.; Tamate, H. Rumen epithelial cell proliferation accelerated by rapid increase in intraruminal butyrate. *J. Dairy Sci.* **1978**, *61*, 1109–1113. [CrossRef] [PubMed]
49. Sakata, T.; Tamate, H. Rumen epithelium cell proliferation accelerated by propionate and acetate. *J. Dairy Sci.* **1979**, *62*, 49–52. [CrossRef] [PubMed]
50. Shen, Z.; Kuhla, S.; Zitnan, R.; Seyfert, H.M.; Schneider, F.; Hagemeister, H.; Chudy, A.; Löhrke, B.; Blum, J.W.; Hammon, H.M. Intraruminal infusion of n-butyric acid induces an increase of ruminal papillae size independent of IGF-1 system in castrated bulls. *Arch. Anim. Nutr.* **2005**, *59*, 213–225. [CrossRef]
51. Smith, P.M.; Howitt, M.R.; Panikov, N.; Michaud, M.; Gallini, C.A.; Bohlooly-Y, M.; Glickman, J.N.; Garrett, W.S. The microbial metabolites, short-chain fatty acids, regulate colonic Treg cell homeostasis. *Science* **2013**, *341*, 569–573. [CrossRef] [PubMed]
52. Li, R.W.; Li, C. Butyrate induces profound changes in gene expression related to multiple signal pathways in bovine kidney epithelial cells. *BMC Genom.* **2006**, *7*, 234. [CrossRef]
53. Li, C.J.; Li, R.W.; Wang, Y.H.; Elsasser, T.H. Pathway analysis identifies perturbation of genetic networks induced by butyrate in a bovine kidney epithelial cell line. *Funct. Integr. Genom.* **2007**, *7*, 193–205. [CrossRef] [PubMed]
54. Marinova, Z.; Leng, Y.; Leeds, P.; Chuang, D.M. Histone deacetylase inhibition alters histone methylation associated with heat shock protein 70 promoter modifications in astrocytes and neurons. *Neuropharmacology* **2011**, *60*, 1109–1115. [CrossRef]
55. Yang, C.M. Response of forage fiber degradation by ruminal microorganisms to branched-chain volatile fatty acids, amino acids, and dipeptides. *J. Dairy Sci.* **2002**, *85*, 1183–1190. [CrossRef]
56. Russell, J.B.; Hespell, R.B. Microbial rumen fermentation. *J. Dairy Sci.* **1981**, *64*, 1153–1169. [CrossRef] [PubMed]
57. Guo, Y.; Yang, R.; Duan, C.; Wang, Y.; Hao, Q.; Ji, S.; Yan, H.; Zhang, Y.; Liu, Y. Effect of dioscorea opposite waste on growth performance, blood parameters, rumen fermentation and rumen bacterial community in weaned lambs. *J. Integr. Agric.* **2023**, *22*, 1833–1846. [CrossRef]
58. Sun, C.; Chen, L.; Shen, Z. Mechanisms of gastrointestinal microflora on drug metabolism in clinical practice. *Saudi Pharm. J.* **2019**, *27*, 1146–1156. [CrossRef]
59. Pokusaeva, K.; Fitzgerald, G.F.; van Sinderen, D. Carbohydrate metabolism in Bifidobacteria. *Genes Nutr.* **2011**, *6*, 285–306. [CrossRef] [PubMed]
60. Bunesova, V.; Vlkova, E.; Rada, V.; Killer, J.; Musilova, S. Bifidobacteria from the gastrointestinal tract of animals: Differences and similarities. *Benef. Microbes* **2014**, *5*, 377–388. [CrossRef] [PubMed]
61. Wu, F.; Lei, H.; Chen, G.; Chen, C.; Song, Y.; Cao, Z.; Zhang, C.; Zhang, C.; Zhou, J.; Lu, Y.; et al. In Vitro and In Vivo Studies Reveal that Hesperetin-7-O-glucoside, a Naturally Occurring Monoglucoside, Exhibits Strong Anti-inflammatory Capacity. *J. Agric. Food Chem.* **2021**, *69*, 12753–12762. [CrossRef]
62. Soo, R.M.; Skennerton, C.T.; Sekiguchi, Y.; Imelfort, M.; Paech, S.J.; Dennis, P.G.; Steen, J.A.; Parks, D.H.; Tyson, G.W.; Hugenholtz, P. An expanded genomic representation of the phylum cyanobacteria. *Genome Biol. Evol.* **2014**, *6*, 1031–1045. [CrossRef]
63. Feng, C.; Li, A.; Yin, C.; Wang, S.; Jin, W.; Liu, Y.; Huo, T.; Jiang, H. Realgar Alleviated Neuroinflammation Induced by High Protein and High Calorie Diet in Rats via the Microbiota-Gut-Brain Axis. *Nutrients* **2022**, *14*, 3958. [CrossRef]
64. Wang, Y.; Nan, X.; Zhao, Y.; Jiang, L.; Wang, M.; Wang, H.; Zhang, F.; Xue, F.; Hua, D.; Liu, J. Rumen microbiome structure and metabolites activity in dairy cows with clinical and subclinical mastitis. *J. Anim. Sci. Biotechnol.* **2021**, *12*, 36. [CrossRef]
65. Bach, A.; Lopez-Garcia, A.; Gonzalez-Recio, O.; Elcoso, G.; Fàbregas, F.; Chaucheyras-Durand, F.; Castex, M. Changes in the rumen and colon microbiota and effects of live yeast dietary supplementation during the transition from the dry period to lactation of dairy cows. *J. Dairy Sci.* **2019**, *102*, 6180–6198. [CrossRef]
66. Perea, K.; Perz, K.; Olivo, S.K.; Williams, A.; Lachman, M.; Ishaq, S.L.; Thomson, J.; Yeoman, C.J. Feed efficiency phenotypes in lambs involve changes in ruminal, colonic, and small-intestine-located microbiota. *J. Anim. Sci.* **2017**, *95*, 2585–2592. [CrossRef] [PubMed]
67. Chen, Z.; Wang, Z.; Li, D.; Zhu, B.; Xia, Y.; Wang, G.; Ai, L.; Zhang, C.; Wang, C. The gut microbiota as a target to improve health conditions in a confined environment. *Front. Microbiol.* **2022**, *13*, 1067756. [CrossRef] [PubMed]
68. Wu, Y.; Jiao, N.; Zhu, R.; Zhang, Y.; Wu, D.; Wang, A.J.; Fang, S.; Tao, L.; Li, Y.; Cheng, S.; et al. Identification of microbial markers across populations in early detection of colorectal cancer. *Nat. Commun.* **2021**, *12*, 3063. [CrossRef] [PubMed]
69. Wang, Y.; Jin, T.; Zhang, N.; Li, J.; Wang, Y.; Kulyar, M.F.; Han, Z.; Li, Y. Effect of stocking density and age on physiological performance and dynamic gut bacterial and fungal communities in Langya hens. *Microb. Cell Factories* **2021**, *20*, 218. [CrossRef]
70. Lun, H.; Yang, W.; Zhao, S.; Jiang, M.; Xu, M.; Liu, F.; Wang, Y. Altered gut microbiota and microbial biomarkers associated with chronic kidney disease. *Microbiol. Open* **2019**, *8*, e678. [CrossRef]
71. Caspi, R.; Billington, R.; Keseler, I.M.; Kothari, A.; Krummenacker, M.; Midford, P.E.; Ong, W.K.; Paley, S.; Subhraveti, P.; Karp, P.D. The MetaCyc database of metabolic pathways and enzymes—A 2019 update. *Nucleic Acids Res.* **2020**, *48*, D445–D453. [CrossRef]